A Mathematical Odyssey

Steven G. Krantz • Harold R. Parks

A Mathematical Odyssey

Journey from the Real to the Complex

 Springer

Steven G. Krantz
Department of Mathematics
Washington University in St. Louis
St. Louis, MO, USA

Harold R. Parks
Department of Mathematics
Oregon State University
Corvalis, OR, USA

ISBN 978-1-4614-8938-2   ISBN 978-1-4614-8939-9 (eBook)
DOI 10.1007/978-1-4614-8939-9
Springer New York Heidelberg Dordrecht London

Library of Congress Control Number: 2014933267

Mathematics Subject Classification (2010): 00-01; 00A05; 00A06; 00A09; 03D15; 03F40; 05C15; 11D41; 11Y11; 91B25; 05D10; 37-01; 42C40; 49Q05; 51M05; 51M10; 57M40; 83A05; 94A60

© Springer Science+Business Media New York 2014
This work is subject to copyright. All rights are reserved by the Publisher, whether the whole or part of the material is concerned, specifically the rights of translation, reprinting, reuse of illustrations, recitation, broadcasting, reproduction on microfilms or in any other physical way, and transmission or information storage and retrieval, electronic adaptation, computer software, or by similar or dissimilar methodology now known or hereafter developed. Exempted from this legal reservation are brief excerpts in connection with reviews or scholarly analysis or material supplied specifically for the purpose of being entered and executed on a computer system, for exclusive use by the purchaser of the work. Duplication of this publication or parts thereof is permitted only under the provisions of the Copyright Law of the Publisher's location, in its current version, and permission for use must always be obtained from Springer. Permissions for use may be obtained through RightsLink at the Copyright Clearance Center. Violations are liable to prosecution under the respective Copyright Law.
The use of general descriptive names, registered names, trademarks, service marks, etc. in this publication does not imply, even in the absence of a specific statement, that such names are exempt from the relevant protective laws and regulations and therefore free for general use.
While the advice and information in this book are believed to be true and accurate at the date of publication, neither the authors nor the editors nor the publisher can accept any legal responsibility for any errors or omissions that may be made. The publisher makes no warranty, express or implied, with respect to the material contained herein.

Cover figure: Mandelbulb generated by Harold Parks with software designed by Krzysztof Marczak

Printed on acid-free paper

Springer is part of Springer Science+Business Media (www.springer.com)

*To the memory of Bill Thurston
(1946–2012)*

Preface

Mathematics is both an art and a science. It is a science because it provides a panoply of analytic tools for understanding the world around us. It is an art because it is elegant and beautiful.

It is a shame that, because of its ostensibly austere nature, mathematics does not appeal to many people. It is true that not only is mathematics is demanding and unforgiving, but it also offers many rewards and insights. The world would be a better place if we could get more people to appreciate the joys of mathematics.

One of the main purposes of this book is to provide the uninitiated reader with some insights into modern mathematics. The past few years have seen some astonishing advances that could offer charms to even the most jaded reader. This book in fact explores several hot-button topics that cannot be found anywhere else in the nontechnical literature. These include the Black–Scholes option pricing scheme, dynamical systems, relativity theory, wavelets, RSA encryption, the **P/NP** problem, primality testing, Fermat's last theorem, and the Poincaré conjecture.

The Black–Scholes theory is Nobel-Prize-winning work that has dramatically changed the financial world. Now finance is greatly influenced by mathematics, and many new mathematics Ph.D.s end up plying their trade on Wall Street.

Dynamical systems, popularly known in the context of fractals, is a new way of looking at the world around us and understanding its structure. Many colorful figures, including Benoît Mandelbrot, have contributed to our growing understanding of dynamical systems.

Of course relativity theory dates back to the days of Albert Einstein. But new studies, especially regarding cosmology and the "grand unified theory," have brought relativity theory into focus. The new theory of strings provides a detailed structure for physics both in the large and in the small.

Wavelets is a new branch of Fourier analysis—the theory of breaking up complicated waves into simple, component waves. Recently a mathematician in New Hampshire won a Grammy Award for using wavelets to clean up the only known live recording of Woody Guthrie. Wavelets have been used to compress images of fingerprints in order to aid the FBI in its data storage activities. Wavelets are used daily in image compression, signal processing, and many other aspects of modern technology. They are a genuine revolution in modern mathematics.

RSA encryption is a new type of coding theory based on ideas from elementary number theory. It makes possible public key encryption and other new types of code paradigms. Much of modern security depends on RSA encryption. And it all hinges on the intractability of factoring large integers. This is a very practical application that depends directly on rather abstract theoretical mathematics.

The **P/NP** problem is considered by many to be the most important unsolved problem in the mathematical sciences. It has to do with the computational complexity of solving certain problems. The prototype problem in this subject is the factoring of large integers (with 150 digits or more). It takes quite a long time to actually *factor* such a number. But only a few moments to verify that a given factorization is correct (by multiplying the factors together). This dialectic is the key to the **P/NP** problem.

Very closely related to RSA encryption is the question of primality testing. In 2004, M. Agrawal and his students in India found a polynomial-time algorithm for testing whether any given integer is prime. It does *not* tell us what the prime factorizaton is, but it tells us whether there is one. This is a dramatic breakthrough and is based on only very elementary mathematical ideas.

Fermat's last theorem was one of the more dramatic mathematical events of the past 20 years. Princeton's Andrew Wiles in 1993 announced that he could solve the 350-year-old enigma of Fermat. He subsequently appeared on the front page of every newspaper in the world. Princeton was overrun with journalists and news cameras. And then Wiles had to announce that there was an error in his proof. It took over a year and the aid of Wiles's student Richard Taylor to finally fix the error and nail down the proof. This is one of the most dramatic stories of modern mathematics.

Finally, there is the saga of the Poincaré conjecture. A hundred-year-old problem about the shape of the universe, the Poincaré conjecture captured the imaginations of scores of mathematicians. Many proofs have been offered, and many have failed. Finally a rather eccentric Russian mathematician, Grigori Perelman from St. Petersburg, put three preprints on an Internet preprint server which claimed to solve this age-old problem. These preprints were problematical because they were so sketchy and enigmatic. But teams of mathematicians jumped into the breach and spent literally years filling in the gaps in Perelman's arguments. And it all worked. The Poincaré conjecture is proved. And now there are new mysteries to fathom.

This book demonstrates that mathematics is an exciting and ongoing enterprise. It shows the neophyte what mathematicians think about, what they care about, and what their goals are. It shares the excitement, the sorrow, and the frustrations of being a modern mathematician. It shows what we are after, what we can achieve, and what we can appreciate in the process.

The book has few prerequisites beyond a good grounding in basic mathematics. And it will put the reader right into the guts of the problems being discussed. We are not afraid to analyze diagrams, create graphs, and manipulate equations. But we do so in a fashion that readers can appreciate and understand. We hope that the reader will continually be tickled into action and hasten to move on to the next idea. This is likely to be a hard book to put down.

Mathematics is a growing and changing enterprise, full of some of the most important and daring ideas of modern times. Our goal here is to help non-mathematicians appreciate this part of the intellectual pie and perhaps to develop some taste for the saga and journey that is mathematics.

It is a pleasure to thank the many and varied referees who offered their wisdom in the development of this book. Our Editor Ann Kostant provided her own edits and contributed a good deal. Finally we thank Lynn Apfel, whose insights and criticisms have proved to be invaluable.

St. Louis, MO, USA   Steven G. Krantz
Corvallis, OR, USA   Harold R. Parks

Contents

Preface		**vii**
1	**The Four-Color Problem**	**1**
	1.1 Humble Beginnings	1
	1.2 Kempe, Heawood, and the Chromatic Number	3
	1.3 Heawood's Estimate Confirmed	12
	1.4 Appel, Haken, and a Computer-Aided Proof	12
	A Look Back	17
	References and Further Reading	19
2	**The Mathematics of Finance**	**21**
	2.1 Ancient Mathematics of Finance	21
	2.2 Loans and Charging Interest	24
	2.3 Compound Interest	26
	2.4 Continuously Compounded Interest	28
	2.5 Raising Capital: Stocks and Bonds	30
	2.6 The Standard Model for Stock Prices	34
	2.7 Parameters in the Standard Model	37
	2.8 Derivatives	38
	2.9 Pricing a Forward	40
	2.10 Arbitrage	42
	2.11 Call Options	43
	2.12 Value of a Call Option at Expiry	45
	2.13 Pricing a Call Option Using a Replicating Portfolio: A Single Time Step	45
	2.14 Pricing a Call Option Using a Replicating Portfolio: Multiple Time Steps	52

	2.15 Black–Scholes Option Pricing	53
	A Look Back	56
	References and Further Reading	57

3 Ramsey Theory — 59
3.1 Introduction 59
3.2 The Pigeonhole Principle 60
3.3 The Happy End Problem 65
3.4 Relationship Tables and Ramsey's Theorem for Pairs 69
3.5 Ramsey's Theorem in General 73
A Look Back 76
References and Further Reading 78

4 Dynamical Systems — 81
4.1 Introduction 81
4.2 Creation of the Mandelbrot Set 82
    4.2.1 Pseudo-Code to Generate the Mandelbrot Set 86
4.3 Staircase Representation of a One-Dimensional Dynamical System 87
    4.3.1 The Use of the Pocket Calculator 89
4.4 A Little Physics 91
    4.4.1 Poincaré and the Three-Body Problem 94
    4.4.2 Lorenz and Chaos 95
4.5 The Cantor Set as a Fractal 97
4.6 Higher-Dimensional Versions of the Cantor Set 105
    4.6.1 The Sierpiński Triangle 105
A Look Back 107
References and Further Reading 109

5 The Plateau Problem — 111
5.1 Paths That Minimize Length 111
5.2 Surfaces That Minimize Area 113
5.3 Curvature of a Plane Curve 118
5.4 Curvature of a Surface 118
5.5 Curvature of Minimal Surfaces 122
5.6 Plateau's Observations 124
5.7 Types of Spanning Surfaces 125

5.8	The Enneper–Weierstrass Formula	127
	5.8.1 Costa's Surface	127
5.9	Solutions by Douglas and Radó	128
5.10	Surfaces Beyond Disc Type	129
5.11	Currents	131
5.12	Regularity Theory	132
5.13	Plateau's Rules	133
A Look Back		133
References and Further Reading		134

6 Euclidean and Non-Euclidean Geometries — 137

- 6.1 The Concept of Euclidean Geometry 137
- 6.2 A Review of the Geometry of Triangles 140
- 6.3 Some Essential Properties of Euclidean Geometry 145
- 6.4 What is Non-Euclidean Geometry? 149
- 6.5 Spherical Geometry 149
- 6.6 Neutral Geometry 152
- 6.7 Hyperbolic Geometry 154
  - 6.7.1 The Question of Consistency 155
  - 6.7.2 Models of Hyperbolic Geometry 156
- A Look Back 160
- References and Further Reading 161

7 Special Relativity — 163

- 7.1 Introduction 163
- 7.2 Principles Underlying Special Relativity 163
- 7.3 Some Consequences of Special Relativity 164
- 7.4 Momentum and Energy 171
  - 7.4.1 Vector Quantities 172
  - 7.4.2 Vectors in Relativity 173
  - 7.4.3 Relativistic Momentum 175
  - 7.4.4 Rest Mass 178
- A Look Back 179
- References and Further Reading 181

8 Wavelets in Our World — 183

- 8.1 Introductory Ideas 183
- 8.2 Fourier's Ideas 184

	8.3 Enter Wavelets	189
	8.4 What Are Wavelets Good for?	191
	8.5 Key Players in the Wavelet Saga	192
	A Look Back	194
	References and Further Reading	196

9 RSA Encryption — 197

- 9.1 Basics and Background 197
- 9.2 Preparation for RSA 200
  - 9.2.1 Background Ideas 200
  - 9.2.2 Modular Arithmetic 200
  - 9.2.3 Euler's Theorem 201
  - 9.2.4 Relatively Prime Integers 203
- 9.3 The RSA System Enunciated 204
- 9.4 The RSA Encryption System Explicated 207
- 9.5 Zero Knowledge Proofs: How to Keep a Secret 208
- A Look Back 213
- References and Further Reading 215

10 The P/NP Problem — 217

- 10.1 Introduction 217
- 10.2 Complexity Theory 219
- 10.3 Automata 221
- 10.4 Turing Machines 230
- 10.5 Examples of Turing Machines 233
- 10.6 Bigger Calculations 241
- 10.7 Nondeterministic Turing Machines 243
- 10.8 $\mathcal{P}$ and $\mathcal{NP}$ 249
- 10.9 NP-Completeness 251
- A Look Back 252
- References and Further Reading 253

11 Primality Testing — 255

- 11.1 Preliminary Concepts 255
- 11.2 Euclid's Theorem 256
- 11.3 The Sieve of Eratosthenes 258
- 11.4 Recognition of Composite Numbers 261
- 11.5 Speed of Algorithms 263

11.6 Pierre de Fermat	264
11.7 Fermat's Little Theorem	266
11.8 The Strong Pseudoprimality Test and Probabilistic Primality Testing	268
11.9 The AKS Primality Test	271
A Look Back	273
References and Further Reading	275

12 The Foundations of Mathematics — 277

- 12.1 The Evolution of the Concept of Proof 277
- 12.2 Kurt Gödel and the Birth of Uncertainty 280
- 12.3 Origins of Kurt Gödel 281
- 12.4 Elements of Formal Logic 285
  - 12.4.1 Definitions of Connectives 288
  - 12.4.2 Logical Syllogisms 294
  - 12.4.3 Quantifiers 296
- 12.5 Truth and Provability 300
- 12.6 Cracks in the Edifice 302
- 12.7 The Gödel Incompleteness Theorem 304
- A Look Back 306
- References and Further Reading 307

13 Fermat's Last Theorem — 309

- 13.1 Introduction 309
- 13.2 Splitting Square Numbers 311
- 13.3 Pythagorean Triples: Another Construction 312
- 13.4 Kummer's Criterion 315
- 13.5 Fields 317
  - 13.5.1 Complex Numbers 318
  - 13.5.2 Polynomial Equations 320
  - 13.5.3 Finite Fields 321
  - 13.5.4 Polynomials Over Finite Fields 323
- 13.6 Algebraic Curves 325
- 13.7 Elliptic Curves 326
  - 13.7.1 Composition 327
  - 13.7.2 Reduction 328
- 13.8 The Modular Group 330
  - 13.8.1 Inversion 330

13.8.2 Translation	331
13.8.3 Modular Transformations	332
13.8.4 The Congruence Subgroup	332
13.9 The Taniyama–Shimura–Weil Conjecture	332
13.9.1 Frey Curves	333
13.9.2 Wiles's Proof of Fermat's Last Theorem	334
A Look Back	335
References and Further Reading	337

14 Ricci Flow and the Poincaré Conjecture — 339

14.1 Introduction	339
14.2 Thurston's Geometrization Program	342
14.3 Ricci Flow	348
14.4 Perelman's Three Papers	351
14.5 Reaction to Perelman's Work	354
14.6 Final Remarks on the Proof	356
A Look Back	361
References and Further Reading	363

Epilogue — 365

Credits for Illustrations — 367

Index — 371

Chapter 1
The Four-Color Problem

1.1 Humble Beginnings

In 1852 Francis W. Guthrie, a graduate of University College London, posed the following question to his brother Frederick:

> Imagine a geographic map on the earth (that is, on a sphere) consisting of countries only—no oceans, lakes, rivers, or other bodies of water. The only rule is that a country must be a single contiguous mass—in one piece, and with no holes—see Fig. 1.1. As cartographers, we wish to *color* the map so that no two adjacent countries will be of the same color (Fig. 1.2). How many colors should the map-maker keep in stock in order to be sure he/she can color any map?

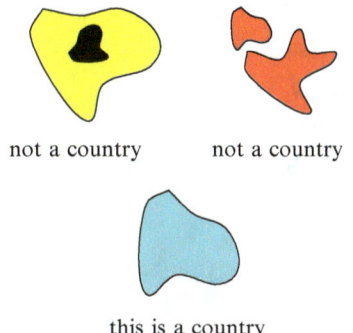

Figure 1.1 What is a country?

In fact, it was already known from experience that four colors suffice to color any map; the point was that no one had given a *mathematical proof* confirming the empirical observation.

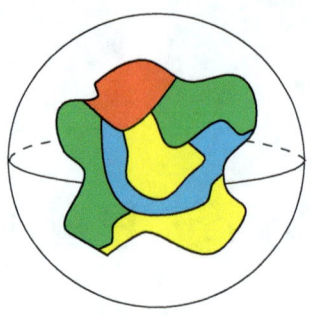

Figure 1.2 A map coloring.

Frederick Guthrie was a student of the mathematician Augustus De Morgan (1806–1871), and he ultimately communicated the problem to his mentor. The problem was passed around among academic mathematicians for a number of years [in fact De Morgan communicated the problem to William Rowan Hamilton (1805–1865)]. The first allusion in print to the problem was by Arthur Cayley (1821–1895) in 1878.

These were all serious mathematicians, and it was rather surprising that such an elementary and accessible problem should prove so resistive to their best efforts. The problem became rather well known, and was dubbed *The Four-Color Problem*.

The eminent mathematician Felix Klein (1849–1925) in Göttingen heard of the problem and declared that the only reason the problem had never been solved is that no capable mathematician had ever worked on it. *He*, Felix Klein, would offer a class, the culmination of which would be a solution to the problem. He failed. Even so, Klein believed (and it was generally believed) that four colors would suffice to color any map. Certainly all the known examples pointed in that direction.

1.2 Kempe, Heawood, and the Chromatic Number

One reason the four-color problem is challenging is that one cannot simply piece together colorings of smaller maps to get a desired coloring of a bigger map. For instance, the map in Fig. 1.3 has been successfully colored with four colors (and all four colors were needed). But if we try to put together two copies of the map side by side as in Fig. 1.4, then we will need to interchange some of the colors to avoid having a border with the same color on both sides. Similarly, if a new country is inserted somewhere in a four-colored map, then some of the existing colors may need to be changed to color the new map with only four colors.

Figure 1.3 A successful coloring.

Figure 1.4 Two four-colored maps needing recoloring before merging.

In 1879, Alfred Kempe (1845–1922) published what was believed to be a solution of the four-color problem. That is, he showed that *any* map on the earth (that is, the sphere) could be colored with four colors. Kempe's proof turned out to be flawed, but it took 11 years for anyone to discover the error.

Kempe's argument involved first removing countries from the map in a carefully chosen order, then adding them back, recoloring the map as needed with each addition. The carefully chosen order was in fact designed to guarantee that each country that is added back borders no more than five countries. No recoloring is required if a reinserted country borders 1, 2, or 3 other countries, but it may be needed when the reinserted country borders 4 or 5 countries.

The first interesting case occurs when the reinserted country borders 4 existing countries and those countries have been colored using all four colors. The situation might be similar to that in Fig. 1.5a. Since the red country near the top of the map in Fig. 1.5a does not border any yellow country, its color can be changed to yellow, and the new country in the center can be colored red as shown in the map in Fig. 1.5b.

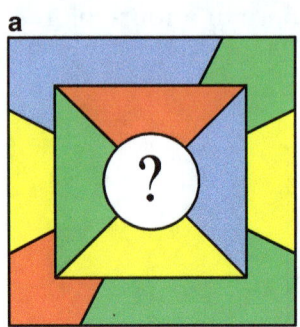 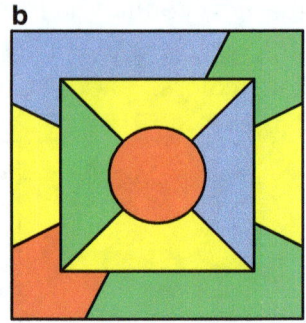

Figure 1.5 The easy case of adding back a country that borders all four colors (**a**) The initial coloring (**b**) After changing one red country to yellow, the new country can be colored red.

Kempe's new idea was to extend this argument to more countries. For example, in Fig. 1.6a the new country is again surrounded by all four colors. None of the four surrounding colors can be changed individually, because each of those four countries also touches all three of the other colors.

Kempe's procedure for coloring the new country relies on planning for many simultaneous color changes. Pick a pair of colors on opposite sides of the new country. We will pick red and yellow. Then, starting with the red country on top, form a larger set containing all the yellow countries it borders. Then include in the set all the red countries that those yellow countries border. Continue this process until no more countries get included. The set of countries just constructed is a *Kempe chain*. (Don't take the name "chain" too literally—a Kempe chain is not necessarily a sequence, it may branch and/or contain contiguous clumps of countries.) Once the chain is formed, as in Fig. 1.6b, it is certainly harmless (i.e., introduces no countries sharing a border that are colored the same) to interchange the colors red and yellow in the Kempe chain.

1.2. Kempe, Heawood, and the Chromatic Number

There are now two cases:

(1) The first possibility is that the Kempe chain does not include the country that is opposite the country that started the chain—that is the case in Fig. 1.6b. Specifically, the yellow country below the new country is not part of the Kempe chain. The colors in the Kempe chain can be interchanged as in Fig. 1.6c, so the red country on the top becomes yellow, and finally the new country can be colored red as in Fig. 1.6d.

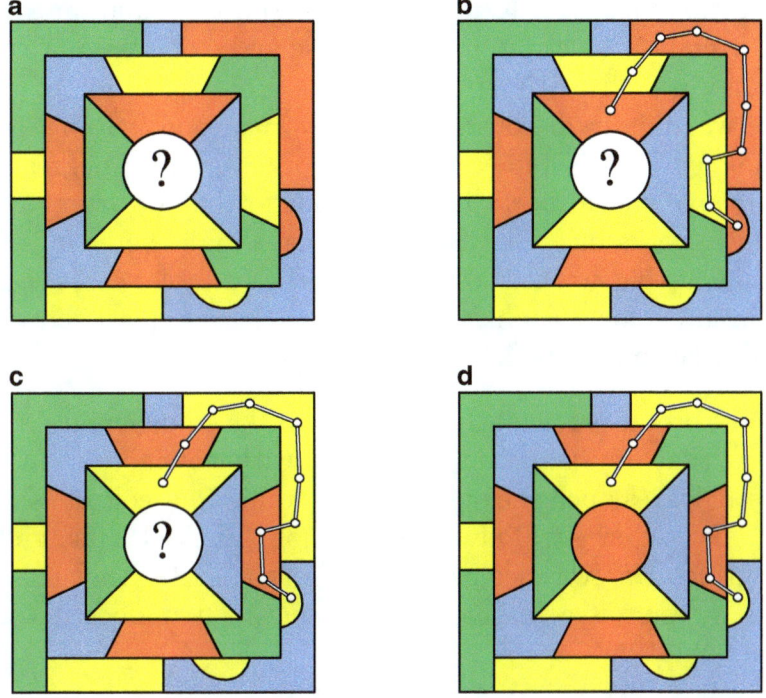

Figure 1.6 Using a Kempe chain to add back a country that borders all four colors when the chain does not close up (**a**) The initial coloring (**b**) Form the Kempe chain (**c**) Change colors in the chain (**d**) Color the new country.

(2) The second possibility is that the Kempe chain *does* include the country that is opposite the country that started the chain. This case is illustrated by the map in Fig. 1.7a and the Kempe chain in Fig. 1.7b. This time the yellow country below the new country *is* part of the Kempe chain, so interchanging red and yellow in the Kempe chain does not facilitate coloring the new country. Fortunately, the fact that the Kempe chain "closes up" guarantees that a Kempe chain starting from either of the other two colors cannot close up. Figure 1.7c shows the Kempe chain starting from the blue country on the right side of the new country. It cannot get around to the green country on the left side of the new country, because it is blocked by the red and yellow countries in the first Kempe chain. In Fig. 1.7d, the colors blue and green have been interchanged in the second Kempe chain. The new country no longer touches any blue country and it can be colored blue as in Fig. 1.7e.

Things are trickier when the country that is added back borders 5 existing countries. In that situation Kempe needed to do two simultaneous color interchanges. Kempe's description of how this can be done is quite convincing, and his error was sufficiently subtle that his proof stood for 11 years. Then the mistake was finally discovered by Percy Heawood.[1] Roughly speaking, Heawood noticed that Kempe's treatment of a vertex of degree five (i.e., a vertex where five edges meet) is too simplistic; his argument of removing vertices, which works nicely in the simpler cases, fails here. It is fascinating how an argument that worked so nicely in earlier, simpler cases can then trick the reader in later, more sophisticated cases. Heawood studied the problem further and came to a number of fascinating conclusions:

- Kempe's proof, particularly his device of "Kempe chains," *does* suffice to show that any map can be colored with not more than five colors. This still leaves open the question of whether four colors—say red, blue, green, and yellow—will always do the job. But red, blue, green, yellow, and purple *will* always color any map on the earth (that is, the sphere). Kempe's result is a real accomplishment, but it does not answer the original question.

- Heawood showed that if the number of countries bordering each country in the map is divisible by 3, then the map is four-colorable.

One of the lovely things about mathematics is that it develops along many parallel tracks that often nurture one another. While some mathematicians were thinking about the four-color problem, others were thinking

---

[1] Yes, mathematicians make mistakes. We all do. And frequently those mistakes are caught by other mathematicians. After all, mathematics is a scholarly process and part of what we do is to validate and, when appropriate, correct each other's work. When the process works correctly, a great many expert mathematicians will check and vouch for the validity of a new proof. Then it is accepted as a valid part of the canon. This is very much like one chemist repeating the experiment of another chemist to check his/her work.

1.2. Kempe, Heawood, and the Chromatic Number

Figure 1.7 Using a Kempe chain to add back a country that borders all four colors when the chain closes up (**a**) The initial coloring (**b**) Form the Kempe chain (**c**) Form a second chain (**d**) Change colors in the second chain (**e**) Color the new country.

about the nature and shape of surfaces. Camille Jordan (1838–1922) and August Möbius (1790–1868) showed that any 2-dimensional surface in space is a sphere with handles attached. See Fig. 1.8. The number of handles is called the *genus*, and we denote it by g. The torus (see Fig. 1.9) is topologically equivalent to a sphere with one handle (that is, one may be deformed to the other with bending and stretching). Thus the torus has genus $g = 1$. A double torus, with two holes, has genus $g = 2$. Heawood used this information to extend the study of the map-coloring problem to any 2-dimensional surface in space.

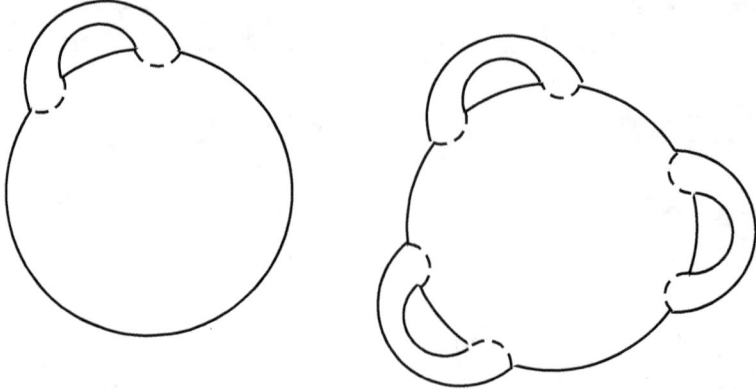

Figure 1.8 Sphere with handles.

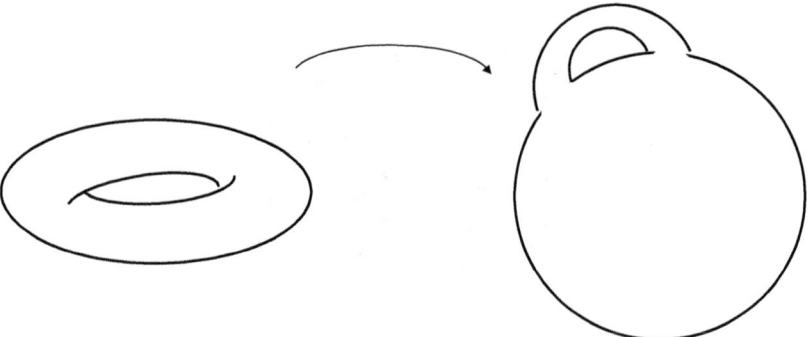

Figure 1.9 Torus as a sphere with one handle.

Heawood found a formula that gives an estimate for the "chromatic number" of any surface. The *chromatic number* of a surface with genus g, written $\chi(g)$, is the least number of colors it will take to color *any* map on that genus-g surface. The following table gives some of Heawood's estimates for the chromatic number of the surface of genus g:

1.2. Kempe, Heawood, and the Chromatic Number

Genus	Chromatic number estimate	Genus	Chromatic number estimate	Genus	Chromatic number estimate
1	7	10	14	22	19
2	8	11	15	⋮	⋮
3	9	12	15	100	38
4	10	13	16	101	38
5	11	14	16	102	38
6	12	15	16	103	38
7	12	⋮	⋮	104	38
8	13	20	19	⋮	⋮
9	13	21	19		

We see that the chromatic numbers start out in a fairly innocent sequence. But then they begin to bunch up: genus 6 and genus 7 have the same chromatic number; genus 13, 14, and 15 have the same chromatic number; genus 20, 21, 22 have the same chromatic number; and genus 100, 101, 102, 103, and 104 have the same chromatic number. As the genus gets larger, the clumping gets more pronounced. In fact there is a formula due to Heawood that gives all his estimates for the chromatic number, and we discuss it in a footnote.[2] Heawood's formula gives the estimate 7 for the chromatic number of the torus. And in fact we can give an example—see Fig. 1.10d—of a map on the torus that requires seven colors.

Figure 1.10 shows how to deform seven countries that wrap around a torus so that you end up in Fig. 1.10d with each of the countries touching the other six. Keep in mind that in Fig. 1.10c, d, while it may look like there

---

[2]Heawood's formula is

$$\chi(g) \leq \left\lfloor \frac{1}{2}\left(7 + \sqrt{48g+1}\right) \right\rfloor$$

as long as $g \geq 1$.

Here is how to read this formula. The Greek letter chi (χ) is the chromatic number of the surface—the least number of colors that it will take to color any map on the surface. Thus $\chi(g)$ is the number of colors that it will take to color any map on a surface that consists of the sphere with g handles. Next, the symbols $\lfloor \ \rfloor$ stand for the "greatest integer function." For example $\lfloor \frac{9}{2} \rfloor = 4$ just because the greatest integer in the number "four and one-half" is 4. Also $\lfloor \pi \rfloor = 3$ because $\pi = 3.14159\ldots$ and the greatest integer in the number pi is 3.

are two countries of each color, the figure gives only the top view: Those pieces are connected on the bottom side of the torus.

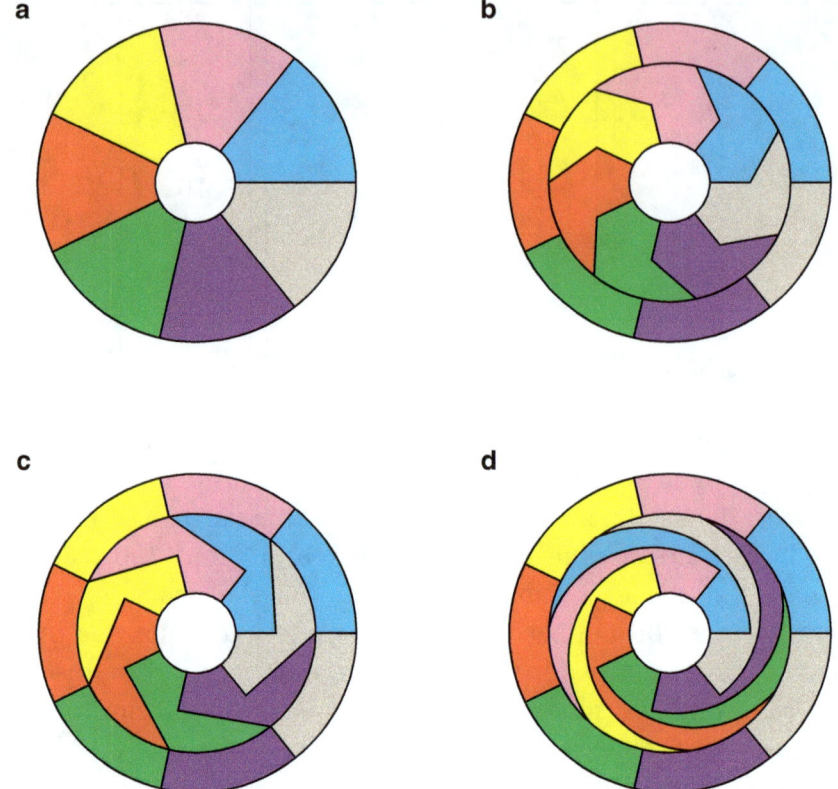

Figure 1.10 A 7-coloring on the torus (**a**) Top view of seven countries wrapping around the torus (**b**) Top view. Part of each country has been twisted counter-clockwise. Pieces of the same color connect on the bottom of the torus (**c**) Top view. Each country has been twisted more. Pieces of the same color connect on the bottom of the torus (**d**) Top view. Each country has been twisted a lot. Pieces of the same color connect on the bottom of the torus.

Since every one of the seven countries in Fig. 1.10d touches all of the other countries, they all must have different colors! This is a map on the torus that requires seven colors; it shows that Heawood's estimate is sharp for this surface.

You may think that we may add more countries to the configuration in Fig. 1.10d and obtain a map with any number of countries in which all the

1.2. Kempe, Heawood, and the Chromatic Number

countries touch each other. But this is not so! Try adding just one country, and you will find that the eighth country cannot touch all the others.

Heawood was able to prove that, for a surface with genus greater than zero, his formula gives an upper bound for the chromatic number of the surface, but he was unable to prove that his formula gives a lower bound. In other words, Heawood told us how many colors would be enough, but he wasn't sure we would need them all.

You will note that if you plug genus $g = 0$ into Heawood's formula in the footnote, it gives $\chi(g) \leq 4$. Had Heawood's proof of the upper bound for the chromatic number been valid when $g = 0$, he would then have solved the four-color problem. But his proof only works when $g \geq 1$. Thus Heawood was unable to decide whether the chromatic number of the sphere is 4 or 5.

Heawood understood, and was able to prove, that the sharp chromatic number for the torus is 7. But he was unable to fully treat the case of higher genus. For the torus with two handles (genus 2), Heawood's formula gives an estimate of 8. Is that the best number? Is there a map on the double torus that actually *requires* eight colors? And so forth: we can ask the same question for every surface of every genus. Heawood could not answer these questions.

The mathematician Tait produced another supposed resolution of the four-color problem in 1880. Julius Petersen pointed out a gap in 1891. Another instance of 11 years lapsing before the error was found!

Percy Heawood

Photo courtesy of Robin J. Wilson

Percy Heawood

Percy John Heawood (1861–1955) was the eldest son of an Anglican priest, and he married the daughter of another clergyman. Heawood was devoutly religious and a classical scholar as well as a mathematician. Throughout his life he was a contributor to theological as well as mathematical journals.

Heawood was educated at Oxford. He became a Lecturer in Mathematics at Durham in 1887. His career at Durham has been described as "spectacularly prominent." He loved committees and administrative tasks, eventually serving as vice-chancellor of the university. A further passion was the saving of Durham Castle, an historical treasure that would have slid into the sea were it not for the efforts of Heawood to raise the money to save it. For this latter work, he was awarded an Honorary Doctorate of Civil Laws by Durham University and created an Officer of the Order of the British Empire.

A total eccentric, Heawood had an enormous moustache, dressed oddly, and brought his dog to his lectures. He set his watch once a year on Christmas day, and rather than adjust it during the interval, he simply tracked how slow or fast it ran. Thus, quite logically, he is quoted as saying "No, it's not two hours fast, it's ten hours slow."

1.3 Heawood's Estimate Confirmed

The four-color problem has a long and curious history. The American mathematician G.D. Birkhoff (1884–1944) did foundational work on the problem that allowed Philip Franklin (1898–1965) in 1922 to prove that the four-color conjecture is true for maps with at most 25 countries. Heinrich Heesch made seminal contributions to the program, and in fact introduced the techniques of reducibility and discharging which were to prove important years later in developing our understanding of the problem. Walter Stromquist proved in his 1975 Harvard Ph.D. thesis [Str 75(a)] that, for any map with 100 or fewer countries, four colors will always suffice. See also [Str 75(b)]. What is particularly baffling is that Ringel and Youngs were able to prove in 1970 that all of Heawood's estimates, for the chromatic number of any surface, are sharp (but remember Heawood's estimate does not apply to the sphere). So the chromatic number of a torus is indeed 7. The chromatic number of a "super torus" with two holes is 8, and so forth. But the Ringel/Youngs proof does not apply to the sphere. They could not improve on Heawood's result that five colors will always suffice.

1.4 Appel, Haken, and a Computer-Aided Proof

Then, in 1974, there was blockbuster news. Using 1,200 hours of computer time on the University of Illinois supercomputer, Kenneth Appel and Wolfgang Haken showed that in fact four colors will always work to color any map on the sphere. Their technique is to identify 633 fundamental configurations

1.4. Appel, Haken, and a Computer-Aided Proof

of maps (to which all others can be reduced) and to prove that each of them is reducible in the sense of Heesch. But the number of "fundamental examples" was very large, and the number of reductions required was beyond the ability of any human to count. And the reasoning is extremely intricate and complicated. Enter the computer.

In those days computing time was expensive and not readily available, and Appel and Haken certainly could not get a 1,200-hour contiguous time slice (50 days) for their work. So the calculations were done late at night, "off the record," during various down times. In fact, Appel and Haken did not know for certain whether the calculation would ever cease. Their point of view was this:

- If the computer finally stopped, then it will have checked all the cases and the four-color problem was solved.
- If the computer never stopped, then they could draw no conclusion.

Well, the computer stopped. But the level of discussion and gossip and disagreement in the mathematical community did not. Was this really a proof? The computer had performed tens of millions of calculations. Nobody could ever check them all. In 1974 our concept of a proof was etched in stone after 2,500 years of development: a proof was a logical sequence of steps that one human being recorded on a piece of paper so that another human being could check them. Some proofs were quite long and difficult (for example, the proof of the celebrated Atiyah–Singer Index Theorem from the mid-1960s was four long papers in the *Annals of Mathematics* and used a great deal of mathematical machinery from other sources). But, nonetheless, they were always checkable by a person or persons. The new "proof" of Appel and Haken was something else again. It required one to place a certain faith in the computer, in the integrity of its central processing unit, and in the algorithm being used. The old IBM adage "Garbage In, Garbage Out" was in the forefront of everyone's mind.

But now the plot thickens. Because in 1975 a mistake was found in the Appel/Haken proof. Specifically, there was something amiss with the algorithm that Appel and Haken fed into the computer. It was later repaired. Their paper [AH 76] was published in 1976. The four-color problem was declared to have been solved.

In fact Oscar Lanford pointed out that in order to justify a computer calculation as part of a proof, one must not only prove that the program is correct, but one must understand how the computer rounds numbers, how the operating system functions, and how the time-sharing system works. It would also help to know how the CPU (central processing unit, or "chip") stores data. There is no evidence thus far that those who use computers heavily in their proofs go beyond the first requirement in this list.

In a 1986 article [AH 86] in the *Mathematical Intelligencer*, Appel and Haken point out that the reader of their seminal 1976 article [AH 76] must face the following:

> 50 pages containing text and diagrams, 85 pages filled with almost 2,500 additional diagrams, and 400 microfiche pages that contain further diagrams and thousands of individual verifications of claims made in the 24 lemmas in the main section of the text.

They go on to acknowledge that their proof required more than 1,200 hours of computer time, and that there were certainly typographical and copying errors in the work. But they offer the reassurance that readers will understand "why the type of errors that crop up in the details do not affect the robustness of the proof." Several errors found subsequent to publication, they record, were "repaired within 2 weeks." By 1981 "about 40 %" of 400 key pages had been independently checked, and 15 errors corrected, by Ulrich Schmidt. In fact, for many years after that, the University of Illinois Mathematics Department had a postmark that appeared on every outgoing letter from their department. It read:

> FOUR COLORS SUFFICE

Quite a triumph for Appel and Haken and their supercomputer.

The mathematical community was slow to accept the new kind of proof that Appel and Haken offered. After the dust cleared and the seminal paper was published, Appel and Haken were invited to give a plenary address at a national meeting of the American Mathematical Society. They presented their ideas to a packed auditorium, and the atmosphere in the room was one of stony silence. At the end of the talk, there was no applause. Mathematicians were not sure of the validity or the value of what they were hearing. They feared that this was a threat to the intellectual monument that had

1.4. Appel, Haken, and a Computer-Aided Proof

been built for 2,500 years. They could not decide about the meaning or the robustness of a computer proof. How does one check such a proof? How does one, with confidence, declare it to be correct?

But Appel and Haken stood their ground. They felt that they had something new and valuable to offer, and they had the patience and the tenacity to explain it to the world. After all, they were both well-established mathematicians of considerable credibility. They deserved to be listened to. What they had done was valid and certifiable. Only the world needed to learn to understand and appreciate it.

But it seems as though there is always trouble in paradise. According to one authority, who prefers to remain nameless, errors continued to be discovered in the Appel/Haken proof. There was considerable confidence that any error that was found could be fixed. And invariably the errors *were* fixed. But the stream of errors never seemed to cease. So is the Appel/Haken work really a proof? Is a proof supposed to be some organic mass that is never quite right, that is constantly being fixed? Not according to the paradigm set down by Euclid 2,500 years ago!

In mathematics there is hardly anything more reassuring than another independent proof. Paul Seymour and his group at Princeton University found another way to attack the problem. In fact they found a new algorithm that seems to be more stable. They also needed to rely on computer assistance. But by the time they did their work computers were *much*, much faster. So they required much less computer time. In any event, this paper appeared in 1994 (see [Sey 95]). It has stood the test of 16 years, with no errors found. And in fact in 2004 G. Gonthier used a computer-driven "mathematical assistant" to check the 1994 proof.[3]

Nonetheless, nobody can check the Seymour proof, in the traditional sense of "check." The computer is still performing many millions of calculations, and it is not humanly possible to do so much checking by hand—nor would anyone want to! The fact is, however, that over the course of 20 years, from the time of the original Appel/Haken proof to the advent of the Seymour proof, we, as a community of scholars, have become much more comfortable with computer-assisted proofs. There are still doubts and concerns, but this new methodology has become part of the furniture. There are enough

---

[3]This process has in fact become an entire industry. A number of significant results have been computer verified; among these is the prime number theorem (about the distribution of primes).

computer-aided proofs around (some of them will be discussed in this book) that a broad cross-section of the community has come to accept them—or at least to tolerate them.

It is still the case that mathematicians are most familiar with, and most comfortable with, a traditional, self-contained proof that consists of a sequence of logical steps recorded on a piece of paper. We still hope that some day there will be such a proof of the four-color theorem. After all, it is only a traditional, Euclidean-style proof that offers the understanding, the insight, and the sense of completion that all scholars seek. For now we live with the computer-aided proof of the four-color theorem.

With the hindsight of 40 years, we can be philosophical about the Appel/Haken proof of the four-color theorem. What is disturbing about it is that this proof lacks the sense of *closure* that we ordinarily associate with mathematical proof. Traditionally, we invest several hours—or perhaps several days or weeks—absorbing and internalizing a new mathematical proof. Our goal in the process is to *learn something*.[4] The end result is new understanding, and a definitive feeling that something has been internalized and accomplished. These new computer proofs do not offer that reward.

The real schism is, as Robert Strichartz [Stri 06] put it, between the quest for knowledge and the quest for certainty. Mathematics has traditionally prided itself on the unshakeable absoluteness of its results. This is the value of our method of proof as established by Euclid. But there are so many new developments that have undercut the foundations of the traditional value system. Also, there are new societal needs: theoretical computer science and engineering and even modern applied mathematics require certain pieces of information and certain techniques. The need for a workable device often far exceeds the need to be *certain* that the technique can stand up to the rigorous rules of logic. The result may be that we will reevaluate the foundations of our subject. The way that mathematics is practiced in the year 2100 may be quite different from the way that it is practiced today.

---

[4] And of course the admittedly selfish motivation is to learn some new techniques that will help the reader with his/her own problems.

A Look Back

The innovative use of computers by Appel and Haken to create a proof of the four color theorem is a mathematical milestone. It was a quite original approach to an age-old problem. And it created quite a stir. Today computer-aided proofs are fairly common, although not always universally accepted. When the theory of wavelets (see our Chap. 8) was young, a key step in its development depended on a computer calculation. Now there is a workaround so that the computer is no longer necessary.

But, among the experts, there have been doubts about the validity of the Appel/Haken proof. Appel and Haken themselves have said:

> The reader of their seminal 1976 article must face
>
> - 50 pages containing text and diagrams,
> - 85 pages filled with almost 2,500 additional diagrams,
> - 400 microfiche pages containing further diagrams and thousands of individual verifications of claims that were made in the 24 lemmas in the main section of the text

They go on to acknowledge that their proof required more than 1,200 hours of computer time, and that there were typographical and copying errors in the work.

In fact, errors in the Appel/Haken work have been discovered regularly over the past 37 years. There have been plenty of reassurances that the method is stable and everything is all right, but this is not a satisfactory situation for mathematics. Enter Paul Seymour and his team.

Seymour is a Professor at Princeton University and a powerful mathematician. He, together with Neil Robertson, Daniel P. Sanders, and Robin Thomas, took it upon themselves to come up with a more reliable proof of the 4-color theorem. Their proof still uses the computer, but in a much more reliable fashion. In their own words[5]:

> Unfortunately, the proof by Appel and Haken (briefly, A&H) has not been fully accepted. There has remained a certain amount of doubt about its validity, basically for two reasons:

---

[5]See [RSST 96].

(i) part of the A&H proof uses a computer, and cannot be verified by hand;

(ii) even the part of the proof that is supposed to be checked by hand is extraordinarily complicated and tedious, and as far as we know, no one has made a complete independent check of it.

Reason **(i)** may be a necessary evil, but reason **(ii)** is more disturbing, particularly since the 4CT (4-color theorem) has a history of incorrect proofs. So in 1993, mainly for our own peace of mind, we resolved to convince ourselves somehow that the 4CT really was true. We began by trying to read the A&H proof, but very soon gave this up. To check that the members of their unavoidable set were all reducible would require a considerable amount of programming, and also would require us to input by hand into the computer descriptions of 1,478 graphs; and this was not even the part of their proof that was most controversial. We decided it would be easier, and more fun, to make up our own proof, using the same general approach as A&H. So we did; it was a year's work, but we were able to convince ourselves that the 4CT is true and provable by this approach.

In addition, our proof turned out to be simpler than that of A&H in several respects. The basic idea of the proof is the same as that of A&H. We exhibit a set of "configurations"; in our case there are 633 of them. We prove that none of these configurations can appear in a minimal counterexample to the 4CT, because if one appeared, it could be replaced by something smaller, to make a smaller counterexample to the 4CT (this is called proving "reducibility"; here we are doing exactly what A&H and several other authors did. ...But every minimal counterexample is an "internally 6-connected triangulation," and in the second part of the proof we prove that at least one of the 633 configurations appears in every internally 6-connected triangulation. (This is called proving "unavoidability.") Consequently, there is no minimal counterexample, and so the 4CT is true. Where our method differs from A&H is in how we prove unavoidability.

Certainly Appel and Haken deserve credit for the ingenuity, determination, and originality. But it is always heartening to have a second, independent proof of any new result. Seymour and his team have provided that for us.

REFERENCES AND FURTHER READING

[AH 76]   Appel, K.I., Haken, W.: A proof of the four color theorem. Discrete Mathematics **16**, 179–180 (1976)

[AH 86]   Appel, K., Haken, W.: The four color proof suffices. Mathematical Intelligencer **8**, 10–20 (1986)

[Dir 63] Dirac, G.A., Heawood, Percy John Heawood: Journal of the London Mathematical Society **38**, 263–277 (1963)

[RSST 96] Robertson, N., Sanders, D.P., Seymour, P.D., Thomas, R.: A new proof of the four-colour theorem. Electronic Research Announcements of the American Mathematical Society **2**, 17–25 (1996)

[Sey 95] Seymour, P.D.:Progress on the four-color theorem. In: Proceedings of the International Congress of Mathematicians (Zürich, 1994), pp. 183–195. Birkhäuser, Basel (1995)

[Stri 06] Strichartz, R.S.: Letter to the editor. Notices of the American Mathematical Society **53**, 406 (2006)

[Str 75(a)] Stromquist, W.R.: Some aspects of the four color problem. Ph.D. thesis, Harvard University (1975)

[Str 75(b)] Stromquist, W.R.: The four-color theorem for small maps. Journal of Combinatorial Theory, Series B **19**, 256–268 (1975)

[Wil 02] Wilson, R.J.: Four Colors Suffice. Princeton University Press, Princeton (2002)

Chapter 2
The Mathematics of Finance

2.1 Ancient Mathematics of Finance

Among the difficulties faced by the earliest emerging civilizations was the need for record keeping. Because we have a written language, record keeping is easy enough for us, but the earliest civilizations did not have that tool. Archaeological evidence indicates that the invention of written language was contemporaneous with the development of civilization. In fact, it is hypothesized that the creation of numerical notation for record keeping was the first step in the process of developing written language (see [Sch 94]).

Middle Eastern artifacts in the form of clay tokens that are believed to have represented units of grain have been dated to as early as 8000 BCE. These tokens are believed to have been used to record amounts of stored grain, and they are believed to be the first mechanism used for that record-keeping task. Figure 2.1 illustrates ancient clay accounting tokens—not the earliest but still dating from before 3100 BCE.

After an amount of grain has been recorded, the next step is to make a record of the ownership of the grain represented by the tokens. Here the archaeological evidence shows that the ownership record was maintained by enclosing the tokens in a clay envelope marked by the owner's seal. Figure 2.2 illustrates a clay envelope and accounting tokens that also date from before 3100 BCE.

A clay envelope is not transparent, so once tokens have been sealed inside the envelope, the record, while safe, is inaccessible. The impossibility of seeing what's inside a clay envelope was overcome by the innovation of making

22   Chapter 2. The Mathematics of Finance

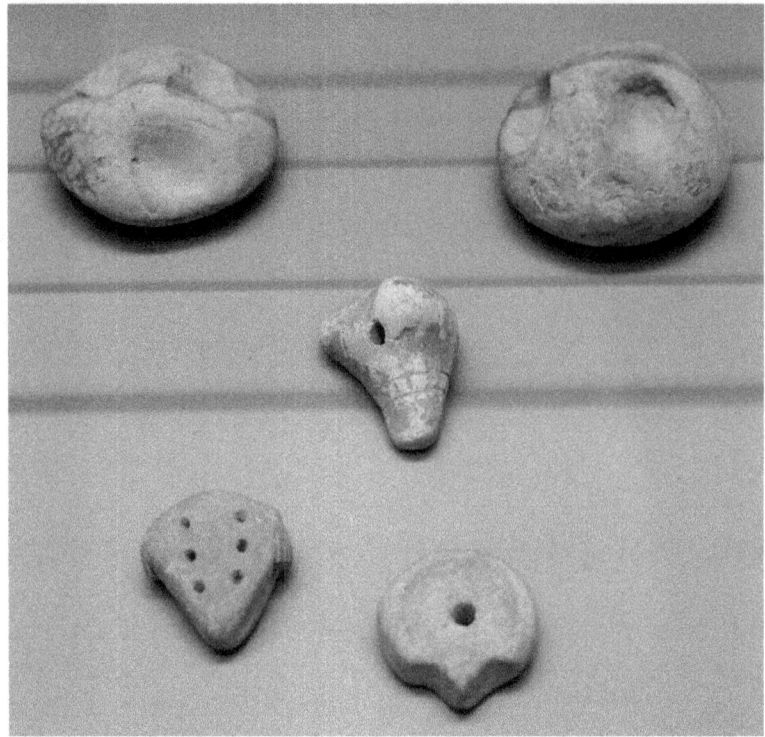

Musée du Louvre, Département des Antiquités Orientales
© Marie-Lan Nguyen / Wikimedia Commons

Figure 2.1 Clay accounting tokens. Susa, Uruk period (4000–3100 BCE).

impressions on the outside of the envelope using the tokens that were to go inside the envelope. In this way, the contents of the envelope could be known without breaking the envelope.

It is apparent to us that, once the contents and ownership are represented on the surface of the envelope, the contents themselves are redundant. The surface of the enclosure itself and the impressed markings thereon provide all the needed information. The next innovation is also apparent to us: Don't bother with the tokens inside the envelope and don't even bother making a spherically shaped clay envelope; simply make the needed marks on a flat surface.

Finally, about 3100 BCE that ultimate step was taken: People began to use a pointed stylus to incise pictures of the tokens in clay tablets instead of impressing the tokens themselves. It is at this point that we can say that a

2.1. Ancient Mathematics of Finance

system of writing had been invented. Figure 2.3 illustrates a clay accounting tablet that dates from 3100–2850 BCE.[1]

Musée du Louvre
© Marie-Lan Nguyen / Wikimedia Commons

Figure 2.2 Clay envelope and accounting tokens. Susa, Uruk period (4000–3100 BCE).

---

[1]Two things that had puzzled archaeologists in the mid-twentieth century were: (1) what was the significance of various bits of clay like those in Fig. 2.1 that were found at nearly every Middle Eastern archaeological site, and (2) why did many early pictographs not look like the things they represented. It was Denise Schmandt-Besserat who recognized that the bits of clay were accounting tokens and that the pictographs represented the tokens.

Musée du Louvre, Département des Antiquités Orientales
Mbzt / Wikimedia Commons

Figure 2.3 Clay accounting tablet. Susa, period III (3100–2850 BCE).

2.2 Loans and Charging Interest

Civilization requires the division of labor and the transfer of goods among various specialized groups of workers. In particular, the producers of food need to be taxed so that some of their production is available to an organizing authority. Not surprisingly, even in ancient times the payment of some taxes was delayed. The records of those delayed payments are the earliest recorded debts. So it is within a few hundred years of the emergence of written language that written records of debt appeared.

There are also early records (circa 2400 BCE) of debts owed by one individual to another. By 1800 BCE there are records of loans requiring the payment of interest. An example of such an early record (see [Sim 78]) reads as follows:

> One and one-sixth shekels silver, to which the standard interest is to be added, Ilshu-bani, the son of Nabi-ilishu, received from Shamash and from Sin-tajjar. At harvest time he will repay the silver and the interest. Before five witnesses. In month seven of the year that Apil-Sin built the temple of Inanna of Elip.

The preceding loan contract is between the borrower Ilshu-bani and the lender Sin-tajjar. Shamash was the name of a god. The shekel is a unit

2.2. Loans and Charging Interest

of measurement equal to approximately 8 grams (although it later became a unit of currency). That there can be a reference to a "standard interest" tells us that the practice of charging interest was commonplace. In fact, the standard interest rate is known to have been 20% for silver and $33\frac{1}{3}$% for grain. Many loans were actually for short time periods and would still be paid with the 20% interest, independent of the time period. Thus it may not be fully appropriate to project our notion of annual interest onto the thinking of those ancient Babylonians making a contract. Nonetheless, it was common to pay 1 shekel interest per lunar month on each 1 mina (which equals 60 shekels) that was borrowed. Since a year usually contains 12 lunar months (but can have 13) we arrive at an interest rate of approximately 20% per annum.

Even though charging interest on loans had been a well-established practice among the Babylonians and had spread over most of the Near East, still the practice of charging interest later came to be seen as, at best, disreputable. For instance, Aristotle said (see [Bar 84]):

> The most hated sort [of wealth getting], and with the greatest reason, is usury, which makes a gain out of money itself, and not from the natural object of it. For money was intended to be used in exchange, but not to increase at interest. And this term interest, which means the birth of money from money, is applied to the breeding of money because the offspring resembles the parent. That is why of all modes of getting wealth this is the most unnatural.

Later the Catholic Church took a dim view of charging interest. Some of the Church's actions against charging interest are as follows: The Council of Nicaea in 325 banned usury among clerics, and the First Council of Carthage in 345 and the Council of Aix in 789 declared it to be reprehensible for laymen to make money by lending with interest. The canonical laws of the Middle Ages absolutely forbade the practice of lending with interest. The Third Council of the Lateran in 1179 and the Second Council of Lyons in 1274 condemn usurers. In the Council of Vienne in the year 1311, it was declared that anyone maintaining that there was no sin in the practice of demanding interest should be punished as a heretic.

The Church's censure notwithstanding, since loans are essential for commerce, people came up with stratagems that could be used to evade the Church's prohibition on charging interest. One common dodge was for the interest to be hidden within the premium (*agio*) charged by bankers for currency conversion. A more blatant evasion was to characterize payments to lenders as discretionary gifts.

2.3 Compound Interest

Compound interest is based on the idea that, after a specific time interval, the outstanding interest also becomes part of the loan and thus from that time on interest must be paid on the interest. In ancient Babylonia most loans were extended for time periods of a year or less, so compound interest was not relevant. Even so, the idea behind compound interest was understood. Our evidence for this is a royal inscription describing a conflict dating to 2400 BCE (see [Coo 86]). The inscription tells us that a portion of the grain harvested from a certain disputed property—1 guru of grain we are told—was to be paid by one city–state to another. But the payment was not made, and after several decades the 1 guru of grain together with its interest had increased to 8.64 million gurus (in modern units about 4.5 trillion liters).

If we apply the principle of compound interest to the above "loan" of 1 guru of grain using the interest rate of $33\frac{1}{3}\%$ that the Babylonians applied to loans of grain, then we see that, after 1 year, $\frac{4}{3} \approx 1.33$ guru of grain is owed. After that the debt grows each year by multiplying by $\frac{4}{3}$. After 2 years, the debt becomes $\frac{4}{3} \times \frac{4}{3} = \frac{16}{9} \approx 1.78$ guru of grain. Continuing in this way, we obtain the following table that shows not only is the amount owed increasing, but the rate of growth of the amount owed is also increasing.

Years	Grain owed (gurus)
1	1.33
2	1.78
3	2.37
4	3.16
5	4.21
10	17.76

Years	Grain owed (gurus)
20	315.34
30	5,599
40	99,437
50	1,765,780
55	7,440,986
56	9,921,315

2.3. Compound Interest

Fibonacci: Leonardo of Pisa

From an engraving by Pelle

Fibonacci

Since the 1970s there has been easy availability of handheld calculators capable of doing many financial calculations which, if done by hand, would be quite laborious. We now take for granted the calculation of compound interest and the calculation of the payments required on a car loan or a mortgage. Centuries before the computer revolution, there was another revolution in computation that allowed the pencil and paper financial calculations needed for commerce. One of the important pioneers in introducing this revolutionary mathematical notation and technique was Leonardo of Pisa (1170–1250), most often referred to as Fibonacci.

Figure 2.4 Location of Béjaïa.

Fibonacci and his family were part of the growing commercial community that arose after the Dark Ages. The Italian city of Pisa had established the colony of Bugia in North Africa (now Béjaïa in Algeria—see Fig. 2.4). Fibonacci's father was an administrative official in that colony and Fibonacci was brought to the colony at his father's request and received training in Arabic mathematical methods.[2] Fibonacci traveled extensively in the Mediterranean. It is believed that he was earning his living as a merchant, but he

---

[2]At that time the Arabs were in many ways on the cutting edge of mathematical development. A number of basic ideas of algebra, and also the arabic system of numerals that we use today, were generated by medieval Arabs. The Arabs learned some of their

was also—and more importantly for us—pursuing mathematical knowledge wherever he went. When he returned from his travels, he wrote the book, *Liber Abaci*.[3] That book is the reason Fibonacci is remembered, while the other merchants of those days have been forgotten.

Liber Abaci was published in 1202 and revised in 1228. Because of its antiquity, copies of *Liber Abaci* were necessarily made by hand. The oldest surviving version dates to the 1290s, omits the preliminary material explaining Arabic numerals and mathematical operations, and is written in Italian rather than Latin. *Liber Abaci* is believed to contain the earliest Arabic numeral multiplication table in Western mathematics.

The specific part of *Liber Abaci* with which modern mathematicians are most familiar is the sequence

$$1, 1, 2, 3, 5, 8, 13, 21, 34, 55, \ldots$$

in which each number, after the first two, is the sum of the preceding two numbers. Now known as the Fibonacci sequence, this sequence provides the solution to a problem in *Liber Abaci* concerning the breeding of rabbits. In fact, much of *Liber Abaci* is devoted to practical problems of finance and commerce. In 1241, the Republic of Pisa granted Fibonacci a pension for "educating its citizens and for his painstaking, dedicated service."

2.4 Continuously Compounded Interest

In our modern world, a financial institution would never wait a full year before compounding the interest on a loan (or on a savings account). In fact, computers allow financial institutions to use compounding periods as short as they wish. Figure 2.5 shows the effect of more frequent compounding. The simple interest rate considered is 64% (per year), a large value chosen so that the effects of compounding will be visible in the graph. The principal is $100, and after 1 year, $100 at 64% simple interest grows to $164, as shown by the bottom graph in each part of Fig. 2.5. Figure 2.5a shows that, by compounding in the middle of the year, the total interest on $100 after 1 year increases from $64 to $74.25; Fig. 2.5b shows that, if the interest is

---

ideas from the East Indians. The intercourse between the Arabs and the Indians came about because of medical needs—with the best doctors traveled the best ideas.

[3]This famous text is about arithmetic, pure and simple. That was a cutting-edge topic at the time. In fact *Liber Abaci* was one of the very first Western books to explain the Hindu-Arabic numerals that we use today.

2.4. Continuously Compounded Interest

compounded after each quarter year, then the total interest becomes $81.06; and Fig. 2.5c shows that, if the interest is compounded after each eighth of a year, then the total interest becomes $85.09.

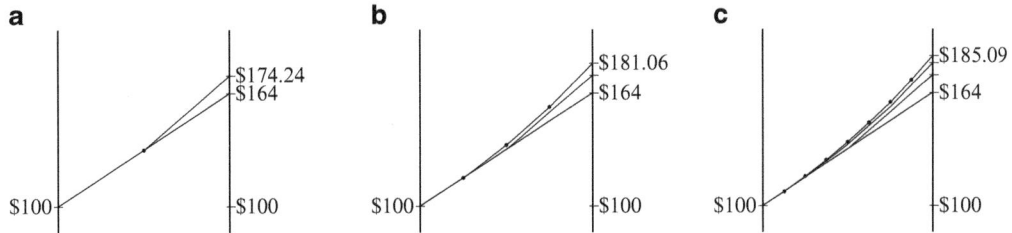

Figure 2.5 Compounding interest. (a) Compounding twice per year. (b) Compounding four times per year. (c) Compounding eight times per year.

Just to be clear about this: When we compound interest in the middle of the year, we calculate half the interest (or 32% applied to the principal) on June 30. Then we calculate the other half of the interest (or 32% applied to the principal plus the first quantity of interest) on December 31. When we compound interest four times a year, we calculate one-fourth of the interest (or 16% applied to the principal) on March 31. Then we calculate another one-fourth of the interest (or 16% applied to the principal plus the first quantity of interest) on June 30. Then we calculate another one-fourth of the interest (or 16% applied to the principal plus the first two quantities of interest) on September 30. Finally, we calculate the last one-fourth of the interest (or 16% applied to the principal plus the first three quantities of interest) on December 31. Other forms of compound interest are calculated in the same way.

So easy have the computations become that it is now commonplace for interest to be compounded continuously. By continuous compounding is meant the limiting value as the number of compounding periods approaches infinity. Figure 2.6 illustrates the effect of continuous compounding. Each graph in Fig. 2.6 shows twice as many compounding periods as the graph immediately below it. As more and more compounding periods are used, the graphs approach the limiting graph that shows continuous compounding and of course is the highest of all the graphs in Fig. 2.6. Even though each increase in the number of compounding periods leads to greater interest, the total interest does not increase to infinity, but instead approaches $89.65.

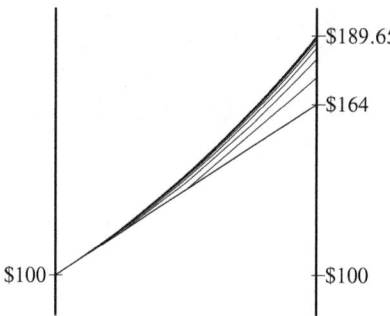

Figure 2.6 The transition from simple interest to continuously compounded interest.

Figure 2.7 compares compound interest to simple interest on a principal amount of $100 at a nominal rate of 10% per annum. The compound interest is computed on the basis of continuous compounding. After 10 years, the compound interest is noticeably more than simple interest, but after 20 years the difference is large and after 30 years it is huge.

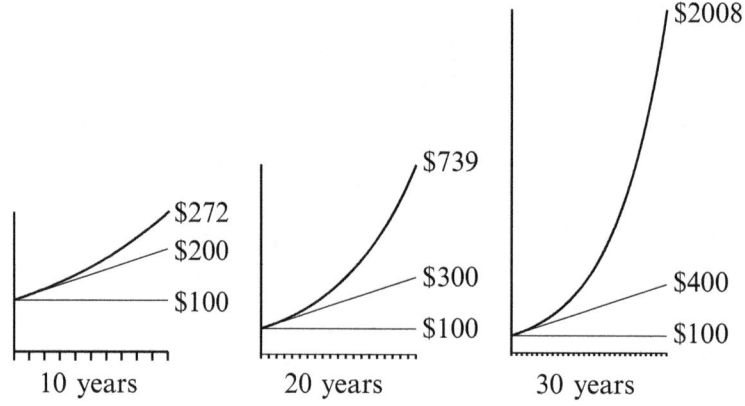

Figure 2.7 Comparison of compound interest and simple interest.

2.5 Raising Capital: Stocks and Bonds

Government endeavors and commercial projects typically require capital far beyond the means of one individual. One way in which a large amount of capital can be raised is by borrowing from many individuals, a process carried

2.5. Raising Capital: Stocks and Bonds

out by issuing bonds. The Italian city–states of Venice, Florence, and Genoa were early leaders in this practice, utilizing both voluntary and forced loans as early as the eleventh century. Some such bonds were non-transferable, but in other cases the bonds were transferable and a secondary market for them consequently emerged.

For a commercial enterprise to raise a large amount of capital, an appropriate and effective legal structure must exist. A "joint contract" among individuals is not well suited to this purpose, because then each individual is fully responsible for the enterprise's liabilities. Even if a partnership has been arranged in such a way that individual liability is limited, there remains the serious problem of resolving how a partner can recover his capital and withdraw from the partnership. One quite successful solution to the problem of raising capital for commercial enterprise is the joint-stock company in which transferable shares representing ownership of a limited liability company are issued. To withdraw and recover his or her capital, an owner can simply sell the shares owned.

The seventeenth century Dutch company Verenigde Nederlandsche Geoctroyeerde Oostindische Compagnie is the prototypical example of the joint-stock company for which a secondary market developed. It is the development of a secondary market in financial instruments that led to, and necessitated, the development of stock exchanges on which shares of companies can be bought and sold. The name "Verenigde Nederlandsche Geoctroyeerde Oostindische Compagnie" translates literally to "United Netherlands Chartered East India Company," but it is commonly referred to as the Dutch East India Company. This company played a crucial role in the history of finance.

The Dutch East India Company was chartered in 1602. The original plan had been to liquidate the company in 1612, but the time required for round trips to Asia makes it obvious to us that a 10 year time frame was far too short; in any event, the Dutch East India Company continued in existence until 1795. The equity capitalization of the company remained essentially fixed throughout the life of the company, but the company also issued bonds. One salubrious side effect of the existence of a secondary market in shares in the Dutch East India Company was that shares owned could then be used as collateral for loans. Since shares could be liquidated easily, the lender was much more secure with shares as collateral as opposed to other less liquid collateral. The effect of using more liquid collateral was a lowering of interest rates, further facilitating commercial enterprise.

In England in 1688 a union of Parliamentarians and an invading army led by the Dutch *stadtholder* William III of Orange-Nassau (William of Orange) overthrew King James II. This is the event called the "Glorious Revolution" in English history. After this revolution England adopted the Dutch model for the joint-stock company, as exemplified by the Dutch East India Company. By using the Dutch model, England gained the same benefits from active secondary markets and lowered interest rates.

The Oldest Live Securities in Modern Capital Markets

Dikes are, and have been, crucial for the existence and survival of the Netherlands. The maintenance of dikes is managed by water boards, of which there may have been as many as 3,500 in the nineteenth century. Mergers have brought the number of extant water boards in the Netherlands down to fewer than 60 at present. Water boards had the power to draft citizens into a "dike army" when needed and were granted taxing authority. It still sometimes happened that expenses would exceed what taxes could raise and bonds would be issued. Some such bonds were issued as perpetuities, i.e., bonds that would not be redeemed, but instead would pay interest forever, or synonymously, in perpetuity.

The management areas of water boards in the Netherlands are dictated by nature and so the water boards are separate from government entities and consequently are often insulated from government upheavals. This independence has led to the perpetuities issued by some water boards continuing to function for centuries. In recent years, at least four seventeenth century bonds issued by the Hoogheemraadschap Lekdijk Bovendams have been presented to the successor organization (namely, the water board of Stichtse Rijnlanden) for payment of interest. The oldest of these bonds dates from 1624. On July 1, 2003, Geert Rouwenhorst, Professor of Finance at Yale University and coeditor of [GR 05], personally collected 26 years back interest on a 1,000 Carolus guilder bond issued at 5% interest on May 15, 1648. The currency in use in 1648, the Carolus guilder worth 20 stuivers ("stuiver" remains a nickname for the 5 euro cent coin in the Netherlands) has been succeeded by the Flemish pound, then the guilder, and finally (so far) the euro. The bond under discussion now pays €11.34 annually, which is the modern equivalent of 25 guilders, the interest rate having been reduced to 2.5% during the eighteenth century.

John Law

From an engraving by Langlois

John Law

John Law (1671–1729) was born into a family of Scottish bankers and goldsmiths. Law initially joined the family business and studied banking especially. After his father died in 1688, Law changed direction and moved to London where he lived extravagantly. Despite being reputed to be brilliant at calculating odds, he lost large sums of money gambling. After killing another man in a duel he was forced to escape to the continent.

Law's contribution to economic theory was the pamphlet "Money and Trade Considered with a Proposal for Supplying the Nation with Money." In modern terms, Law proposed that economic activity could be spurred by increasing the money supply, and that the increasing material production would then be sufficient to prevent inflation. Having left the continent for Scotland, Law unsuccessfully attempted to get the Scots to adopt his proposed monetary policies. But after the union of Scotland and England in 1707, he again had to flee to the continent.

In France, after the death of Louis XIV (1638–1715), Law found fertile ground for his financial innovations. At that time, France was bankrupt— having been drained by continuous warfare. Money was in short supply. The opportunity that presented itself to Law was the conversion of the huge government debt into equity. In 1716, Law established the General Bank (Banque Générale Privé) which developed the use of paper money. But the main tool for the conversion of government debt into equity was to be a large trading company along the lines of the Dutch East India Company and the similar British East India Company. In 1717, Law took over the Company of the West (Compagnie d'Occident), which owned the trading rights to Louisiana (at that time a territory far larger than the present day state of Louisiana);[4] this in exchange for also taking over France's short-term debt. That short-term debt was then converted into long-term debt

---

[4] In fact one of the important events in the development of the United States was the so-called "Louisiana Purchase" by President Thomas Jefferson in 1803. This added considerably to the land mass of the country. The acquisition was celebrated by the 1903 World's Fair held in St. Louis. In fact many of the Washington University campus buildings were originally erected as administration buildings for the fair. The fair introduced to the world, among other things, cotton candy and hot dogs.

at a lower interest rate. The effect was to ease France's debt problem while simultaneously establishing an income stream for the Company of the West.

The Company of the West issued shares and embarked on a series of acquisitions. The resulting company is generally known as the Mississippi Company. Another step in building Law's financial empire was the conversion of the General Bank mentioned above into the Royal Bank. As part of the conversion from General to Royal Bank, the king's Council was entrusted with the power to determine the bank's note issue. The result was the printing of an excess of money, and that served to drive up the price of shares in the Mississippi Company. In 1720, the Royal Bank and the Mississippi Company were united and Law was appointed Controller General of Finances for France.

By 1720, France was awash in liquidity, but investors began to lose confidence in the Mississippi Company. The King of France, who was himself a large shareholder, was one of the first to lose confidence. He sold his entire holdings, and the market value of shares in the Mississippi Company collapsed. Law was dismissed from his post, and he fled France. He died in Venice, a poor man.

2.6 The Standard Model for Stock Prices

As mentioned above, the advent of computers has made possible many computations that, say 100 years ago, a person could only dream of doing. Not surprisingly, it was military needs (such as the calculation of artillery trajectories) that spurred the development of the earliest electromechanical and electronic computers during the 1940s. But it was not too long afterwards that electronic computers entered the civilian realm and major calculations were done to satisfy intellectual curiosity. One such computation was performed by the British statistician Maurice Kendall. Kendall's goal was to find the underlying price cycles that were presumed to exist in stock prices. Instead of finding such underlying price cycles, Kendall concluded that stock prices move randomly (see [Ken 53]). This work was published in 1953, and at that time Kendall's conclusion that stock prices move randomly was difficult to accept. Part of the difficulty was the absence of any theoretical explanation.

2.6. The Standard Model for Stock Prices

About a decade later a theoretical explanation for the random movement of stock prices emerged. The nascent justification was the *efficient market theory* first described in Eugene Fama's 1964 University of Chicago Ph.D. thesis.[5] This efficient market theory postulates that prices on any widely traded asset already reflect all known information. A prototypical example of a widely traded asset would be shares of stock trading on one of the world's major exchanges. Since insider trading and market manipulation via rumor are (now) explicitly illegal, there is reason to believe that the efficient market theory may hold a measure of truth.

If all available information is already reflected in stock prices, then we may wonder what it is that stock prices are doing, since clearly stock prices are neither constant nor steadily increasing or decreasing. One possible explanation for the randomness in stock prices is that, since all known information is already included in the existing price of a stock, the only thing that can contribute to changing the prices is the emergence of previously unknown (and hence unpredictable) information. Tautologically, the unpredictable is unpredictable, so the effect is that stock prices change randomly.

Let's look at randomness as a model of stock price movements. A common example of a random phenomenon is the outcome of a coin toss. Figure 2.8 illustrates the outcome of 500 computer simulated coin tosses. Reading from left to right, each outcome is indicated by a line segment that is above the midline (for a head) or a line segment that is below the midline (for a tail).

Figure 2.8 500 coin tosses.

A random walk is the motion in one dimension generated by taking one step up or down in response to the outcome of a coin toss. Figure 2.9 illustrates the random walk that is based on the coin tosses shown in Fig. 2.8.

The standard model for a stock price is that it is a geometric random walk. In a geometric random walk the outcome of the coin toss determines the ratio of the stock price after the time step to the stock price before the

---

[5]Fama was one of the receipients of the 2013 Nobel Prize in Economics (the others were Richard Shiller and Lars Peter Hansen).

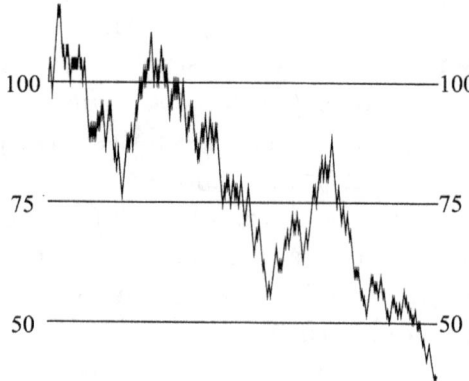

Figure 2.9 Random walked based on 500 coin tosses.

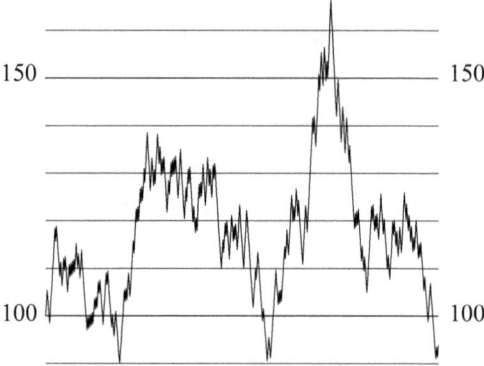

Figure 2.10 Geometric random walk based on 500 coin tosses.

time step. Since it seems to be an observable fact that stock prices rise gradually over time, the standard model also includes a growth factor. To simulate a geometric random walk using the coin-tosses from Fig. 2.8, we multiply or divide the share price by a number greater than 1, the choice of multiplication or division depending on the outcome of the coin toss. Figure 2.10 illustrates the result of this process with a beginning share price of 100. The vertical scales are different in Figs. 2.9 and 2.10, but the ups and downs occur at the same places. The random walk in Fig. 2.9 could become negative if continued, but the geometric random walk can never become negative (because we multiply and divide instead of adding and subtracting).

The geometric random walk model can be converted into a continuous model by letting the length of the time interval decrease to 0 while the step size also decreases appropriately. The resulting continuous process is called geometric Brownian motion. The mathematics involved in describing a geometric Brownian motion is significantly more technical than that required to describe a geometric random walk, so we will be content to model stock prices using a geometric random walk.

Louis Jean-Baptiste Alphonse Bachelier

The French mathematician Louis Bachelier (1870–1946) spent the majority of his career on the faculty at the university in Besançon, the small, isolated capital of the French province Franche-Comté. His relatively obscure academic position contrasts starkly with the fact that his 1900 Ph.D. thesis developed and applied the mathematical theory of Brownian motion 5 years in advance of Einstein's renowned work on the subject.

Various events in Bachelier's life seem to have contributed to the obscurity of his academic career. Bachelier's parents died early, so he was forced to take over the family business. That event hampered his educational plans, so he did not get on the proper track to ultimately obtain a prestigious academic position. The direction in which his own interests led him may have been a problem as well. Bachelier's thesis was concerned with applying probability to the stock market. In 1900, the subject of probability was not considered to be a proper part of mathematics. Further, there is a distinct likelihood that studying something as "sordid" as the stock market would have been considered déclassé. A final problem is that Bachelier may have had a "difficult" personality.

With the advantage of hindsight, Bachelier's work is now recognized as pioneering. It was Bachelier who first used Brownian motion to investigate stock options. His work was not appreciated and not advanced by economists until the 1960s, well after Bachelier's death.

2.7 Parameters in the Standard Model

The general upward drift of the price of a stock is called the *average return*. The average return is an important parameter in the geometric random walk model of a stock price. Likewise, the average return must figure into the continuous model of stock prices based on geometric Brownian motion. The second important parameter describes the magnitude of the fluctuations in the stock price. This parameter is called *volatility*. In fact the average return and the volatility are the only parameters in the standard model for stock prices.

Given their importance, one would most certainly want to know and use actual values for the average return and the volatility. Unfortunately, and disappointingly, statistical theory tells us that the average return cannot be accurately computed from data. The difficulty is that additional data does not narrow the margin of error. Consequently, the average return of a stock is effectively unknowable.

In contrast to the average return, the volatility can be estimated from data, and the accuracy and reliability of the estimate can be quantified in much the way the accuracy and reliability of an opinion poll is quantified.

It is appropriate to note here that, over the long term, actual companies and operating conditions for them do change. As a consequence, the parameters associated with the price of the stock of any specific company will be subject to gradual change. Over a period of decades the change can be enormous. For example, early in the twentieth century the industrialist, investor, and art collector Henry Clay Frick (1849–1919) declared that railroad stocks were the "Rembrandts of investments." Contrary to Frick's pronouncement, in 1970 the Penn Central Railroad[6] was forced into bankruptcy. In 1970, Penn Central had assets of $6.5 billion, making its bankruptcy the largest up until that time. Over the same period Rembrandt's work appreciated in value—and continues to appreciate.

2.8 Derivatives

Financial markets deal in two types of instruments. The first is called an *underlying asset*; these are company shares, bonds, foreign currencies, and commodities. The second type of instrument is a *derivative*. A derivative involves a future payment for, or future delivery of, an underlying asset. If there were no underlying assets, then derivatives would not exist.

One might wonder why derivatives *do* exist. Wouldn't it make sense to simply buy or sell the underlying asset rather than fool around with some future contract?

---

[6]Penn Central was formed by the merger of the Pennsylvania Railroad and the New York Central Railroad, once the two largest and most renowned of U.S. railroads.

2.8. Derivatives

One reason for the existence of future contracts is that, in some instances, it is literally impossible to transfer the underlying asset at a particular time: the underlying asset may still be growing in the field. A second reason for the existence of future contracts is to decrease (or increase) risk. For instance, if you know you will need a particular commodity at a particular time, then you can use a future contract to guarantee the availability of the commodity at a known price.

Derivatives emerge when the underlying assets can be readily traded, are available in sufficient quantity, and are subject to price changes. Since the prices of the various types of underlying assets are variable, the prices of the derivatives based on them also vary. One interesting and important problem is to determine an appropriate price for a derivative.

The simplest derivative is a *forward contract*, or more simply a *forward*, in which two parties agree on a price for the underlying asset at a specific future time. The forward is a very old type of derivative; there exists evidence of forward contracts in early records written in cuneiform script.

A more sophisticated type of derivative is an *option*. An option gives the right, but not the obligation, to buy, or sell, the underlying asset at some future time. Trade in derivatives is facilitated by certain clearing institutions, one of the main functions of which is canceling offsetting claims, a process called *netting*. Clearing institutions may also provide coverage against counterparty risk, that is, the risk of one of the parties to a contract defaulting on the contract.

Tulipmania

The classic example of a speculative bubble is provided by the trade in tulip bulbs in Holland in the 1630s. Tulips are very slow to grow from seeds, so they are usually propagated from bulbs. According to Charles Mackay's *Extraordinary Popular Delusions and the Madness of Crowds*, originally published in the mid 1800s, the price of tulip bulbs reached a phenomenal peak in January 1637. This phenomenal peak was followed by a disastrous collapse in February 1637. In fact, this market frenzy and speculative bubble—called "tulipmania"— would not have involved physically transferring actual tulip bulbs, because bulbs are not uprooted and moved during the winter. What was being traded were derivatives for which the underlying asset was tulips. Nonetheless, when the price of an asset seems to be inordinately high, it is often wondered if people are "buying tulips."

2.9 Pricing a Forward

The simplest derivative is a forward contract in which two parties agree on a price for the underlying asset at a specific future time; that is, two parties, say A and B, agree that at a particular time in the future A will deliver the underlying asset to B and B will pay the agreed upon price to A. Note that the forward contract described in this example is *not* a case of paying now for future delivery of the asset; both the asset and the payment will be exchanged at the specified future time.

To determine a price for the forward, we need to assume that the underlying asset is widely traded, can be bought and sold without transaction costs, does not degrade with time, and can be borrowed without cost: shares of stock in a company come close enough to meeting these requirements.

The possibility of borrowing the underlying asset without cost is a non-obvious assumption, but it is very powerful. This assumption makes it possible to "buy negative shares" through the mechanism of selling shares that have been borrowed; this is called *selling short*.[7] The possibility of selling short makes the circumstances of the buyer and the seller equivalent but with plus and minus signs interchanged.

Let's consider the pricing for a forward contract in which A will deliver 1 share of a stock to B 1 year in the future, at which time B will pay the price of \$$P$ to A. Suppose for simplicity that the current price per share of stock is \$100. Our task is to determine \$$P$.

You might reasonably think that the appropriate price for the forward is closely related to the probability distribution for the price of the underlying stock 1 year in the future. For instance, suppose the stock that is currently priced at \$100 per share will, 1 year in the future, have either the price \$200 or \$50, each with equal likelihood. You might think that the appropriate price to set for the forward contract is $\frac{1}{2} \times \$200 + \frac{1}{2} \times \$50 = \$125$.

---

[7]Those who invest in the stock market today speak of the "bull market" and the "bear market." The bull market is when stock prices go up. You obviously make money in a bull market by buying stock today and selling it later at a higher price. The bear market is different. There you anticipate that the price of a stock will go down. So you sell shares of the stock (which you do not own) today, but do not actually buy the shares until, say, three days later, at which time you deliver them to the buyer. Since the stock market went down, you paid less for the shares than the price that the buyer gave you. So you made money. The just-described process of taking advantage of a falling stock market is called "selling short."

2.9. Pricing a Forward

Even with the specific price information in the preceding paragraph, the price of $125 for 1 share, 1 year in the future, will almost surely be the wrong price. In fact, the correct price for the future has almost nothing to do with what is going to happen to the stock price in the future. It turns out that the correct price in the forward contract is determined by a crucial additional ingredient that we have not yet mentioned; that is, the risk-free interest rate.

The *risk-free interest rate*, which we will denote by r, tells us that if X dollars are invested in a money market account (money market accounts are considered risk-free) at the present time, then the account will grow to $(1+r)X$ dollars in 1 year. Similarly, X dollars can be borrowed, but $(1+r)X$ dollars must be repaid in 1 year. We are assuming that the same interest rate applies to saving and borrowing. We are also simplifying the interest rate model by assuming that the saving and borrowing will be for exactly 1 year. Thus we are using r for the *effective annual interest rate*.

To illustrate the correct way to set the price in a forward contract, let's make the specific assumption that the risk-free interest rate is 10%. In that case, the correct price for the forward contract is the current price per share, $100, plus 1 year's interest on that price, $10 per share; that is, the correct price in the forward contract is $110 per share.

The price of $110 per share for the forward contract is correct even if we are given the information that, 1 year in the future, the share price will be $200 or $50 with equal likelihood. Here's why we are confident that $110 is the correct price:

- If the price of the forward contract were more than $110, then we assert that everyone would want to be a seller of the forward, because a risk-free profit could be made on each share sold forward. To make this risk-free profit, at the start of the contract, say $T = 0$, the seller of the forward, A, would borrow $100 at 10% interest and buy 1 share of stock. Then, 1 year later, at $T = 1$, A would have the 1 share of stock available to fulfill the contract, A could pay back the loan, including the interest, using $110 of the money received under the forward contract, and there would still be money left over as profit. To make the example concrete, observe that if the forward price is $125 per share, then the seller could make $15 profit per share—risk-free.

- If the price of the forward were less than $110, then we assert that everyone would want to be a buyer of the forward, because a risk-free profit could be made on every share bought forward. To make this risk-free profit, at $T = 0$, the buyer of the forward, B, would borrow 1 share of stock, sell it for $100, and deposit the $100 at 10% interest. Then, 1 year later, at $T = 1$, B would use the forward contract to buy 1 share of stock for less than $110, that 1 share would be used to pay back the borrowed share of stock, and there would still be money left over as profit. To make the example concrete, observe that if the forward price is $105 per share, then the buyer of the forward makes $5 profit per share—risk-free.

2.10 Arbitrage

It was possible for us to determine the correct price for a forward contract because we made the fundamental assumption that it is impossible to make a risk-free profit, at least not without tying up some of your own money in the proposed investment. That is, we assumed that there is no arbitrage. *Arbitrage* is trading simultaneously, or nearly so, in different markets to take advantage of price differences. The idea is that the low and high prices may exist simultaneously, but in different markets. Provided the markets are large enough, the arbitrageur can buy at the low price and immediately sell at the high price, making a profit with no risk.

You seldom see opportunities to make risk-free profit, and any such opportunities that do happen to arise in the financial markets tend to disappear quickly, because prices change in response. When we determine the correct price for a derivative, we make the assumption that an arbitrage opportunity is not merely a rare thing, but that it is non-existent. The price for the forward contract that we constructed in the preceding section is thus called the *no-arbitrage price* or *arbitrage-free price*.

An Arbitrage Opportunity: Hoover's Free Travel Offer

In the summer of 1992, Maytag-UK had built up an excess inventory of washing machines and Hoover vacuum cleaners. To clear the backlog, the company started a promotion in which the purchase of £100 or more of merchandise entitled the purchaser to 2 roundtrip tickets to key cities on the continent. The promotion was successful, so the company improved it by offering 2 roundtrip tickets to the US.

Two roundtrip tickets from the UK to the continent were already worth more than the £100 purchase required, but two roundtrip flights to the US were worth a lot more. The public recognized an arbitrage opportunity and bought vacuum cleaners simply to get the airline tickets.

2.11. Call Options

In any contract, there is always counterparty risk, that is, the risk that the other party to the contract will default. Maytag–UK attempted to default on the contract with consumers by not honoring the airline vouchers that consumers had obtained. Litigation resulted, and the company lost tens of millions of pounds sterling and its reputation.

2.11 Call Options

The forward contract, such as the example discussed above (1 share of stock 1 year in the future), is a particularly simple derivative and the appropriate arbitrage-free price for our example was relatively easy to determine. In finance a *call option*, or simply a *call*, is a contract that gives the buyer of the option the right, but not the obligation, to buy the underlying asset, at a specified time, called the *expiry date* or simply the *expiry*, and at a specified price, called the *strike price*. Again, a typical example of an underlying asset would be shares of a stock.

The second type of option is a *put option* which gives the buyer of the option the right, but not the obligation, to sell the underlying asset, at a specified time, the *expiry date*, and at a specified price, the *strike price*. For simplicity, we will consider only call options.

In reality, stock options trade in units that apply to 100 shares of the underlying stock, but instead of exactly modeling reality, we will consider options for 1 share of stock. By considering options on 1 share of stock, we will eliminate a factor of 100 that we would otherwise need to include.

As with derivatives in general, you might think the whole idea of options is silly. If a person thinks the stock will go up, why not simply buy the stock at the current low price and sell it later for more? The point is that options cost far less than the underlying stock, so a person may be able to capture a large profit from an upward movement in price without investing as much money.

A *European call option* is an option to buy the underlying asset that can only be exercised at the expiry date, while an *American call option* allows the purchase of the underlying asset at the strike price at any time up to and including the expiry date. Our goal is to compute the no-arbitrage price for an option. Because a European option can be exercised at only one time, the explanation of the no-arbitrage price will be simpler if we limit our attention to European options, and we shall do so. It is reasonable to assert that the no-arbitrage price is the correct price for an option.

Alfred Winslow Jones

The first modern hedge fund, A.W. Jones & Co., was the 1949 creation of Alfred Winslow Jones (1900–1989). Jones was born in Australia, grew up in the United States, and obtained his undergraduate education at Harvard. After working various jobs unrelated to finance, he enrolled in Columbia University where he earned a doctorate in sociology in 1941.

In the 1940s, Jones worked for *Fortune* magazine, but he wrote mainly on non-financial topics. However, for the March 1949 issue of *Fortune*, Jones wrote an article entitled "Forecasting fashions" about stock market forecasting and forecasters. Apparently Jones learned quite a bit while doing research for this article, because 2 months before it appeared, he formed an investment partnership. Jones directed the investment strategy for that partnership with the goals of decreasing risk and increasing returns by employing short selling, options, and leverage.[8]

"Hedge" is a gambling term describing betting both for and against an outcome. A hedged investment is one in which a profit will be made whether the asset goes up or down. For example, if the market is valuing a call option incorrectly, say the price is too low, then you could short the stock and simultaneously buy the option. Of course, there is the difficulty of knowing what the price of the call option should be. That question will be the main topic of the remainder of this chapter.

A.W. Jones & Co. was a spectacularly successful hedge fund. The company made large profits, apparently with little or no risk. On the other hand, not every company that has called itself a hedge fund has been successful. For example, the ironically named hedge fund Long-Term Capital Management opened in 1994, performed very well for a couple of years, but lost heavily in 1998. Because it was highly leveraged, the company owed a vast amount of money to many banks, so much money that the failure of the company would likely trigger major bank failures. The New York Federal Reserve Bank was forced by those circumstances to orchestrate a bailout—though government money was not used. Long-Term Capital Management finally dissolved in 2000 [Low 00].

---

[8] "Leverage" is obtained by borrowing money to invest more than could be done otherwise.

2.12 Value of a Call Option at Expiry

At the expiry date of a European option, the decision to be made by the holder of that option is whether to use the option to buy the stock at the strike price. The correct decision at that time may be clear, but we state it explicitly here:

- If at expiry the price of the underlying stock is higher than the strike price of the option, then the option should be used to buy the stock at the strike price. Note that, in this situation, an immediate profit can be made by buying the stock at the strike price, then selling it at the market price. We also conclude that when the market price at expiry exceeds the strike price, then the value of the option at expiry is the market price minus the strike price.

- If at expiry the price of the underlying stock is lower than the strike price, then the option should not be used, because the stock can be purchased more cheaply at the market price. We also conclude that when the market price at expiry is less than the strike price, then the value of the option at expiry is $0.

If at expiry the price of the underlying stock is exactly equal to the market price, then using the option has no effect on the purchase price of the stock. The value of the option at expiry is again $0.

2.13 Pricing a Call Option Using a Replicating Portfolio: A Single Time Step

In the preceding section we determined the value of a European option at expiry. A more difficult problem is determining the value of an option before expiry. The full theory of options pricing is technical, so we will address the problem in a simplified setting.

We will find the initial no-arbitrage price for the option under the following simplifying assumptions:

- The underlying stock only trades at the time when the option is purchased and at expiry.

- The stock price at expiry can only be one of two values, one larger than the strike price and the other lower than the strike price. Which of these two values will happen, we don't know.

- The stock price at expiry is determined by some random experiment such as a coin toss—possibly with an unfair coin.

The assumption that the stock price either goes up one "step" or down one "step" is motivated by the geometric random walk model for stock prices. We don't know what the odds are for the random experiment that determines the stock price at expiry, but it will turn out that the no-arbitrage price for the option does not depend on those odds. This surprising independence of the option price and the odds of the stock price going up or down is crucial to the usefulness of the theory.

To find the arbitrage-free initial price of a European option we will use the *replication method*. In the replication method, we determine a portfolio consisting of a mixture of the underlying stock and cash that produces exactly the same result at expiry as the option produces. Then the initial value of that portfolio must be equal to the initial value of the option that it models.

Rather than using formulas and equations, we will simplify further by making specific choices for the prices of the underlying stock and the strike price. After working through that specific example, we will show how to use graphical methods to construct the replicating portfolio.

Our specific example uses the following (arbitrarily chosen) values:

(1) the underlying stock trades only at time $T = 0$ and $T = 1$,

(2) the price of the underlying stock at time $T = 0$ is $300,

(3) at time $T = 1$, the underlying stock can only take one of two values, a high value of $700 or a low value of $400,

(4) the strike price in the call option is $550.

As was true for finding the no-arbitrage price of a forward in Sect. 2.9, the risk-free interest rate, r, is an essential additional ingredient. So that the effect of the risk-free interest rate will show clearly, we will assume the very high interest rate of 100%. Thus we have one more assumption

(5) the effective risk-free interest rate over one time period is 100%.

2.13. Pricing a Call Option: Single Time Step

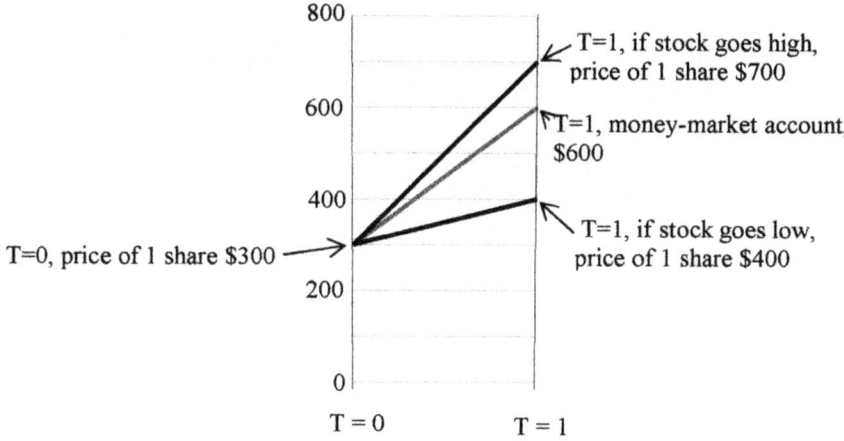

Figure 2.11 Possible stock price movements vs. money market.

Figure 2.11 illustrates two things that you could do with $300 at $T = 0$ and the possible consequences of your choice. You could simply put the $300 in the risk-free money market account and have $600 at $T = 1$. This possibility is shown by the green line in Fig. 2.11. A second possibility is to buy 1 share of stock with the $300. If you buy the stock at $T = 0$ and the stock price at $T = 1$ is the low value of $400, then you end up with $200 less than if you had put your $300 in the money market account. On the other hand, if you buy the stock at $T = 0$ and the stock price at $T = 1$ is the high value of $700, then you end up with $100 more than if you had put your $300 in the money market account.

Notice that if we were to change assumption (3) above by making the high value of the stock at $T = 1$ equal to $600 or less, then buying the stock would be a bad choice no matter whether the stock went to its high or low value. Similarly, if we were to change assumption (3) above by making the low value of the stock at $T = 1$ equal to $600 or more, then putting money in the money market account would be a bad choice no matter whether the stock went to its high or low value.

It is convenient to define the value spread of 1 share of the stock to be the difference between the high and low values that the stock could assume at $T = 1$; that is, the value spread equals $700 − $400 = $300. The black arrow pointing up in Fig. 2.12 illustrates that the value spread of the underlying stock is +$300.

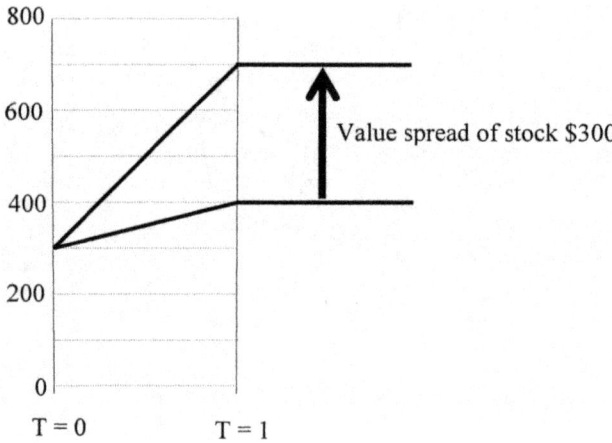

Figure 2.12 The value spread of one share of the underlying stock.

Any derivative based on the underlying stock also will have a value spread at $T = 1$. The value spread for the underlying stock is by definition positive (high minus low). We can make no such general assertion about whether the value spread of a derivative is positive or negative. We also have not yet shown how to go about finding the value spread of a derivative.

We next illustrate how to find the value spread for the European call option in our example. For the particular case of a European call option at expiry, we know how to find the value of the option from the value of the underlying stock and the strike price.

- If the underlying stock price at $T = 1$ assumes the high value $700, then the option is worth the difference between the stock price and the strike price; that is the value of the option equals $700 - $550 = $150.

- On the other hand, if the stock price at $T = 1$ assumes the low value $400, then the option is worthless, because if you want to buy a share of stock at $T = 1$, it can be bought for $400 which is less than the $550 that the option entitles you to pay.

We see that the value of the call option could be as high as $150 and as low as $0 and that gives the call option a value spread of $150 - $0 = $150. The red arrow pointing up in Fig. 2.13 illustrates the fact that the value spread of the call option is +$150.

Again we note that in general it is possible for the value spread of something to be negative. The value spread of anything is calculated by first

2.13. Pricing a Call Option: Single Time Step

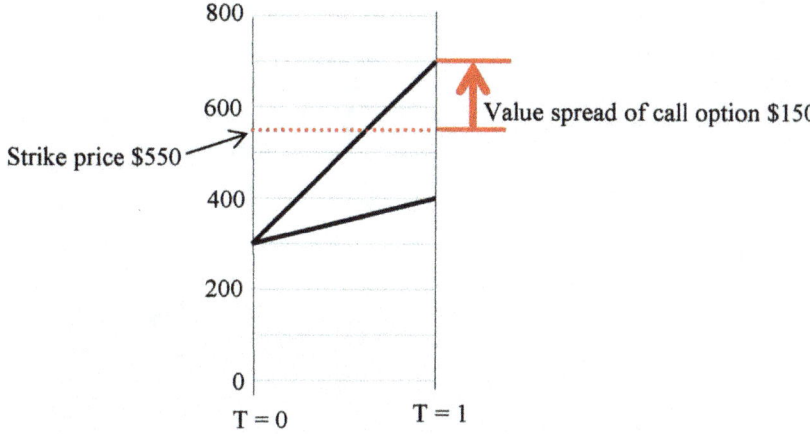

Figure 2.13 The value spread of the derivative: European call option.

computing its value when the stock is assigned its high value and then subtracting its value when the stock is assign its low value. In particular, the value spread for a negative number of shares of the underlying stock must be negative. Recall that a negative number of shares is obtained by selling short, i.e., borrowing the shares and selling them.

In order to determine the no-arbitrage price of the call option, we construct a portfolio consisting of the underlying stock and cash that exactly mimics the behavior of the option at $T = 1$, that is, we construct a replicating portfolio. Since there is no arbitrage, the value of the option at $T = 0$ must equal the value of the replicating portfolio at $T = 0$; were that not the case you would sell whichever is overpriced to buy the other and make a risk-free profit at $T = 1$.

We will construct the replicating portfolio in two steps. First, we construct a preliminary portfolio that matches the value spread of the call option. Because there is no value spread for cash, all of the value spread in the preliminary portfolio must come from the underlying stock. Since the value spread of the option is $150 and the value spread of 1 share of stock is $300, we see that $\frac{1}{2}$ share of the stock will produce the desired value spread of $150. We begin by putting $\frac{1}{2}$ share of stock in the preliminary portfolio. The behavior of the preliminary portfolio is illustrated by the blue arrow in Fig. 2.13.

Figure 2.14 also includes a red arrow that corresponds to the value spread of the call option at $T = 1$. The top and bottom of the red arrow are located

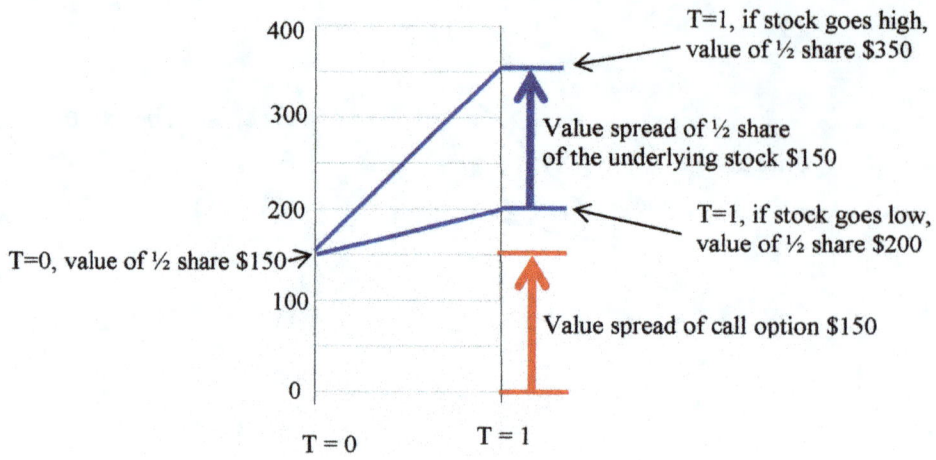

Figure 2.14 The value spread of $\frac{1}{2}$ share of the underlying stock.

at positions that correspond to the possible values of the call option. Likewise, the top and bottom of the blue arrow are located at the positions that correspond to the possible values of the $\frac{1}{2}$ share of stock in the preliminary portfolio.

By construction, the blue arrow in Fig. 2.14 is the same length and points in the same direction as the red arrow in Fig. 2.14. But the blue arrow is positioned higher than the red arrow. In fact, the blue arrow is higher than the red arrow by a distance that corresponds to $200.

The replicating portfolio must exactly match the behavior of the call option when $T = 1$, so we must add cash to the preliminary portfolio so as to leave the value spread unchanged while lowering the ending values by $200. To make the replicating portfolio worth $200 less at $T = 1$, we add to the preliminary portfolio a loan on which we owe $200 at $T = 1$. Since the risk-free interest rate is assumed to be 100%, the loan must be for $100 at $T = 0$. This loan is illustrated by the green line in Fig. 2.15.

Figure 2.15 also illustrates the total value of the replicating portfolio with the blue lines. Notice that the value of the replicating portfolio is $50 at $T = 0$. From this total value at $T = 0$, we conclude that the no-arbitrage price for the call option at $T = 0$ equals $50. The behavior of the option is illustrated in Fig. 2.16.

2.13. Pricing a Call Option: Single Time Step

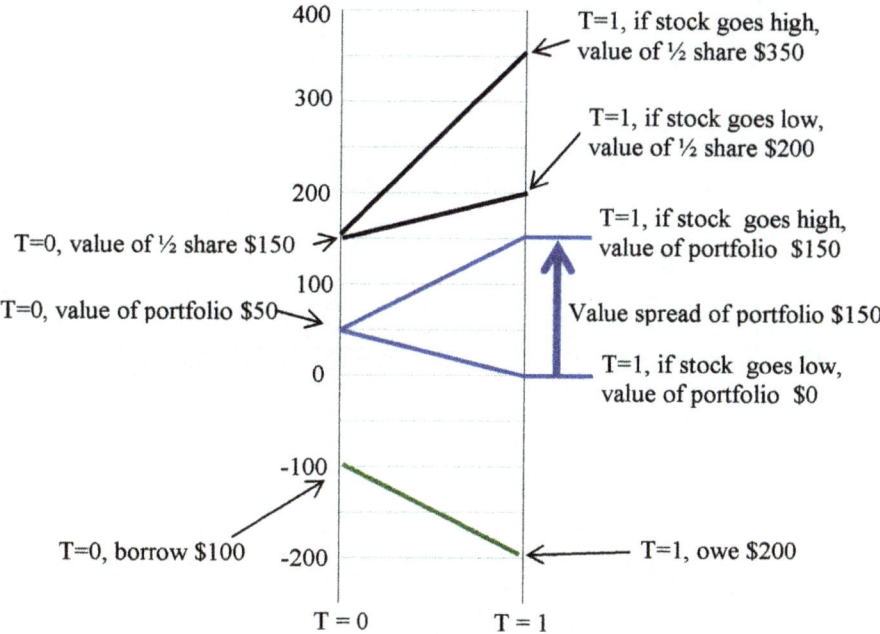

Figure 2.15 Behavior of the replicating portfolio.

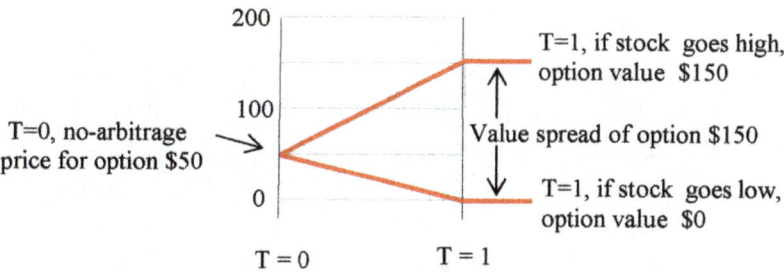

Figure 2.16 Behavior of the value of the European call option.

In review, we found the correct amount of stock for the replicating portfolio by matching the value spread of the option with the appropriate amount of the underlying stock, then we matched the $T = 1$ values themselves by including the appropriate amount of cash at $T = 1$. Once we know how much stock and cash is in the portfolio at $T = 1$, we can then use the stock price at $T = 0$ (which is known) to calculate the value at $T = 0$ of the stock in the portfolio, and we can use the risk-free interest rate (which is known) to calculate the amount of cash in the portfolio at $T = 0$. The total value of the portfolio at $T = 0$ is the no-arbitrage price for the call option at $T = 0$, namely $50. This value of $50 is unaffected by the probability that the stock price goes up or down; it does depend on the value spread of the stock and the risk-free interest rate.

The same process,

(1) match the value spread with stock,

(2) shift the final values with cash,

(3) calculate the value at $T = 0$ from the stock price and the interest rate,

can be applied to find the no-arbitrage value of any derivative based on the underlying stock. It is not necessary to know the probability of an upward or downward movement of the stock price. Even if the probability is known, it has no effect on the no-arbitrage valuation for the derivative.

2.14 Pricing a Call Option Using a Replicating Portfolio: Multiple Time Steps

The geometric random walk model for stock prices suggests that we might be able to calculate the no-arbitrage value of a derivative by repeated applications of the method illustrated in the preceding subsection. Figure 2.17 illustrates the tree of stock price movements implied by the geometric random walk model. Each dot corresponds to a specific price for the stock at a specific time. The left-hand side of the figure corresponds to the present, at which time the stock price is known (so there is only one dot). With each time step the stock price can go to a higher value or a lower value. As we move through more time-steps the number of possible stock prices increases. Also, notice that some pairs of dots are connected by more than one path, because there can be more than one way to get from one price to another.

To determine the value of a European call option, we look at a tree like that in Fig. 2.17. The dots on the right-hand side correspond to the stock price at the expiry date, $T = 4$ in the figure. At the expiry date, $T = 4$, the value of an option depends on the strike price, but the strike price is a known quantity. In Fig. 2.18, a strike price has been superimposed over the stock prices. For the two expiry date stock prices that are above the strike price, the option has the values represented by the red arrows. The values

2.15. Black–Scholes Option Pricing

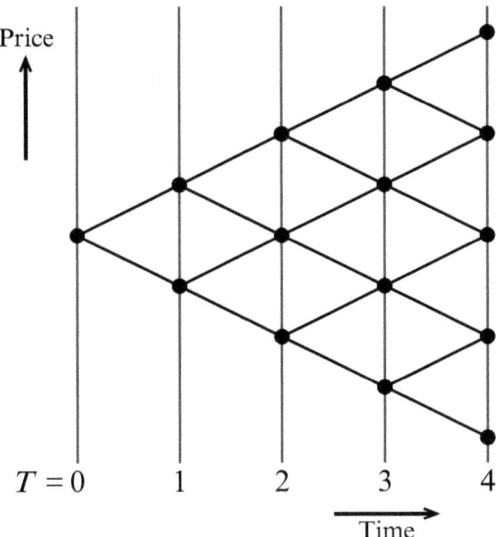

Figure 2.17 The tree structure corresponding to the geometric random walk model for stock prices.

are positive, so the arrows point up. For the three expiry date stock prices that are below the strike price, the value of the option is 0.

For each dot corresponding to a stock price at one time step before the expiry date, that is, at $T = 3$, there are 2 paths leading to 2 of the dots representing stock prices at expiry. For each such set of 3 dots, that is, one dot at $T = 3$ connected to 2 dots at $T = 4$, we can calculate the no-arbitrage value of the option at $T = 3$ using the method described in the preceding section. There are four such no-arbitrage option values at $T = 3$ that must be computed.

Once the no-arbitrage values at $T = 3$ have been computed, we work back one time step to $T = 2$. After the 3 no-arbitrage values of the option at $T = 2$ have been calculated, we work back to $T = 1$, and then finally we compute the no-arbitrage value for the option at $T = 0$.

2.15 Black–Scholes Option Pricing

The multiple time step method for option pricing that we described in the preceding subsection is based on the geometric random walk model for stock prices. The geometric random walk can be extended to its continuous limit,

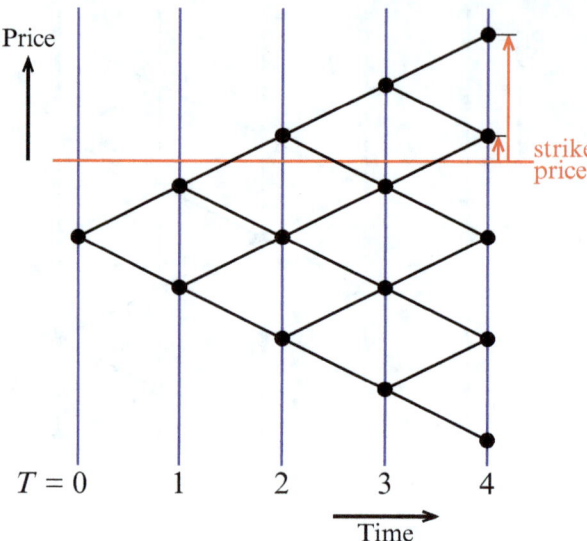

Figure 2.18 Calculate the value of the call option starting at the expiry date. Work backward in time until reaching the present.

that is, to a geometric Brownian motion. The no-arbitrage option pricing obtained in this continuous limit was developed by Fischer Black and Myron Scholes in the 1970s.

The risk-free interest rate clearly plays a crucial role in determining the no-arbitrage price of an option. Because we have been building our option prices without using equations, it is not clear how the option price is affected by the two stock-price parameters, average return and volatility. When the full analysis is carried out, one finds that of the two stock-price parameters, only the volatility enters into the no-arbitrage price for a European call option. Because the average return cannot be precisely estimated, but the volatility can be, the fact that the no-arbitrage price for an option depends only on the volatility is a tremendously important discovery. This insight is one of the main reasons that the work of Black and Scholes is considered to have revolutionized financial mathematics.

2.15. Black–Scholes Option Pricing

Fischer Sheffey Black, Myron Samuel Scholes, and Robert Cox Merton

Photograph courtesy of Alethea Black

Fischer S. Black

2008, Nobel laureates photographer

Myron Scholes

The names of Fischer Black (1938–1995) and Myron Scholes (b. 1941) are often linked in reference to the Black–Scholes equation. The Black–Scholes equation is a stochastic partial differential equation[9] which is used to model the behavior of the price of an option. Black and Scholes's discovery of a rational basis for option pricing began a revolution in finance.

Scholes and Robert C. Merton (b. 1944) extended the initial work of Black and Scholes. It was Scholes and Merton who received the 1997 Nobel Prize in Economics for their part in starting the financial revolution. Black was not so recognized, because he succumbed to throat cancer in 1995 and Nobel Prizes are not awarded posthumously.[10]

Had he lived long enough, Fischer Black would have been one of the few mathematicians to win a Nobel Prize. Of course, we associate Black's name with economics and finance, and his most famous work was indeed in economics, but his interests and background were broader than that. His undergraduate degree from Harvard was in physics and he started his doctoral work in that field. He then switched to mathematics. Ultimately his thesis, *A Deductive Question Answering System*, was on artificial intelligence.[11]

---

[9] A partial differential equation involves an unknown function and that function's rate of change with respect to two or more variables. If a random variable is also involved, then it is a stochastic partial differential equation.

[10] The matter of posthumous Nobel prizes is complicated. In fact, the 1931 Nobel Prize in Literature was awarded posthumously to Erik Axel Karlfeldt. Since 1974 the Nobel Foundation statutes have prohibited posthumous prizes, with the proviso that if a prize winner is announced but dies before the presentation ceremony, then the award will remain valid. In 2011, Dr. Ralph Steinman died hours before the Nobel Committee selected him for the prize in Medicine (along with two other men). In this case, it was decided that since the committee made the award on the good faith assumption that Dr. Steinman was alive, the award was valid.

[11] A slightly edited version of Black's thesis appears as a chapter in *Semantic Information Processing*, edited by Marvin Minsky.

2006, Digarnick
Robert C. Merton

One of Fischer Black's children is Alethea Black—author of *I Knew You'd Be Lovely*. The name Alethea is from the Greek $\alpha\lambda\eta\theta\epsilon\iota\alpha$ meaning *truth*. We would imagine, as she long did, that the name was chosen because of her mathematician father's love of and quest for truth. In fact, her parents got her name from the character Alethea Staunton played by Karen Black in a 1968 episode of *Judd for the Defense*.

A Look Back

Fisher Black was a Professor of Finance at Harvard University and Myron Scholes a Professor at Stanford. Scholes was awarded the Nobel Prize in Economics in 1997. (Black was deceased at the time, so he received no tribute.) The Black/Scholes option pricing model had really turned the world of finance on its ear, and this recognition seemed ever so appropriate.

Like most universities, Stanford has a monthly, in-house newsletter whose purpose is to inform the members of the university community about recent news and accomplishments of their colleagues. You can imagine that quite a big deal was made of Scholes winning the Nobel Prize, and a substantial article was written about the event.

Unfortunately, the editor assigned to handle the Scholes article carelessly applied a spell checker to the piece. He ended up replacing every occurrence of "Myron Scholes" with "moron schools." And that is how the article went into print. More's the pity for all of us.

The Black/Scholes model for option pricing has had a considerable impact on Wall Street. There are now a good many mathematics Ph.D.s who get jobs in the finance world. At Washington University we had a recent Ph.D. whose first position was in a Wall Street investment firm for a starting salary of $200,000.

But sometimes there is trouble in paradise. The concept of stochastic volatility—fundamental for the Black/Scholes model—has now become a mainstay of the investment world. But not everyone subscribes to that model. The stock market crash of 1987 lent weight to the nay-sayers. There have been other crashes since then.

Some say that the Black/Scholes model has too many simplifying assumptions. Others have asserted that volatility must be modeled in a different way in order to make the Black/Scholes model work effectively.

Any good idea continues to grow and develop. Mathematician James H. Simons founded and ran the historically most successful hedge fund (named Renaissance Technologies). Edward O. Thorp, after writing the famous book *Beat the Dealer* about how to win at blackjack in Las Vegas, wrote another book called *Beat the Market* in which he applied similar strategies to investments. He went on to become a multi-billionaire and is now happily retired (see [Pou 05]). Both men have proprietary mathematical investment schemes, and both certainly took into account the stochastic analysis that goes into the Black/Scholes model.

REFERENCES AND FURTHER READING

[Bac 06]   Bachelier, L.: Louis Bachelier's Theory of Speculation: The Origins of Modern Finance, translated and with commentary by Mark Davis and Alison Etheridge. Princeton University Press, Princeton (2006)

[Bar 84]   Barnes, J. (ed.) Politics. In: The Complete Works of Aristotle, vol. 2, p. 1997. Princeton University Press, Princeton (1984)

[BM 97]   Baskin, J.B., Miranti, P.J. Jr.: A History of Corporate Finance. Cambridge University Press, Cambridge (1997)

[BR 96]   Baxter, M., Rennie, A.: Financial Calculus: An Introduction to Derivative Pricing. Cambridge University Press, Cambridge (1996)

[BS 73]   Black, F., Scholes, M.: The pricing of options and corporate liabilities. Journal of Political Economy **81**, 637–654 (1973)

[BM 84]   Brealey, R., Myers, S.: Principles of Corporate Finance. McGraw-Hill, New York (1984)

[Buc 06]   Buchanan, J.R.: An Undergraduate Introduction to Financial Mathematics. World Scientific Press, Singapore (2006)

[Coo 86] Cooper, J.S.: Sumerian and Akkadian Royal Inscriptions I. The American Oriental Society, New Haven (1986)

[Dur 96] Durrett, R.: Stochastic Calculus: A Practical Introduction. CRC Press, Boca Raton (1996)

[Eth 02] Etheridge, A.: Financial Calculus. Cambridge University Press, Cambridge (2002)

[Gar 00] Garber, P.M.: Famous First Bubbles: The Fundamentals of Early Manias. MIT Press, Cambridge (2000)

[GR 05] Goetzmann, W.N., Rouwenhorst, K.G. (eds.) The Origins of Value: The Financial Innovations That Created Modern Capital Markets. Oxford University Press, Oxford (2005)

[Ken 53] Kendall, M.: The Analysis of Economic Time-Series-Part I: Prices. Journal of the Royal Statistical Society **116**, 11–34 (1953)

[Low 00] Lowenstein, R.: When Genius Failed: The Rise and Fall of Long-Term Capital Management. Random House, New York (2000)

[Mal 99] Malkiel, B.G.: A Random Walk Down Wall Street. W.W. Norton & Company, New York (1999)

[Pou 05] Poundstone, W.: Fortune's Formula: The Untold Story of the Scientific Betting System that Beat the Casinos and Wall Street. Hill and Wang, New York (2005)

[Pro 08] Protter, P.: Review of Louis Bachelier's theory of speculation: the origins of modern finance (Mark Davis and Alison Etheridge). Bulletin of the American Mathematical Society **45**, 657–660 (2008)

[Sch 94] Schmandt-Besserat, D.: Oneness, twoness, threeness: how ancient accountants invented numbers. In: Swetz, F.J. (ed.) From Five Fingers to Infinity. Open Court, Chicago (1994)

[Shr 05] Shreve, S.E.: Stochastic Calculus for Finance I: The Binomial Asset Pricing Model. Springer, New York (2005)

[Sim 78] Simmons, S.: Early Old Babylonian Documents, Yale Oriental Series, vol. 14. Yale University Press, New Haven (1978)

Chapter 3
Ramsey Theory

3.1 Introduction

Counting is a big part of modern mathematics. Many mathematical problems necessitate the estimation of a particular, precisely specified number having a certain technical description. Certainly questions of airline scheduling, Internet routing, queueing theory, and crystalline structure are of this nature. For instance, how many different airline routes are there from San Francisco to Boston with not more than two stops along the way? This is a nontrivial question with a meaningful and useful answer. Along with Lejeune Dirichlet, Frank Ramsey was one of the pioneers of counting theory. His Ramsey's theorem pervades large parts of mathematics.

In fact Ramsey taught us that absolute chaos does not exist; there is always a well-structured subsystem. The purpose of this chapter is to explain this contention. That is what Ramsey theory is about.

A simple example is this: imagine that each point in the plane is colored either red or blue. We claim that there are two points that are distance 1 apart and have the same color. So we see structure in apparent chaos. The proof? Take an ideal (that is, perfectly shaped) equilateral triangle of side 1 and toss it onto the plane. Each vertex sits on a point of some color. Two of the vertices have the same color. End of discussion.

The simplest example of a problem in pure Ramsey theory is the following brain-teaser: "Show that, in any group of six people, there are at least three people who all know each other or at least three people who all do not know each other." (We will address this problem in detail below.) What if we want to be sure that there are four people who all know each other or who all do not know each other? What if we want 40 people who all know each other or who all do not know each other? Will a group of a million people be enough to ensure there are forty who all know each other or who all do not know each other? Maybe we need a billion people to get such a special group of forty people. Ron Graham asked a question similar to the brain-teaser above. Graham's question can be expressed in terms of forming committees. The question is, "How many people does it take to guarantee that if all the pairs of committees formed from those people are divided into two categories of pairs of committees, then there will be four committees such that all pairs from among those four committees fall in the same one of the two categories and such that each member of any of those four committees is a member of an even number of committees?" In [GR 71], Graham and Rothschild were able to show that there is in fact a large enough number of people to guarantee the existence of the four desired committees. Their upper bound has come to be known as the *Graham number* and it was cited in the 1980 *Guinness Book of World Records* as the largest number (i.e., positive integer) created by man.

3.2 The Pigeonhole Principle

The comedian Rodney Dangerfield used to say, "I have three kids, one of each." He could get a laugh with this absurdity because when you heard him say he had three children your mind instantly prepared you for "two" of one gender and "one" of the other.

Figure 3.1 Pigeonholes for mail delivery.

The unconscious reasoning that Dangerfield relied on to make his joke work is the elegant and powerful mathematical tool called the *pigeonhole*

3.2. The Pigeonhole Principle

principle. In this context, "pigeonholes" are the small compartments in an old fashioned desk or the mail delivery slots in an office (Fig. 3.1). The pigeonhole principle says that if you have more letters to distribute than there are pigeonholes, then after you've put all the letters into the pigeonholes, one of the pigeonholes will hold at least two letters.

One solution of the brain-teaser with which we began this section starts with an application of the pigeonhole principle. Recall that the problem is to confirm the following assertion: In any group of six people, either

  (a) there are three people who all know each other

  or

  (b) there are three people no two of whom know each other.

To see that this claim is true, we reason as follows: Pick any one of the people. Let's say you have chosen Ann. Now consider two metaphorical pigeonholes: In one pigeonhole you put all the people (other than Ann) in the group that Ann knows, and in the other pigeonhole you put all the people in the group that Ann does not know. Since there are five people other than Ann and they have been put into two pigeonholes, one of those pigeonholes must contain at least three people. We now want to look at the three (or more) people who wound up in the same pigeonhole. At this point the argument breaks into two cases below:

Case 1: The group of people that Ann knows contains three or more people.

If any two of the people that Ann knows also know each other, then those two together with Ann make a set of three people who all know each other. On the other hand, if among the people that Ann knows, no two know each other, then the people that Ann knows form a set of at least three people none of whom knows the other. Either way, in Case 1 we have shown that there are three people who all know each other or there are three people no two of whom know each other.

Case 2: The group of people that Ann does not know contains three or more people.

If any two of the people that Ann does not know also do not know each other, then those two together with Ann make a set of three people, no two of whom know each other. On the other hand, if among the people that Ann does not know, every two do know each other, then the people that Ann does not know form a set of at least three people, all of whom know each other. Either way, in Case 2 we have shown that there are three people who all know each other, or there are three people, no two of whom know each other.

The fact that in any group of six people, there are three who are mutually acquainted or three who are mutually unacquainted is a special case of Ramsey's theorem. The more general result that Frank Ramsey proved is as follows:

Ramsey's Theorem. *Pick any three positive integers K, L, and M with M at least as large as K. Then* **there is an N large enough** *that the following is true for any N or more objects (the "objects" could be people or points or anything else):*

- *Whenever all the K-element sets of objects are classified into L categories (above we considered "pairs of people" and the categories "know each other" and "do not know each other"),*[1]

  then

- *there will be a subset of M of the objects with the property that every K-element set chosen from those M falls in the same category (above we sought $M = 3$ people).*

What we showed above was that if $K = 2$, $L = 2$, and $M = 3$, then $N = 6$ is large enough. The smallest number N that always works for specific L, K, and M is called the *Ramsey number* for that problem. In fact, 6 is the Ramsey number for $K = 2$, $L = 2$, and $M = 3$. To show that 6 is the smallest N that works for $K = 2$, $L = 2$, and $M = 3$, we need to construct an example of five people with no group of three satisfying either of the conditions (a) and (b). To do this, imagine two women who know each other from work, but have never met each other's husbands and their husbands have never met. Then add to those four people a third man who happens to know each of the husbands, but has never met the wives.

Here is an alternative, interesting way to see that five people will not solve the problem just described (in other words, 5 is not the Ramsey number for

---

[1] Technically a "category" of K-element sets is simply a set whose elements are K-element sets of objects.

3.2. The Pigeonhole Principle

$K = 2$, $L = 2$, $M = 3$). We represent the five people as points in the plane. We connect two points (or people) with a green line if those two people are acquainted. And we connect them with a red line if they are unacquainted. As Fig. 3.2 shows, there are no three people in this set of five who are mutually acquainted and there are no three people who are mutually unacquainted. These two assertions are clear because there is no triangle with all three sides green lines, and there is no triangle with all three sides red lines.

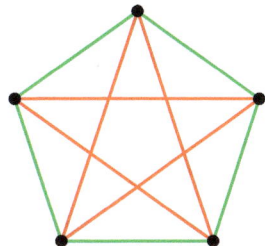

Figure 3.2 Five is not the Ramsey number for $K = 2$, $L = 2$, $M = 3$.

Figure 3.2 suggests that you can think about pairs of objects by identifying the objects with points in the plane (no three of which are collinear) and the pairs with line segments connecting the two objects. Each category can then be identified with a color and the line segments colored accordingly. Similarly triples of objects can be treated using points in space (no four of which are coplanar) to represent the objects and triangles to represent the triples. Again each category can be identified with a color and the triangles colored accordingly. Our ability to visualize the situation breaks down with quadruples, quintuples, and so on, but by thinking about pairs and triples we can then try to reason by analogy.

Next we list in a table the known information about how many people N will be needed at a party in order to guarantee either M acquaintances or M unacquaintances. Any two people are either acquainted or not, so when $M = 2$, then $N = 2$. Our discussion above showed that when $M = 3$, then $N = 6$. The case $M = 4$ is very difficult, but it is known that $N = 18$. The cases $M = 5$ and higher are currently unresolved, so for $M \geq 5$, we indicate a range of values within which N must lie. Paul Erdős (see the next section) used to say that if the devil challenged you to solve the case $M = 5$ then you should harness all the computing power in the world and you might stand a chance. But if the devil asked you to solve $M = 6$, then you better make a deal with the devil.

M	N
2	2
3	6
4	18
5	$[43, 49]$
6	$[102, 165]$
7	$[205, 540]$

M	N
8	$[282, 1\,870]$
9	$[565, 6\,588]$
10	$[798, 23\,556]$
11	$[1\,597, 184\,755]$
12	$[1\,637, 705\,431]$

Table 3.1 Values of and estimates for Ramsey numbers: $K = 2$, $L = 2$.

Frank Ramsey

Frank Ramsey, aged 18

Frank Plumpton Ramsey (1903–1930) was a brilliant polymath (a person conversant with many subjects) who made significant contributions to mathematics, economics, and philosophy despite dying shortly before his 27th birthday. Frank received his secondary education at Winchester College (founded 1382), one of Britain's ancient "public schools." He won a scholarship to Trinity College, Cambridge, where he graduated as a wrangler (i.e., with first class honors in mathematics) in 1923. In fact, he was the top mathematics student in his class—the Senior Wrangler.

Ramsey undertook the task of making the first translation from German into English of Ludwig Wittgenstein's difficult *Tractatus Logico-Philosphicus* (see [Wit 83]). The translation of the *Tractatus* was a serious project in the furtherance of which Ramsey traveled to Austria in September, 1923. He went to the village of Puchberg am Schneeberg where Wittgenstein was teaching primary school. Wittgenstein had given away his substantial inheritance and was living on a meager teacher's salary. During this trip Ramsey was able to clarify a number of issues related to the *Tractatus*.

In 1924, Ramsey was elected to a Fellowship at King's College, Cambridge. In 1925, he married Lettice Baker. Nonetheless, Ramsey was not as conventional as it might seem: Lettice was 5 years older than Frank and felt that they should live together out of wedlock. Frank was fearful of the disapprobation that might result, and they married—but in a Registry Office, because

3.3. The Happy End Problem

Frank was a militant atheist. It is interesting to note that Frank's younger brother Arthur Michael Ramsey was ordained in the Anglican Church in 1928 and later served as Archbishop of Canterbury.

Frank and Lettice Ramsey seemed to have had what is now called an "open marriage." Frank found that arrangement less satisfactory when it was Lettice who fell in love with someone else. We can only wonder what difficulties their relationship might have later endured, because their time together was cut short. Frank had been suffering from chronic liver trouble for some time. He was the victim of a severe attack of jaundice in 1930 and that attack necessitated exploratory surgery from which he never recovered. He died on 19 January 1930. Obituaries appeared in the *Journal of the London Mathematical Society*, *The Economic Journal*, and in *The Times*. His obituary in *The Economic Journal* was written by his friend, the economist John Maynard Keynes.

3.3 The Happy End Problem

Ramsey's theorem might have remained a fairly obscure mathematical result, but instead it has developed a devoted following and has spawned huge amounts of further research into what is now called Ramsey Theory. It is likely that the enormous interest in, and growth of, Ramsey Theory can be attributed to Paul Erdős and George Szekeres, who rediscovered Ramsey's theorem. Erdős was a peripatetic mathematician who spent his long life traveling the globe, collaborating with mathematicians, spreading his ideas, and infecting others with his interests. More will be said about him later.

The Ramsey-type result that first intrigued Paul Erdős was initially discovered by Esther Klein. At that time neither Klein nor Erdős, nor anyone in their mathematical circle, had ever heard of Ramsey's theorem, so Klein's discovery and Erdős and Szekeres's subsequent work generalizing it were all independent developments.

Klein's observation was that, for any five points in the plane, no three of which are collinear, there are four of those points that form a convex quadrilateral, that is, a four-sided figure in which all the corners are outward pointing (Fig. 3.3).

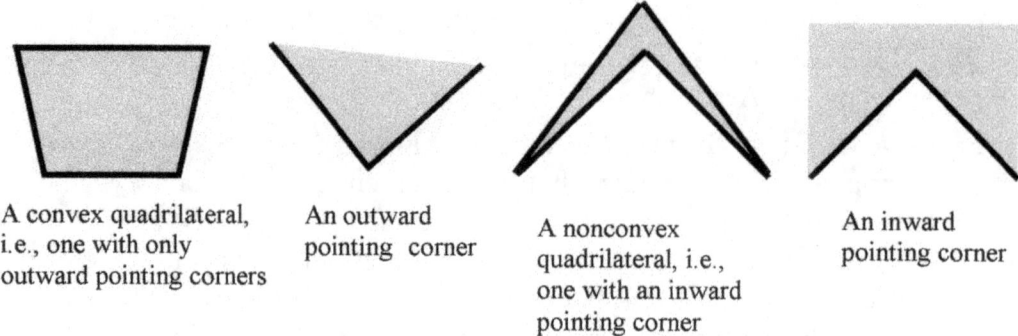

A convex quadrilateral, i.e., one with only outward pointing corners

An outward pointing corner

A nonconvex quadrilateral, i.e., one with an inward pointing corner

An inward pointing corner

Figure 3.3 Quadrilaterals: convex and nonconvex.

To see that Klein's observation is true, pick a direction that is not parallel to the direction of any line determined by any two of the five points. For convenience, let us suppose that the vertical direction is one such direction. Start with a vertical line that is to the left of all five points and move that line to the right until it just touches one of the points. Rotating the line clockwise, we will find a line that just touches a second point, and rotating the line counterclockwise, we will find a line that just touches another point. The situation will look like that pictured in Fig. 3.4.

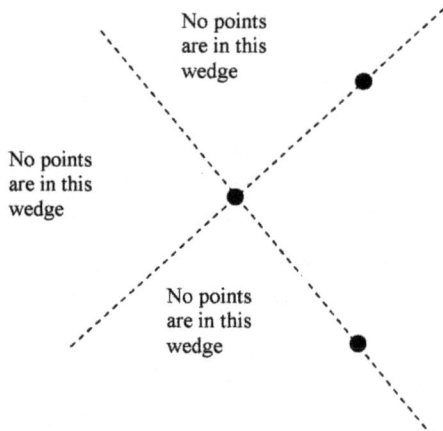

Figure 3.4 Trapping the points in a wedge.

There are two remaining points not accounted for in Fig. 3.4. The three points in Fig. 3.4 form a triangle, and each of those two remaining points will either be inside that triangle or outside of it (Fig. 3.5).

3.3. The Happy End Problem

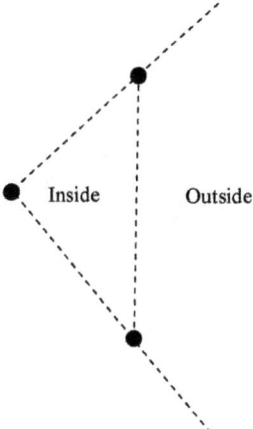

Figure 3.5 The triangle formed by the first three points.

The proof now splits into two cases:

Case 1: If either of the two remaining points is in the region labeled "Outside" in Fig. 3.5, then that point together with the three points forming the triangle will form the required quadrilateral with outward pointing corners.

Case 2: If neither of the two points is in the region labeled "Outside," then the line through those two points will split the plane into two halfplanes, and one of those halfplanes will contain two of the three points that formed the triangle (see Fig. 3.6). Those two vertices of the triangle, together with the two points inside the triangle, form the required convex quadrilateral.

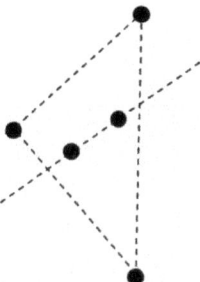

Figure 3.6 Case 2: the two remaining points are inside the triangle.

The fact that five points in the plane, no three of which are collinear, always contain four points that determine a convex quadrilateral inspires the more general question, "Is there a number of points N in the plane that is large enough to guarantee that M of the points will form a convex M-gon, that is, a figure with M sides that has only outward pointing corners?" One can also ask, "If there is such a number N, what is the smallest possible N?"

Klein told both Erdős and Szekeres about the fact that five points in the plane, no three of which are collinear, always contain four points that determine a convex quadrilateral and Erdős and Szekeres then pursued answers to the more general questions. Szekeres came up with the first proof that for any M there is an N such that N points in the plane, no three of which are collinear, always contain M points that form a convex M-gon. Klein and Szekeres became a couple and ultimately married. Erdős felt that Szekeres's cleverness in solving the problem triggered the romance, so Erdős called it the *happy end problem*.

Paul Erdős

1992, K.M. Heckrodt
Paul ErdÖs

Paul Erdős (1913–1996) was the most prolific mathematician of the twentieth century (1,475 academic research articles). He was history's most collaborative mathematician: 485 people (including one of the present authors) coauthored at least one paper with Erdős. He was also one of the twentieth century's more eccentric mathematicians.

He was born March 26, 1913, in Budapest. He was the son of two high school math teachers. This would seem a prosaic enough beginning, but unfortunately, while Erdős's mother was in the hospital giving birth to Paul, his two older sisters contracted septic scarlet fever and died. Then, little more than a year later, World War I began and shortly thereafter Paul's father was captured by the Russians and held prisoner in Siberia for 6 years. These events had a profound effect on Paul's mother. Because of the fear of contagion, Paul was kept out of school until high school and even then he went only every other year as his mother's fears waxed and waned. At 17, he entered the University of Budapest from which he graduated with a Ph.D. in 1934. He was protected and pampered by his mother until he went to England to study in 1934. It is not surprising that he grew up to be unworldly, but his unworldliness became legend. He tied his own shoes for the first time at age 11. He buttered his first piece of bread at age 21, and then only because he was in England without his mother or a servant to do it for him.

Erdős had no wife, no children, no hobbies, no job, and no home. He went from place to place searching for interesting mathematical problems and mathematicians with whom to solve those problems. Erdős's *modus operandi* was to show up on the doorstep of a friend or collaborator and declare "My brain is open." He would then expect, as a matter of course, to be fed, housed, and clothed. In fact, Erdős's nomadic existence would have been impossible were it not for mathematicians the world over who took care of him.

As a Hungarian Jew coming of age in the 1930s, geopolitics had a huge impact on Erdős's life. The Nazis brought disaster but, even after the Nazis were defeated, Erdős continued to have problems. As Erdős would say, he had problems with "Joe" (for Joseph Stalin and the Soviet Union) and he had problems with "Sam" (for Uncle Sam and the United States). Erdős did not want to go back to Hungary because of Joe, so he lived in the United States. His problems with Sam started in 1954 when he left the U.S. to go to a conference in Amsterdam. When he attempted to reenter the United States, the immigration officials asked Erdős for his opinion of Marx. He naively replied that he was not competent to judge, but he had no doubt that Marx was a great philosopher. That ill-advised answer kept Erdős out of the United States until the 1960s.

3.4 Relationship Tables and Ramsey's Theorem for Pairs

We now return to the instance of Ramsey's theorem in which you want to show that, among any N people, there are three people who all know each other, or there are three people no two of whom know each other. We already know that $N = 6$ works. Since we want to give an argument that we can generalize to larger numbers of acquaintances and unacquaintances, we will show instead the much weaker alternative result that $N = 32$ works. Our method will rely on representing the relationships among people using a table, called a *relationship table*, such as is shown in Fig. 3.7.

Figure 3.7 shows a table that represents whether each pair of five people are acquainted. The people are identified by the letters A, B, C, D, and E. Green indicates that the two people know each other and red indicates that they do not know each other. If we look at the part of Fig. 3.7 that is shown in Fig. 3.8, the colors there tell us that A knows B and C, but A does not

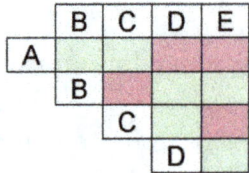

Figure 3.7 A relationship table for five people.

know D and E. The labels for the rows in Fig. 3.7 are indented to the right as you read down the table, because A knows B if and only if B knows A, so we do not need to explicitly include the information that B knows A since the first row has already told us that A knows B.

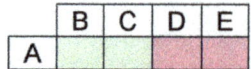

Figure 3.8 The first row of the relationship table.

To show that any group of 32 people includes a subgroup of three who all know each other or who all do not know each other, we will build the relationship table for the 32 people in a special way that will be more helpful.

Suppose we have $N = 32$ people. Pick a person, and call him or her A. Then, among the pairs formed by A and the 31 other people, at least 16 pairs fall into the same relationship (acquainted or unacquainted). Denote that more prevalent relationship with green. In this table, green will represent either "knowing each other" or "not knowing each other" depending on how many people A knows. Thus the first row of the relationship table looks like that shown in Fig. 3.9, where the gray square indicates that we do not know or care what the relationships are.

Figure 3.9 Either A knows at least 16 of the others or A does not know at least 16 of the others.

Next, pick any one of the people in the green relationship with A and call that person B. We consider the 15 or more people, other than B, who are in the green relationship with A. For those 15 or more people, we consider their relationships with B. Since there are two possible relationships and 15

3.4. Relationship Tables

people, we see that 8 or more must be in the same relationship with B. We use turquoise to represent that relationship. Thus the first two rows of the relationship table look like the ones in Fig. 3.10.

Figure 3.10 Either B knows at least 8 of the 16 or B does not know at least 8 of the 16.

Next, pick any one of the eight or more people corresponding to the turquoise squares in the second row of the table in Fig. 3.10. Note that those people automatically correspond to green squares in the first row of the table in Fig. 3.10. Call this newly chosen person C. We consider the seven or more people, other than C, who are in the turquoise relationship with B. Since there are two possible relationships and seven people, we see that 4 or more must be in the same relationship with C. We use tan to represent that relationship. Thus the first three rows of the relationship table look like the ones in Fig. 3.11.

Figure 3.11 Either C knows at least 4 of the 8 or C does not know at least 4 of the 8.

Next, pick any one of the four or more people corresponding to the tan squares in the third row of the table in Fig. 3.11. Note that those people automatically correspond to green squares in the first row of the table in Fig. 3.11 and to turquoise squares in the second row of the table in Fig. 3.10. Call this newly chosen person D. We consider the three or more people, other than D, who are in the tan relationship with C. Since there are two possible relationships and three people, we see that 2 or more must be in the same relationship with D. We use purple to represent that relationship. Thus the first four rows of the relationship table look like those shown in Fig. 3.12.

Finally we call the two people corresponding to the purple squares E and F. We fill in one more row of the table to include the relationship between E and F. We use blue to represent that relationship. Thus the first five rows of the relationship table look like those shown in Fig. 3.13.

Figure 3.12 Either D knows at least 2 of the 4 or D does not know at least 2 of the 4.

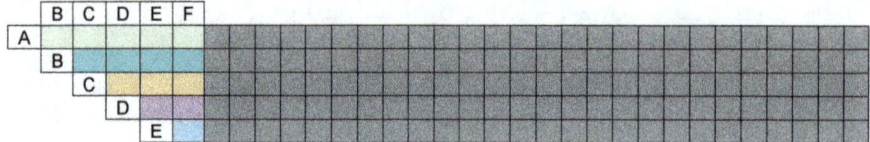

Figure 3.13 Relationship table for A, B, C, D, E, and F.

Each row in the table in Fig. 3.13 has a distinct color, but the five colors green, turquoise, tan, purple, and blue represent only the two possibilities (1) "know each other" or (2) "don't know each other." Really the five rows should be marked with only two colors. Since there are five rows to mark with two colors, we conclude that will be three rows that are marked with the same color. The three people corresponding to those three rows (with the same color) form the group of three people who all know each other or no two of whom know each other.

We can summarize the argument as follows: Each row is associated with a person and that person has the same relationship with all the people associated with lower rows. To obtain three people who are all pairwise in the same relationship, we need $5 = 1 + 2 \cdot 2$ rows, because there are two types of relationships and, once we have used two rows for each, one more row puts a third row into one of the two relationships. At the bottom of the table, we have one colored square in the fifth row. We work back up from fifth row with 1 square, to a fourth row with 2 squares, to a third row with $4 = 2^2$ squares, to a second row with $8 = 2^3$ squares, to a first row with $16 = 2^4$ squares, and to obtain those 16 squares all of the same color, we need $32 = 2^5$ people.

Now, suppose our goal is to end up with M people instead of 3 who all know each other or no two of whom know each other. Then we will need more rows. Specifically, we need $2M - 1$ rows, to ensure that at least M rows will represent the same relationship: (1) "know each other" or (2) "don't know each other." Following the above procedure of working up the table, doubling the number of squares at each step, starting from one square in the bottom row, and with a final doubling in the top row, we conclude that to build a table of the desired type with $2M - 1$ rows requires that we start with at least 2^{2M-1} people.

As special cases of this last formula, we conclude that $128 = 2^{2 \cdot 4 - 1}$ people is a large enough group so that it will always contain a subgroup of four people who all know each other or all do not know each other. Just as $32 = 2^{2 \cdot 3 - 1}$ is larger than the minimum number of people (i.e., 6) that guarantees that there is a subgroup of three people who all know each other or all do not know each other, likewise 128 is more than the minimum size that guarantees the existence of a subgroup of four people who all know each other or all do not know each other. In fact, we saw in Table 3.1 that the minimum size is 18 (but we will not prove that here). Applying our formula with $M = 5$, we see that $512 = 2^{2 \cdot 5 - 1}$ people is a large enough group so that it will always contain a subgroup of five people who all know each other or all do not know each other. You may find it remarkable that the minimum number for this problem is still not known. Table 3.1 shows that our estimate of 512 is too large by a factor of 10.

3.5 Ramsey's Theorem in General

In his paper [Ram 30], Ramsey generalized the facts described above. Recall from Sect. 3.2 that what Ramsey proved was that if

(1) you look at K-element sets of objects,

(2) you put those K-element sets into L categories, and

(3) you seek a set of M objects with the property that every K-element set chosen from those M falls in the same category,

then there is some large enough number of objects N such that you will always be able to find a subset of M objects with the property (3). The smallest such number is the Ramsey number for that problem and it is denoted by

$$R_K(M; L).$$

If K or L is not explicitly shown, then it is assumed to be 2. So for instance,

$$R(M; L) = R_2(M; L)$$

and
$$R(M) = R_2(M; 2).$$

If we work with pairs (i.e., $K = 2$), and there are L categories into which the pairs have been divided, then the relationship table will require L colors. When we go through this process as we did in the preceding section, we will need to work our way down to a table with $1 + L(M - 1)$ rows. And to obtain each row from the row above, we will need to have L times as many squares of the same color in the row above. To get the needed squares of the same color in the top row, we will need another factor of L people. Thus $N = L^{1+L(M-1)}$ people will suffice.

For instance, if the pairs are put into three categories and we ask to have a subgroup of at least five people such that all pairs chosen from those five people fall into the same category, then it will suffice to start with

$$N = 3^{1+3(5-1)} = 3^{13} = 1,594,323$$

people.

You might suspect that it does not really require a million-and-a-half people to guarantee that there is a subgroup of 5 who either (a) know each other and first met on the job, (b) know each other and first met away from work, or (c) do not know each other. In fact, you might suspect that the number is much lower. The smallest number of people is denoted by $R(5; 3)$, but nobody knows exactly what that number is.

In the general version of Ramsey's theorem, when one considers not pairs, but triples and even larger sets, it is no longer possible to use a pictorial proof with a two-dimensional relationship table to show that a large enough N exists. Nonetheless, the idea we used with the tables can be generalized inductively. Suppose that we consider unordered $(K + 1)$-tuples[2] (here K is at least 2) and suppose also that we have already proved Ramsey's theorem for unordered K-tuples. Subsequently, it is to be understood that K-tuples are unordered.

Pick an arbitrary element and label it A_1. We will now establish L categories of K-tuples using the same labels for these categories as the labels we were given for the categories of $(K + 1)$-tuples. We do this as follows: Given

---

[2] The word "unordered" is included to emphasize that, for instance, in the case of triples $\{a, b, c\}$ and $\{c, a, b\}$ are both the same triple.

3.5. Ramsey's Theorem in General

a K-tuple of elements that does not include A_1 we define that K-tuple to be in the same category as the $(K+1)$-tuple obtained when A_1 is added to the K-tuple.

Since we have proved Ramsey's theorem for K-tuples we know that, by starting with N objects, we can arrange to have a given number M_1 of objects such that all K-tuples of those M_1 objects belong to the same category. Whatever category that is, we will say that A_1 is associated with it.

Next we work with the M_1 objects obtained in the preceding paragraph. Pick one and label it A_2. Define the categories of K-tuples as before, but with A_2 as the $(K+1)$st entry. If we have arranged for M_1 to be large enough, then we can arrange to have a given number M_2 of objects such that all K-tuples of those M_2 objects belong to the same category. Again we say that A_2 is associated with that category.

Keep applying the above procedure until you have labeled Q objects, $A_1, A_2, \ldots, A_Q$. Choose Q large enough so that it forces M of the objects A_i to be associated with the same category.

Now, given that there are L categories, you can specify how large a Q is needed to obtain the desired number, M, of A_is that are all associated with the same category. Next, we want to see how many objects are needed to ensure that we can apply the above procedure the required Q number of times. When A_Q, is selected, the only requirement it needs to satisfy is that it is one element from among $K+1$, so that it will be associated with a category. So now you know that $M_Q = K+1$. Knowing M_Q tells you how to choose M_{Q-1}. Work your way up the chain of M_is all the way to M_1 at the top and you know how to choose N.

The argument just given produces a number N large enough that we can be sure it will work. Doubtless the N obtained will be larger than needed, possibly vastly larger than needed. Similarly, the Graham number described at the beginning of this chapter is large enough that

> if all the pairs of committees formed from that many people are divided into two categories of pairs of committees, then there will be four committees such that all pairs from among those four committees fall in the same one of the two categories and such that each member of any of those four committees is a member of an even number of committees,

but the Graham number is probably vastly larger than needed. (Recall that the Graham number was cited in the 1980 *Guinness Book of World Records* as the largest number, i.e., positive integer, created by man.) Graham himself thought 6 might be large enough to solve the problem, but in [Exo 03] it is shown that the number must be at least 11. There is still much room for improvement.

Here is a brief description of Donald Knuth's special notation for describing the famous Graham number. We define a new type of exponentiation. Let $3 \uparrow k$ denote 3 to the power k. This is a familiar idea. Now let $3 \uparrow\uparrow 3$ denote $3 \uparrow (3 \uparrow 3)$. In general

$$3 \underbrace{\uparrow\uparrow \cdots \uparrow\uparrow}_{k \text{ arrows}} 3$$

denotes

$$3 \underbrace{\uparrow\uparrow \cdots \uparrow\uparrow}_{(k-1) \text{ arrows}} (3 \underbrace{\uparrow\uparrow \cdots \uparrow\uparrow}_{(k-1) \text{ arrows}} 3).$$

Now let

$$M = 3 \underbrace{\uparrow\uparrow \cdots \uparrow\uparrow}_{3\uparrow\uparrow\uparrow\uparrow 3 \text{ arrows}} 3.$$

Next define

$$P = 3 \underbrace{\uparrow\uparrow \cdots \uparrow\uparrow}_{M \text{ arrows}} 3.$$

Now iterate this construction 61 more times. You will then obtain Graham's number G.

Ramsey theory gives a lovely and dynamic example of profound mathematics that gives rise to meaningful questions that are accessible to a broad audience. Ramsey theory comes up in the theory of queueing, in Internet routing, in design theory, in microchip structure and creation, and in many other parts of modern technology. It is an important part of combinatorics, and combinatorics is in turn one of the most active and productive fields of modern mathematics.

A Look Back

The problem that we used to motivate this chapter has received considerable attention. Paul Erdős gives it a mathematical formulation as follows:

A Look Back

Let k be a positive integer. How many people $N = N(k)$ do you need in a room to be sure that at least k people are mutually acquainted or at least k people are mutually unacquainted?

When $k = 2$, the answer is trivially $N = 3$. When $k = 3$, the answer is 6. The table below shows what we know for values of k up to 7:

Value of k	Value of N
2	2
3	6
4	18
5	Between 43 and 49
6	Between 102 and 165
7	Between 205 and 540

Paul Erdős liked to describe the situation as follows: If an evil spirit would appear and say, "Tell me the value of N when k equals 5, or I will exterminate the human race," it would be best to get all the computers in the world to try to solve the problem. But if the evil spirit would ask for the value when k equals 6, it would be best to try to exterminate the evil spirit.

He goes on to say, "And if we could get the right answer just by thinking, we wouldn't have to be afraid of him [the evil spirit], because we would be so clever that he couldn't do us any harm."

Of course the case $k = 7$ is beyond impossible.

The subject of Ramsey theory is centered around the notion of a "Ramsey number." This number $\mathcal{R}(m,n)$ is the least number N such that if every pair of N points in the plane is connected with either a red or a blue line, then either m of them are red or n of them are blue. We think here of a red line as indicating that the two points are acquainted and a blue line as indicating that the two points are unacquainted.

In the table above we recorded that

$$\mathcal{R}(2,2) = 2, \quad 42 < \mathcal{R}(5,5) < 50,$$
$$\mathcal{R}(3,3) = 6, \quad 101 < \mathcal{R}(6,6) < 166,$$
$$\mathcal{R}(4,4) = 18, \quad 204 < \mathcal{R}(7,7) < 541.$$

Work continues on these fascinating problems.

REFERENCES AND FURTHER READING

[**AZH 09**] Aigner, M., Ziegler, G.M., Hoffman, K.H.: Proofs from the Book, 4th edn. Springer, New York (2009)

[**Exo 03**] Exoo, G.: A Euclidean Ramsey problem. Discrete and Computational Geometry **29**, 223–227 (2003)

[**GR 71**] Graham, R.L., Rothschild, B.L.: Ramsey's theorem for n-parameter sets. Transactions of the American Mathematical Society **159**, 257–292 (1971)

[**GRS 90**] Graham, R.L., Rothschild, B.L., Spencer, J.H.: Spencer Ramsey Theory. Wiley, New York (1990)

[**GG 55**] Greenwood, R.E. Jr, Gleason, A.M.: Combinatorial relations and chromatic graphs. Canadian Journal of Mathematics **7**, 1–7 (1955)

[**Hof 98**] Hoffman, P.: The Man Who Loved Only Numbers. Hyperion, New York (1998)

[**Kra 02**] Krantz, S.G.: Mathematical Apocrypha. Mathematical Association of America, Washington, DC (2002)

[**Kra 05**] Krantz, S.G.: Mathematical Apocrypha Redux. Mathematical Association of America, Washington, DC (2005)

[**Ram 30**] Ramsey, F.P.: On a problem of formal logic. Proceedings of the London Mathematical Society **30**, 264–286 (1930)

[**Ram 50**] Ramsey, F.P.: The foundations of mathematics and other logical essays (edited by Richard Bevan Braithwaite, with a preface by George Edward Moore). The Humanities Press, New York (1950)

[**Sah 90**] Sahlin, N.-E.: The Philosophy of F. P. Ramsey. Cambridge University Press, Cambridge (1990)

[**Tay 06**] Taylor, G.: Frank Ramsey—a biographical sketch. In: Galavotti, M.C. (ed.) Cambridge and Vienna—Frank P. Ramsey and the Vienna Circle. Springer, New York (2006)

References and Further Reading

[**Wit 83**] Wittgenstein, L.: Letters to C. K. Ogden with Comments on the English Translation of the "Tractatus Logico-Philosophicus"; Edited with an Introduction by Georg Henrik von Wright, and with an Appendix of Letters by Frank Plumpton Ramsey. Basil Blackwell/Routledge K. Paul, Oxford/Boston (1983)

Chapter 4
Dynamical Systems

4.1 Introduction

Many processes in life arise from the iteration of some simple rule or function. As an instance, the amount of money in your savings account is determined by the fact that the bank pays r percent interest per year. So if the initial deposit is some principal amount P, then

- after 1 year the account holds $(1 + r \cdot 0.01) \cdot P$ dollars;
- after 2 years the account holds $(1 + r \cdot 0.01)^2 \cdot P$ dollars;
- after 3 years the account holds $(1 + r \cdot 0.01)^3 \cdot P$ dollars;

and so on.

This is an instance of a *discrete dynamical system*, since the process develops in distinct steps. Many dynamical systems are in fact *continuous dynamical systems*. The function involved is defined on a line or interval, and the process develops continuously. As an example, consider a petri dish full of bacteria. It is known that a typical bacterium reproduces every 10 hours. But there are millions of bacteria in the dish, and they are all reproducing at different times. So the bacteria population is in effect continuously evolving. We can certainly say that in 10 hours, the bacteria population doubles. And in another 10 hours it doubles again. At intermediate times the population is growing as well.

Many physical systems can be thought of in terms of the iteration of a process or function. The oscillation of a pendulum and the vibration of a spring are just two examples of dynamical systems in our world. The book [Dev 81] provides considerable context and background—especially from the point of view of classical physics—for the theory of dynamical systems.

Dynamical systems theory began in the late nineteenth century with the work of Henri Poincaré (1854–1912). He solved an important problem in celestial mechanics by using techniques of dynamical systems. In modern times, Stephen Smale, Adrien Douady, John Milnor, Lennart Carleson, and many other notable mathematicians have helped to develop this theory into a powerful mathematical tool. The subject is exciting in that it integrates analysis, geometry, and computer graphics into a whole that is greater than the sum of its parts.

One of the interesting features of a dynamical system is that just a small change in the "rule" by which the system iterates can result in a large deviation in global behavior. This phenomenon is sometimes termed *chaos*. Popular books on dynamical systems, fractals, and chaos are fond of posing the question, "Can the flap of a butterfly wing in Brazil cause a tsunami in Indonesia?" If it were so, then that would perhaps be an instance of the sort of chaotic behavior we have just been describing.

4.2 Creation of the Mandelbrot Set

The Mandelbrot set is one of the most famous objects in all of the mathematical sciences. Named after Benoît Mandelbrot (1924–2010), the set is a beautiful instance of an artifact of a dynamical system. And the pictures of the Mandelbrot set are fascinating. Figure 4.1 shows the full Mandelbrot set, and Fig. 4.2 shows portions of the Mandelbrot set blown up.

Here we give an informal description of how the Mandelbrot set is created.

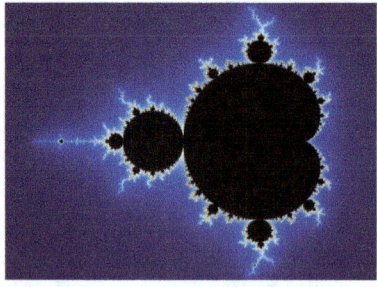

Wolfgang Beyer

Figure 4.1 The Mandelbrot set.

4.2. Creation of the Mandelbrot Set

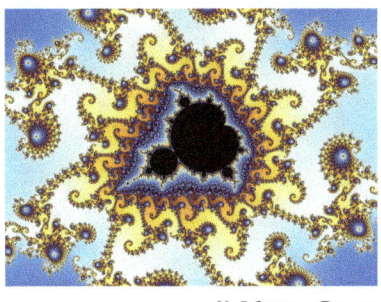 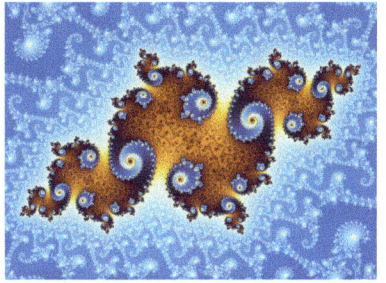

Wolfgang Beyer   Wolfgang Beyer

Figure 4.2 Blown up portions of the Mandelbrot set.

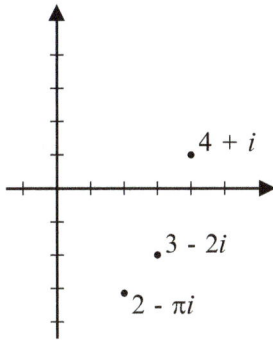

Figure 4.3 An Argand diagram.

If you are adept at computer programming, you may wish to program your computer to generate a picture of the Mandelbrot set. At the end of this discussion, we provide some sample pseudo-code that will serve that purpose.

We begin by considering the function

$$f(z) = z^2 + c.$$

Here z is a complex variable. Thus $z = x+iy$, where x and y are real numbers, and i is the square root of -1; we call x the *real part* of the complex number z and y is called the *imaginary part*. Examples of complex numbers are $3 - 2i$, $4 + i$, and $2 - \pi i$. We plot a complex number in the plane (in a picture called an *Argand diagram*) just as we plot ordered pairs of real numbers, but with the real part of the complex number determining the corresponding point's location in the horizontal direction and the imaginary part determining the location in the vertical direction. See Fig. 4.3.

Now the dynamical system induced by f is this: Take a complex number z and calculate

- $f(z)$
- $f(f(z))$
- $f(f(f(z)))$
- and so forth.

This gives a string of points in the plane. This string will either run out to infinity or it will not. In the second instance we take c to be an element of the Mandelbrot set. In the first instance we do not.

For convenience of calculation, we generally take $z = 0$. Now, as an example, take $c = 1$. Then $f(z) = z^2 + 1$ and our string of points is

- $f(0) = 0^2 + 1 = 1$
- $f(f(0)) = f(1) = 1^2 + 1 = 2$
- $f(f(f(0))) = f(2) = 2^2 + 1 = 5$
- $f(f(f(f(0)))) = f(5) = 5^2 + 1 = 26$
- and so forth.

This string of points plainly runs out to infinity. So 1 is *not* in the Mandelbrot set.

Now consider the value $c = i$. The function is now $f(z) = z^2 + i$ and our string of points is

- $f(0) = 0^2 + i = i$
- $f(f(0)) = i^2 + i = -1 + i$
- $f(f(f(0))) = (-1 + i)^2 + i = (1 - 2i - 1) + i = -i$
- $f(f(f(f(0)))) = (-i)^2 + i = -1 + i$
- and so forth.

4.2. Creation of the Mandelbrot Set

It is clear that in this case further iterations will just repeat the values $-1+i$ and $-i$. All of these complex numbers lie within distance 3 of the origin. So they stay in a bounded subset of the plane; they do *not* run out to infinity. So the number i is *in* the Mandelbrot set.

It is great fun to calculate elements of the Mandelbrot set[1] and to plot them. The resulting set is endlessly complicated.

There are various geometric questions that one can ask about the Mandelbrot set. For instance, is it connected—that is, is it just of one piece? This turns out to be true, and requires a sophisticated mathematical proof (see [DH 84]).

There is a three-dimensional version of the Mandelbrot set, called the *Mandelbulb*, that was created by Daniel White and Paul Nylander. A spectacular image of the Mandelbulb created by Ondřej Karlík is exhibited in Fig. 4.4.

Ondřej Karlík

Figure 4.4 The Mandelbulb.

---

[1] To be fair to history, it should be noted that Brooks and Matelski [BM 81] described and produced the Mandelbrot set a couple of years before Mandelbrot studied it. But Mandelbrot and his followers did a lot more with it.

4.2.1 Pseudo-Code to Generate the Mandelbrot Set

Pseudo-code is a device for showing the logic of a program, without actually implementing the particular syntax of a specific programming language. If you are adept at programming, then you can easily translate the pseudo-code below into C++, Python, JAVA, or any other language.

For each pixel on the screen perform this operation:

```
{
x0 = x co-ordinate of pixel
y0 = y co-ordinate of pixel

x = 0
y = 0

iteration = 0
max_iteration = 1000

while ( x*x + y*y <= (2*2) AND iteration < max_iteration )
{
   xtemp = x*x - y*y + x0
   y = 2*x*y + y0

   x = xtemp

   iteration = iteration + 1
}

if ( iteration == max_iteration )
then
   color = black
plot(x0,y0,color)
else
   color = iteration

plot(x0,y0,color)
}
```

4.3 Staircase Representation of a One-Dimensional Dynamical System

If f is a function and p_0 is a point, then the *orbit* of p_0 is the collection of points

$$\mathcal{O}(p_0) = \left\{ p_0, f(p_0), f(f(p_0)), f(f(f(p_0))), \ldots \right\}.$$

In other words, the orbit is the string of points that is the result of the action of the dynamical system. In practice it is useful to let

$$f^j(p) = \underbrace{f(f(f(\cdots f(p) \cdots)))}_{j \text{ times}}.$$

But it should be noted here that j is *not* an exponent in the usual sense. It does *not* mean that f is to be multiplied by itself j times. Rather, f is composed with itself j times. With this notation, the orbit of p_0 is

$$\mathcal{O}(p_0) = \left\{ p_0, f(p_0), f^2(p_0), f^3(p_0), \ldots \right\}.$$

One of the basic ideas in one-dimensional dynamical systems theory is that of the "staircase representation" for orbits. Let $f : I \to I$ be a mapping of an interval[2] I to itself and fix a point $\mathbf{p}_0 \in I$. Draw the graphs of $y = f(x)$ and $y = x$ on the same set of axes in the x-y plane. See Fig. 4.5. The staircase representation of the orbit $\mathbf{p}_j = f^j(\mathbf{p}_0)$ is generated as follows. Draw a vertical line from the point $Q_0 = (\mathbf{p}_0, \mathbf{p}_0)$ on the line $y = x$ to the graph of $y = f(x)$; the intersection will be at the point $P_1 = (\mathbf{p}_0, \mathbf{p}_1)$. Next, draw a horizontal line from P_1 to the graph of $y = x$; the intersection will be at the point $Q_1 = (\mathbf{p}_1, \mathbf{p}_1)$. We repeat this process. Draw a vertical line from Q_1 to the graph of $y = f(x)$; the intersection will be at $P_2 = (\mathbf{p}_1, \mathbf{p}_2)$. Next, draw a horizontal line from P_2 to $Q_2 = (\mathbf{p}_2, \mathbf{p}_2)$ on the line $y = x$. And so forth. Refer to Fig. 4.6, and convince yourself that this process may be continued indefinitely and that the y-coordinates of the points generated form the orbit of $\mathbf{p}_0$.

---

[2] An interval is a contiguous set of numbers in the real line, for example, the unit interval $[0, 1]$ consisting of the numbers 0, 1, and all numbers between them.

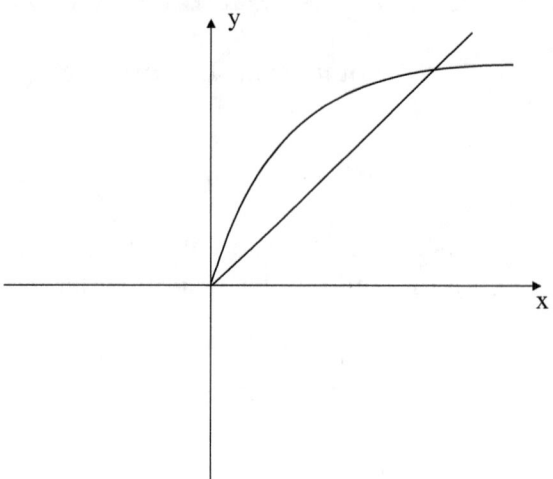

Figure 4.5 Start of the staircase.

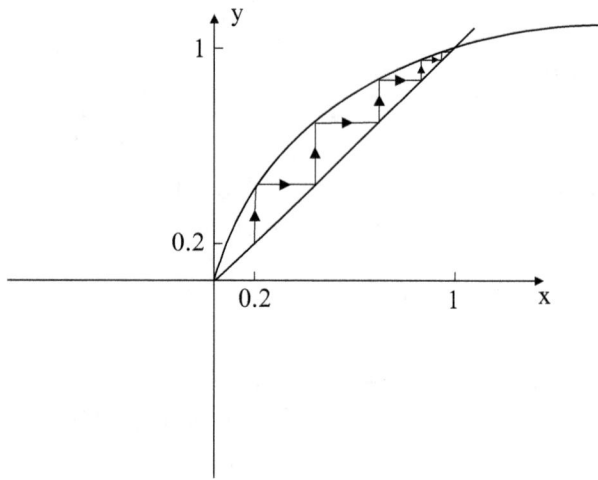

Figure 4.6 The staircase representation.

4.3. Staircase Representation of a Dynamical System

For You To Try: Sketch the staircase representation for each of the following functions:

1. $f(x) = \cos x + \dfrac{1}{3}\cos \pi x$, $\mathbf{p}_0 = 1/8$

2. $g(x) = \dfrac{1}{2}(x^2 + 2x + 1)$, $\mathbf{p}_0 = 1$

3. $h(x) = \dfrac{1}{2}(x^2 - 2x + 1)$, $\mathbf{p}_0 = -2$

4. $f(x) = \dfrac{1}{2}(x^2 + x + 1)$, $\mathbf{p}_0 = 3$

5. $g(x) = \dfrac{1}{3}(x^2 + 2x - 1)$, $\mathbf{p}_0 = -1$

6. $h(x) = 3x(1-x)$, $\mathbf{p}_0 = 0.4$

4.3.1 The Use of the Pocket Calculator

One of the genuinely fun and also very concrete realizations of a dynamical system can be seen with your pocket calculator.

As a first instance, enter any positive number on your calculator and then repeatedly push the $\sqrt{}$ button (i.e., the "square root" button). Thus the function that we are iterating is simply the square root function. If the number you began with is less then one, then you will see the orbit (i.e., the string of your answers) increasing to 1. If instead the number you began with is greater than one, then you will see the orbit (i.e., the string of your answers) decreasing to 1. Thus 1 is the limit point *no matter what the initial value in the system.*[3]

A more interesting, and certainly less obvious, situation, is this. Begin with any number on your calculator. Put the calculator into radian mode. Now press the ⟨cos⟩ button (i.e., the "cosine" button). Do so repeatedly. The limit number, no matter where you began, will be 0.739085.... If instead

---

[3]If you begin with a negative number, then the first time you hit the square root button you will get an error message. If you begin with 0, then the string of your answers will all be 0s. Thus the phenomenon we are describing only applies to positive numbers.

you press the ⟨sin⟩ button (i.e, the "sine" button) repeatedly, then the limit value (which is reached *very slowly*) is 0. Why is this? What is going on here?

It is easiest to give a mathematical analysis of the first example—the square root function. If we let x_0 be the initial value entered, $x(j)$ be the value at the jth push of the button, and $f(x) = \sqrt{x}$, then we may set

$$\begin{aligned} x(0) &= x_0 \\ x(1) &= f(x(0)) = \sqrt{x(0)} = \sqrt{x_0} = x_0^{1/2} \\ x(2) &= f(x(1)) = \sqrt{x(1)} = \sqrt{x_0^{1/2}} = x_0^{1/4} \\ x(3) &= f(x(2)) = \sqrt{x(2)} = \sqrt{x_0^{1/4}} = x_0^{1/8} \\ x(4) &= f(x(3)) = \sqrt{x(3)} = \sqrt{x_0^{1/8}} = x_0^{1/16} \end{aligned}$$

and so forth. Thus

$$x(j) = x_0^{1/2^j}.$$

Since the exponent tends to 0, $x(j)$ must tend to 1.

We may "guess" the answer to the cosine example as follows. What is the first fixed point of the function $f(x) = \cos x$? That is, for what positive value of x do we have $\cos x = x$? You may find such a number x by drawing a very accurate graph, or by experimenting with your calculator. It turns out that the answer is $x^* = 0.739085\ldots$. Try punching this number into your calculator—in radian mode of course—and press the ⟨cos⟩ button. The answer you get will be $0.739085\ldots$. Now what does this have to do with our dynamical system? Let us take it for granted that the dynamical system has some limit point y. Thus

$$\lim_{j \to \infty} x(j) = y.$$

Applying the cosine function to both sides of this equation gives

$$\cos\left[\lim_{j \to \infty} x(j)\right] = \cos(y).$$

Now the cosine function is continuous, so we may pass it inside the limit. Thus we have

$$\lim_{j \to \infty} \cos(x(j)) = \cos(y).$$

This may be rewritten as

$$\lim_{j \to \infty} (x(j+1)) = \cos(y).$$

But the left-hand side is just y itself. We have therefore derived the identity

$$y = \cos(y).$$

In other words the limit point y of the dynamical system is also its fixed point.

You might enjoy applying the staircase representation to the dynamical system generated by the cosine function to obtain a graphical reason for this last result.

The same reasoning may be applied to the dynamical system generated by the sine function. The first fixed point of sine is 0. Thus the dynamical system will converge to 0. Again, try applying the staircase representation to obtain a graphical analysis of this situation.

If you are adept with spreadsheet software (Microsoft `Excel`, for example), you can easily generate values for a dynamical system. You can enter initial values $x(0)$ for the dynamical system in the first row, and program the spreadsheet to produce $x(1)$ in the second row. Then you can use the copy and paste function to automatically get $x(2)$ in the third row, and so forth.

4.4 A Little Physics

Imagine a vibrating spring with the following configuration. There is a weight on a cart that can travel on a horizontal surface with no friction. The cart is attached to a spring, and the spring is in turn attached to a wall. See Fig. 4.7.

The spring obeys Hooke's law:

> **Hooke's Law.** *The force exerted by the spring is proportional to the displacement of the cart from the rest position.*

How can we describe the resulting motion?

It is productive to view this matter as a dynamical system. Of course this is a continuous, not a discrete, dynamical system. At any given position $x(t)$, the function being applied is the force $-cx(t)$, where c is a positive constant

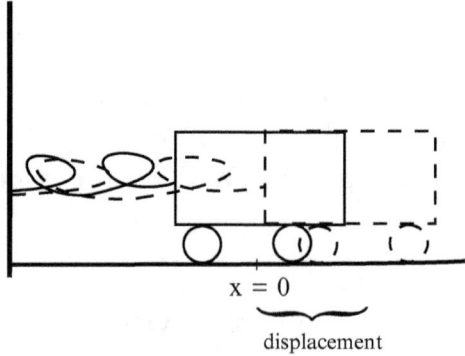

Figure 4.7 A vibrating spring.

of proportionality. This is Hooke's law and the constant of proportionality is called the *spring constant*. Clearly the force of the spring is in the direction *opposite* to the displacement $x(t)$. Using Newton's second law, this fact can be expressed as a differential equation, that is, as an equation involving the displacement of the cart, $x(t)$, as well as the acceleration of the cart. We shall not do so here.

Suffice it to say that this differential equation can be solved, and the solution has the form

$$x(t) = A\cos(\omega t) + B\sin(\omega t). \qquad (*)$$

Here A, B, and ω are constants. This is not a great surprise. For the motion of the cart should be periodic (that is, repeating after a specified period of time), and sine and cosine are our most basic periodic functions.

The value of the constant ω depends on the ratio of the spring constant and the weight of the cart[4] (and the choice of units used when measuring them). A stronger spring leads to more rapid oscillation, and a heavier cart leads to less rapid oscillation.

If we take the initial position of the cart to be $x = 0$, then A must be zero and our motion is the sine function

$$x(t) = B\sin(\omega t).$$

If instead we assume that the spring has some initial displacement $A > 0$ but is not moving initially, then the motion of the cart is given by

$$x(t) = A\cos(\omega t).$$

---

[4]If we wished to be more pedantic we would say the "mass of the cart."

4.4. A Little Physics

Depending on the initial position and velocity of the cart, the motion will have the general form (∗). Plotting displacement as a function of time, we find that, whatever the initial conditions, the graph is shaped like Fig. 4.8. The properties that can vary are:

(1) the amplitude, i.e., how high and low the curve goes,

(2) the frequency, i.e., how close together the peaks are, and

(3) the phase, i.e., how the curve is shifted from left to right.

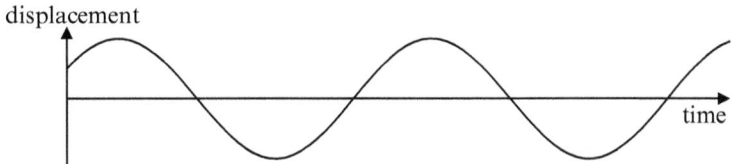

Figure 4.8 Position of the cart as a function of time.

It must be confessed that this mathematical model is not entirely realistic. A typical mass on a spring will not oscillate with constant amplitude forever. Friction will cause the cart to slow down, and the amplitude of its motion will decrease. Eventually the oscillation will come to a halt. Another deficiency of our model is that Hooke's law is only an approximation to the behavior of an actual spring.

Nonetheless the model is useful despite its known shortcomings. The model is meaningful because small errors in the model lead to only small errors in its predictions. Thus, if the cart rolls with only a little friction, then we can safely ignore friction, at least for a short time period. Similarly, for small deformations of a spring, Hooke's law is a good approximation to the behavior of an actual spring. Consequently, as long as the friction involved in moving the cart is small and the spring is not stretched too far, the model will give a good approximation to the actual motion.

Assuming that there is little friction and that the spring is not stretched too far, then each cycle of the cart's motion will come back to nearly where it started in nearly the time predicted by the model. After that each cycle will be only a little different from the previous cycle. If we can also add some energy from an external source in a well-timed way, then we can make up for the energy being lost to friction and keep all the cycles nearly the same. Finally, if we were to add a counter to keep track of the cycles, we could even use the cart/spring system to make a clock.

4.4.1 Poincaré and the Three-Body Problem

The paradigm used above may be familiar: Some basic physical laws are used to create a mathematical model for a particular system. Characteristics of the actual objects are measured. The mathematical model predicts what will happen, but there is a margin of error in the prediction because some parts of the model are approximations and because none of the measurements can be made with infinite precision or accuracy. Often classical examples emphasize regular repeating behavior, such as is shown by orbiting celestial bodies. Indeed, the solar system exhibits regular periodic behavior, and that behavior shaped people's expectations of what mathematical models of the solar system should and would predict.

The more bodies in a mathematical model of the solar system, the more difficult it is to solve the equations. In a basic physic course, one follows the path laid out by Newton in his great work *Philosophiæ Naturalis Principia Mathematica* and models the motion of two celestial bodies, but with one of the bodies so massive that it can be considered stationary. The solution of this restricted two-body problem applies to a planet orbiting the sun or a moon orbiting a planet. Kepler's laws can be obtained as a consequence of the model. This work must be considered one of the great triumphs of science.

The general two-body problem was solved by Johann Bernoulli (1667–1748) in 1710, a little more than 20 years after the publication of Newton's *Principia*. The next step should have been the solution of the three-body problem. But for 175 years no solution was forthcoming.

To motivate mathematicians to work harder on the three-body problem, or on the general n-body problem, the prominent mathematician Gösta Mittag-Leffler (1846–1927) convinced King Oscar II of Sweden and Norway (1829–1907) to offer a prize for its solution. The prize was awarded to Henri Poincaré. The resulting 270 page paper, "Sur le problème des trois corps et les équations de la dynamique," was published in the 1890 volume of *Acta Mathematica*. In this paper, Poincaré showed that, contrary to expectations, the solution of the three-body problem does not always exhibit nice regu-

4.4. A Little Physics

lar behavior. Instead three bodies moving under the force of gravity may exhibit unstable behavior. To show this Poincaré used the qualitative topological method that he had invented.

So compelling was the widely held belief in the stability of the solar system that in his original version of his paper Poincaré claimed, and believed, that he had proved that a particular class of three-body problem has a stable solution. It was only during the process of preparing the paper for publication that Lars Edvard Phragmén (1863–1937) raised questions that ultimately led Poincaré to find a major error in his paper.

The revisions the paper needed were also major. One crude indicator of the magnitude of the revision needed is that the length of the paper increased from 158 pages to 270 pages. The psychological pressure on Poincaré must have been enormous, but also Mittag-Leffler was in an awkward position with the prestige of the prize and his journal in the balance. Ultimately, it was necessary for the initial printing of that volume of *Acta Mathematica* to be destroyed (though a few copies survived).

4.4.2 Lorenz and Chaos

One might hope that the instability of the three-body problem was an exceptional situation. That is not the case, and a central role in discovering this fact was played not by a mathematician, but by a meteorologist.

No doubt you have noticed that weather forecasts are not particularly reliable. Were the phenomenon not so commonplace it would be shocking, because elaborate computer models running on fast computers analyze enormous amounts of data to produce those inaccurate forecasts. So what is going wrong?

One of the earliest pioneers of applying computer technology to meteorology was Edward Lorenz (1917–2008). Working in 1960 using what we now would consider to be a primitive computer, Lorenz needed to simplify the equations governing weather to a basic—essentially schematic—form involving but a few variables. Specifically, the equations Lorenz considered were

$$\frac{dx}{dt} = \sigma(y-x),$$
$$\frac{dy}{dt} = x(\rho-z)-y,$$
$$\frac{dz}{dt} = xy - \beta z.$$

Here σ, ρ, and β are positive constants; x, y, and z are dependent variables; and t is the independent variable, which we think of as representing time. The special values of $\sigma = 10$, $\rho = 28$, and $\beta = 8/3$ led to Lorenz's remarkable discovery (see below). It is important that the second and third equations involve products of the dependent variables on their right-hand sides. This means that the equations are *nonlinear*. In linear equations the right-hand sides would only involve sums and differences of constant multiples of the variables.

The above equations cannot be solved using formulas. Solutions must be approximated numerically. That numerical approximation is best done using a computer, which is what Lorenz did. Of course, Lorenz found the results of his computations fascinating, but sometime during the winter of 1961, Lorenz came upon an output that he felt was especially interesting and that should be examined for differing values of t. Going all the way back to the beginning of the computation seemed like a waste of time, so Lorenz took the shortcut of entering a midway state of the previous run as the initial state of the new run. What he found was that the new run *did not* proceed as the old run had. At first the difference between them was nil as far as he could tell—because he had entered apparently identical data—but as t increased the two runs diverged from each other so that after a while they seemed totally unrelated.

The behavior that Lorenz observed seems at first paradoxical, since digital computers are deterministic and, if functioning properly, must produce the identical output when given identical input. On the other hand, Lorenz did not in fact provide identical input for his new run. The computer was keeping more digits in its internal memory (six decimal places) than it was printing out (three decimal places[5]) and Lorenz was entering the previous output as the new input, thus making changes on the order of 10^{-4}.

Lorenz's bad news was that an almost infinitesimal change in the input can lead to a large difference in the output at a later time. This phenomenon is summed up in the *Butterfly Effect*: A butterfly flapping its wings on one side of the globe can eventually lead to a storm on the other side of the globe.

Note that the Butterfly Effect does not apply to every nonlinear set of equations. Sometimes a nonlinear system will exhibit periodic behavior, so

---

[5]Three decimal places of output seems crude nowadays, but in 1961 multiplication was often done with the aid of a slide rule and two decimal places was one's typical precision with a slide rule.

4.5. The Cantor Set as a Fractal

that it does have a tendency to stabilize itself. Even the most frantic efforts of the largest whale will not disrupt the tides.

The type of system subject to the Butterfly Effect is nonlinear and non-periodic. But, even in such a system, there may exist a type of orderliness in the midst of the seeming disorder with the trajectories approaching a set called an *attractor*. In a beginning course on systems of ordinary differential equations one considers attractors that are points or curves, but for the Lorenz equations the attractor is a fractal with dimension strictly between 2 and 3. Such an attractor is called, somewhat fancifully, a *strange attractor* (Fig. 4.9).

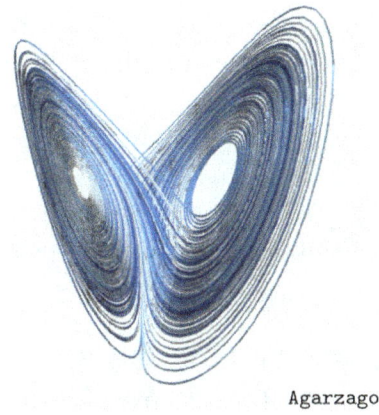

Figure 4.9 The Lorenz attractor.

4.5 The Cantor Set as a Fractal

The Cantor set is an important artifact of classical analysis. It is also an interesting example of a one-dimensional fractal. We examine it here.

Imagine the unit interval $I = [0, 1]$. For convenience we call this set S_0. Now we extract from S_0 the open interval which consists of the middle third. This produces the set S_1 (Fig. 4.10).

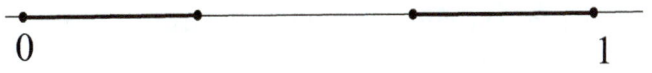

Figure 4.10 Construction of S_1 for the Cantor set.

Now we construct S_2 from S_1 by extracting from each of its two intervals the middle third (Fig. 4.11):

$$S_2 = [0, 1/9] \cup [2/9, 3/9] \cup [6/9, 7/9] \cup [8/9, 1].$$

Figure 4.11 Construction of S_2 for the Cantor set.

Continuing in this fashion, we construct S_{j+1} from S_j by extracting the middle third from each of its component subintervals. We define the *Cantor set C* to be

$$C = \bigcap_{j=1}^{\infty} S_j.$$

It can be shown—and we discuss this point below—that the set C is nonempty; in fact it has a great many points in it.

Proposition 4.5.1 *The Cantor set C has zero length, in the sense that the complementary set $[0, 1] \setminus C$ (that is, the set of elements not in the given set) has length 1.*

The notation $\setminus$ that appears in the proposition is for the operation on sets called the "set-theoretic difference." The set

$$A \setminus B$$

consists of those elements of A that are not elements of B. Another way to describe $A \setminus B$ is that it is the intersection of A with the complement of B.

Proof In the construction of S_1, we removed from the unit interval one interval of length 3^{-1}. In constructing S_2, we further removed two intervals of length 3^{-2}. In constructing S_j, we removed 2^{j-1} intervals of length 3^{-j}. Thus the total length of the intervals removed from the unit interval is

$$\sum_{j=1}^{\infty} 2^{j-1} \cdot 3^{-j}.$$

4.5. The Cantor Set as a Fractal

This last series equals

$$\frac{1}{3}\sum_{j=0}^{\infty}\left(\frac{2}{3}\right)^j.$$

The geometric series sums easily[6] and we find that the total length of the intervals removed is

$$\frac{1}{3}\left(\frac{1}{1-2/3}\right)=1.$$

Thus the Cantor set has length zero because its complement in the unit interval has length one. □

Proposition 4.5.2 *The Cantor set is infinite.*

Proof Since only open middle thirds are removed, the endpoints 0 and 1 of S_0 are never removed in forming C. So there are at least those two points in C. Now look at S_1 which consists of 2 intervals.

The endpoints of the intervals in S_1 are 0, 1/3, 2/3, and 3/3 = 1. Since only open middle thirds are removed, the endpoints of the intervals in S_1 are never removed in forming C. So there are at least those $2^2 = 4$ points in C. Next look at S_2 which consists of $2^2 = 4$ intervals.

The endpoints of the intervals in S_2 are 0, 1/9, 2/9, ..., 8/9, and 9/9 = 1. Since only open middle thirds are removed, the endpoints of the intervals in S_2 are never removed in forming C. So there are at least those $2^3 = 8$ points in C. In general, we look at S_n which consists of 2^n intervals. The endpoints of the intervals in S_n are 0, $1/3^n$, ..., $(3^n-1)/3^n$, and $3^n/3^n = 1$.

Since only open middle thirds are removed, the endpoints of the intervals in S_n are never removed in forming C. So for any n there are at least 2^n points in C. Thus the Cantor set is infinite. □

---

[6] We use the formula $\sum_{j=0}^{\infty} r^j = 1/(1-r)$, valid for $-1 < r < 1$.

The Cantor set is quite thin (it has zero length) but it is large in the sense that it has infinitely many elements. The next result reveals a surprising, and not generally well known, property of this "thin" set:

Theorem 4.5.3 *Let C be the Cantor set and define*

$$S = \{x + y : x \in C,\ y \in C\}.$$

Then $S = [0, 2]$.

Proof We sketch the proof.

Since $C \subseteq [0, 1]$ it is clear that $S \subseteq [0, 2]$. For the reverse inclusion, fix an element $t \in [0, 2]$. Our job is to find two elements c and d in C such that $c + d = t$.

First, observe that $\{x + y : x \in S_1,\ y \in S_1\} = [0, 2]$. Therefore there exist $x_1 \in S_1$ and $y_1 \in S_1$ such that $x_1 + y_1 = t$.

Similarly, $\{x + y : x \in S_2,\ y \in S_2\} = [0, 2]$. Therefore there exist $x_2 \in S_2$ and $y_2 \in S_2$ such that $x_2 + y_2 = t$.

Continuing in this fashion we may find for each j numbers x_j and y_j such that $x_j, y_j \in S_j$ and $x_j + y_j = t$. Of course $\{x_j\} \subseteq C$ and $\{y_j\} \subseteq C$. Since C has the technical property known as compactness, there are subsequences $\{x_{j_k}\}$ and $\{y_{j_k}\}$ which converge to real numbers $c \in C$ and $d \in D$. The operation of addition respects limits, thus we may pass to the limit as $k \to \infty$ in the equation

$$x_{j_k} + y_{j_k} = t$$

to obtain

$$c + d = t.$$

Therefore $[0, 2] \subseteq \{x + y : x \in C\}$. This completes the proof. □

If we do arithmetic in base 3, then we can come to an understanding of which points lie in the Cantor set (or *Cantor ternary set*, as it is sometimes called). Recall that base 3 works like base 10, except that we only work with the digits 0, 1, 2. A typical base 3 number is

$$21011.$$

4.5. The Cantor Set as a Fractal

This is (reading from the right)

- a 1 in the 1s place,
- a 1 in the 3s place,
- a 0 in the 9s place,
- a 1 in the 27s place,
- a 2 in the 81s place.

Thus our number is

$$1 \cdot 1 + 1 \cdot 3 + 0 \cdot 9 + 1 \cdot 27 + 2 \cdot 81 = 193$$

in base 10 notation.

We can also express numbers smaller than 1 in base 3 notation. The number

$$0.110221$$

has (reading from left to right)

- a 1 in the 1/3s place,
- a 1 in the 1/9s place,
- a 0 in the 1/27s place,
- a 2 in the 1/81s place,
- a 2 in the 1/243s place,
- a 1 in the 1/729s place.

Thus our number is

$$\frac{1}{3} + \frac{1}{9} + \frac{0}{27} + \frac{2}{81} + \frac{2}{243} + \frac{1}{729} = \frac{349}{729},$$

expressed as a rational fraction in base 10 notation.

Now let us begin our analysis by examining some points that are obviously in the Cantor set. Look at S_1. It has among its endpoints 1/3 and 2/3. These

two points are never removed at subsequent stages, so they are certainly in the Cantor set C. And
$$\frac{1}{3} = 0.1_3 \,,$$
$$\frac{2}{3} = 0.2_3 \,.$$

Note that we add here a subscript 3 to emphasize that (on the right) this is base 3 notation. In fact it is convenient for our purposes here to exploit the ambiguity built into base-arithmetic notation. Just as in base 10, we can write the number 82 also as $81.\overline{9}$ (where the overbar means that the 9 is repeated infinitely often), so in base 3 notation we can write 0.1_3 as $0.0\overline{2}_3$.

The four endpoints of the intervals in S_1 are thus 0.0_3, $0.0\overline{2}_3$, 0.2_3, and $0.\overline{2}_3$. At the level of S_2, we add the endpoints $1/9$, $2/9$, $7/9$, $8/9$ (in base 10 notation). In base 3 notation these new endpoints are

$$0.00\overline{2}_3 \,,$$
$$0.02_3 \,,$$
$$0.20\overline{2}_3 \,,$$
$$0.22_3 \,.$$

At the level of S_3, we add the endpoints $1/27$, $2/27$, $7/27$, $8/27$, $19/27$, $20/27$, $25/27$, $26/27$ (in base 10 notation). In base 3 notation these new endpoints are

$$0.000\overline{2}_3 \,,$$
$$0.002_3 \,,$$
$$0.020\overline{2}_3 \,,$$
$$0.022_3 \,,$$
$$0.200\overline{2}_3 \,,$$
$$0.202_3 \,,$$
$$0.220\overline{2}_3 \,,$$
$$0.222_3 \,.$$

And now a pattern begins to emerge. Namely, a point is in the Cantor ternary set C if and only if its ternary representation (using the extended notation of infinitely many 2s whenever there would ordinarily be a terminal 1) contains

4.5. The Cantor Set as a Fractal

only 0s and 2s. We invite the reader to compare this result with the use of 0s and 2s in our proof of Proposition 4.5.2 in which we "counted" the elements of the Cantor set.

Yet another way to think about Cantor-like sets is in the language that we used to describe and construct the Mandelbrot set. Consider the function

$$f_c(x) = f(x) = x^2 + c.$$

In a slight deviation from our approach to the Mandelbrot set, we now fix a value for c and determine for which x the orbit under f remains bounded. We give the set of such x the name B.

A convenient value of c to study is $c = -2.64$. Set $f(x) = f_c(x)$. One sees right away that $f(2.2) = 2.2$. So 2.2 is a fixed point for f, and it must lie in B. Also $f(-2.2) = 2.2$. So iterates of f applied to -2.2 converge to 2.2. Hence -2.2 lies in B. The curious feature about f is that its iterates exhibit chaotic behavior. For example, 2.3 is quite close to 2.2 but $f(2.3) = 2.65$. And $f(2.65) = 4.3825$. Clearly the iterates of f applied to 2.3 march off to infinity. Also 2.1 is close to 2.2, but

- $f(2.1) = 1.77$
- $f(1.77) = 0.4929$
- $f(0.4929) = -2.3970$
- $f(-2.3970) = 3.1121$
- $f(3.1121) = 7.0452$

and it is clear that the orbit of 2.1 wanders off to infinity. So 2.1 does *not* lie in B.

It requires considerable calculation—using a computer of course—to determine the composition of B. We make only a few remarks here.

We note first that points outside of the interval $[-2.2, 2.2]$ have orbits which tend to infinity. Our calculations already indicate this fact. If we solve the inequality $f(x) < -2.2$, then we find that

$$x^2 - 2.64 < -2.2$$

so that

$$x^2 < 0.44$$

and therefore
$$|x| < 0.6633.$$
Let $G_1 = (-0.6633, 0.6633)$. For any number x in this interval, $|f(x)| > 2.2$, so the orbit of x will tend to infinity. Thus no number in G_1 will lie in B. To put it more explicitly,
$$B \subseteq [-2.2, 2.2] \setminus G_1.$$
Recall that $\setminus$ is the notation for the set-theoretic difference, so the set
$$[-2.2, 2.2] \setminus G_1$$
consists of those numbers which lie in $[-2.2, 2.2]$ but not in G_1. We note that the endpoints -0.6633 and 0.6633 *do* lie in B, as simple calculations show—because $f(\pm 0.6633) = -2.2$.

Now G_1 breaks the interval $[-2.2, 2.2]$ into two pieces:
$$L = [-2.2, -0.6633]$$
and
$$R = [0.6633, 2.2].$$
We see that $B \subseteq L \cup R$. It is easy to calculate that the image of L under f is the full interval $[-2.2, 2.2]$ and the image of R under f is the full interval $[-2.2, 2.2]$.

Now here is the main point. Within each of the intervals L and R there are subintervals where $f^2(x) < -2.2$. Call these subintervals $G_{2,1}$ and $G_{2,2}$. The intervals $G_{2,1}$ and $G_{2,2}$ further subdivide L and R into four new closed intervals which we call LL, LR, RL, and RR. We can continue in this fashion with an analysis of f^3. And so forth.

The upshot is that we end up with a decreasing sequence of sets:

$$L \quad \cup \quad R$$
$$LL \quad \cup \quad LR \quad \cup \quad RL \quad \cup \quad RR$$
$$LLL \quad \cup \quad LLR \quad \cup \quad LRR \quad \cup \quad RRR$$
$$LLLL \quad \cup \quad LLLR \quad \cup \quad LLRR \quad \cup \quad LRRR \quad \cup \quad RRRR$$

and so forth. The intersection of these sets is plainly a Cantor-like set (not precisely the Cantor ternary set, but something very similar). So we now have a dynamical way to think about the Cantor set.

Self-Similarity

One of the striking features of the Cantor set is its *self-similarity properties*. In this manner it is very much like a fractal (although most familiar examples of fractals are subsets of the plane). For instance, when we look at the portion of the Cantor ternary set C that lies between $2/81$ and $3/81$, and blow it up, it looks exactly like the original Cantor set C. And one can continue in this fashion, focusing in on "smaller" and smaller parts of the Cantor set. A similar statement can be made about the Mandelbrot set, or about any other fractal.

4.6 Higher-Dimensional Versions of the Cantor Set

Historically speaking, the Cantor set inspired the modern theory of fractals. There are many higher-dimensional versions of the Cantor ternary set. We briefly describe two of them here.

4.6.1 The Sierpiński Triangle

Begin with an equilateral triangle T_0, as shown in Fig. 4.12. Now extract from T_0 an open equilateral triangle whose vertices are the midpoints of the sides of T_0. See Fig. 4.13. This is the first step in the construction of the *Sierpiński triangle* (also known as the *Sierpiński gasket*).

Figure 4.12 Starting triangle for construction of the Sierpiński triangle.

What we see now in our figure is three congruent triangles, $T_{1,1}$, $T_{1,2}$, $T_{1,3}$, each of which is similar to the triangle T_0 with which we began. We can perform the same operation on each of these smaller triangles: extract from the triangle an open equilateral triangle whose vertices are the midpoints of the sides. See Fig. 4.14. This is the second step in the construction of the Sierpiński triangle.

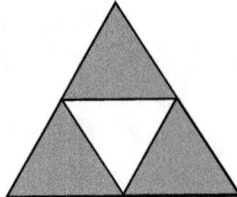

Figure 4.13 First step in the construction of the Sierpiński triangle.

Now continue. The result of the second step is nine congruent subtriangles, each of which is similar to the triangle T_0 with which we began. And we may perform our operation on each of those triangles.

Figure 4.14 Second step in the construction of the Sierpiński triangle.

We continue indefinitely. The intersection of all these regions is the Sierpiński triangle. See Fig. 4.15.

Figure 4.15 The Sierpiński triangle.

A Look Back

An alternative construction involves adding triangles instead of removing them. We begin again with an equilateral triangle as shown in Fig. 4.16.

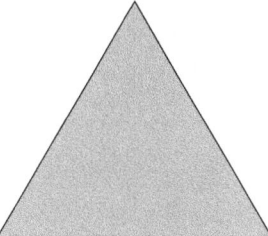

Figure 4.16 Starting triangle for construction of the von Koch snowflake.

But now, at each step we push out the middle third of each side of the triangle to form a new equilateral triangle which is one-third the size of, but similar to, the original triangle. See Fig. 4.17.

Figure 4.17 First step in the construction of the von Koch snowflake.

We continue this procedure indefinitely. The next step is shown in Fig. 4.18. The union of these infinitely many constructions is called the *von Koch snowflake*. See Fig. 4.19.

A Look Back

Dynamical systems is one of the most exciting and active areas of modern mathematical research. The reasons are that **(i)** the subject involves an interesting mix of mathematics, physics, and computer experimentation, **(ii)** many parts of the subject are broadly accessible—even to non-experts,

Figure 4.18 Second step in the construction of the von Koch snowflake.

Figure 4.19 The von Koch snowflake.

and **(iii)** Benoît Mandelbrot has served as a charismatic and powerful leader and expositor of many of the key ideas. Mandelbrot certainly promoted the idea of fractional dimension, of fractal dimension, of self-replicating sets, and of the Mandelbrot set. He helped to popularize stories about the flap of a butterfly wing causing a typhoon in Indonesia, and about the length of the coastline of England. There has rarely been a mathematician as magnetic and powerful as Mandlebrot. His followers are multitudinous.

There are those who have decried fractal geometry as too much of a descriptive language, with not enough substance. But it is difficult to argue with success. The language of the Mandelbrot set, of strange attractors, and of fractal dimension have become nearly universal in the subject. Not only have mathematicians embraced many of these ideas, but so have physicists and many other physical scientists.

Part of what is going on here is that those in the experimental sciences are always looking for new paradigms to hang their hats on. Certainly the ideas behind fractal geometry—that the basic geometry of nature is not smooth and regular but rather is rough and jagged and irregular—is a new way of looking at things. This is certainly quite different from Isaac Newton's point of view, and it has led to many new insights.

The wonderful color pictures of fractals suggest a whole new world of scientific—and even artistic—phenomena. It has been suggested—by Stephen Wolfram and others—that virtually all the phenomena of nature can be described or explained using the language of fractals (and the related idea of cellular automata). Everything from snowflakes to the coloration of panthers can, at least in principle, be explained or modeled using some fractal design.

It is not often that a major paradigm shift takes place in the language of science. Having a completely new way to see and explain nature is both exciting and challenging. Replacing the age-old models of Newton with new mathematical descriptions of the world around us opens up new possibilities for all aspects of modeling and calculation. One of Mandelbrot's remarkable traits was that he not only gave a new way to describe mathematics and physics, but also the financial markets and many other aspects of our lives. Also, Mandelbrot was a powerful and talented writer.[7] He wrote convincingly and well of his new way of doing mathematics. He rejected the old rigorous paradigm of Euclid and the other hallowed heroes of mathematics and forged a new path of experimental and phenomenological mathematics. He was a true revolutionary, and his ideas will live on for some time.

REFERENCES AND FURTHER READING

[**BM 81**]   Brooks, R., Matelski, J.P.: The dynamics of 2-generator subgroups of $PSL(2,\mathbb{C})$. In: Kra, I., Maskit, B. (eds.) Riemann Surfaces and Related Topics. Annals of Mathematics Studies, vol. 97, pp. 65–71. Princeton University Press, Princeton (1981)

[**Dev 81**]   Devaney, R.: Classical Mechanics and Dynamical Systems. CRC Press, Boca Raton (1981)

[**Dev 89**]   Devaney, R.: An Introduction to Chaotic Dynamical Systems, 2nd edn. Addison-Wesley, Redwood City (1989)

---

[7] This is not to say that Mandelbrot was right and Newton was wrong. Rather, Mandelbrot offered a new way to look at things.

[**DH 96**] Diacu, F., Holmes, P.: Celestial Encounters: The Origins of Chaos and Stability. Princeton University Press, Princeton (1996)

[**DH 84**] Douady, A., Hubbard, J.H.: Étude dynamique des polynômes complexes. Prépublications mathématiques d'Orsay **2/4** (1984/1985)

[**Fal 90**] Falconer, K.: Fractal Geometry: Mathematical Foundations and Applications. Wiley, New York (1990)

[**Fle 87**] Fleick, J.: Chaos: Making a New Science. Viking, New York (1987)

[**Man 82**] Mandelbrot, B.B.: The Fractal Geometry of Nature. W.H. Freeman, San Francisco (1982)

Chapter 5
The Plateau Problem

5.1 Paths That Minimize Length

Thanks to David Brewster's 1824 translation of Legendre's *Éléments de Géométrie*, even children in school know that a straight line is the shortest distance between two points.[1] But what if we want to solve the following problem?

Problem: Given three points, find the shortest path connecting them.
(5.1)

A typical set of three points that we might wish to connect is shown in Fig. 5.1a.

Figure 5.1 Two ways of connecting three points. (**a**) Three given points. (**b**) Shortest path using only the given points. (**c**) Shortest path when new points are allowed.

---

[1]We are considering only Euclidean space—the familiar geometric space that we live and breathe in every day.

If we are only allowed to use paths with no forks in them, then the shortest path connecting the three points is shown in Fig. 5.1b. On the other hand, if we are willing to allow a fork in the path, then the three line segments shown in Fig. 5.1c form the shortest path connecting the three points.

We can further generalize the shortest path problem by increasing the number of points that are to be connected. We can also allow any number of forks in the path or even more complicated intersections. As it turns out, the path of shortest length will always consist of (a possibly empty) set of new points and line segments connecting the various new and given points. It will also be the case that, at any of the new points, there will be exactly three line segments coming together and they will meet at equal angles of 120° (Fig. 5.2). At most three segments will end at any of the originally given points, and for any two segments ending at one of the given points, the angle they form will be at least 120°. Even if we consider sets of points in three-dimensional space instead of restricting the points to lie in a plane, the same conclusions will apply. In particular, it will still be the case that, at any of the new points, there will be exactly three line segments coming together, the three segments will lie in a plane, and they will meet at equal angles of 120°.

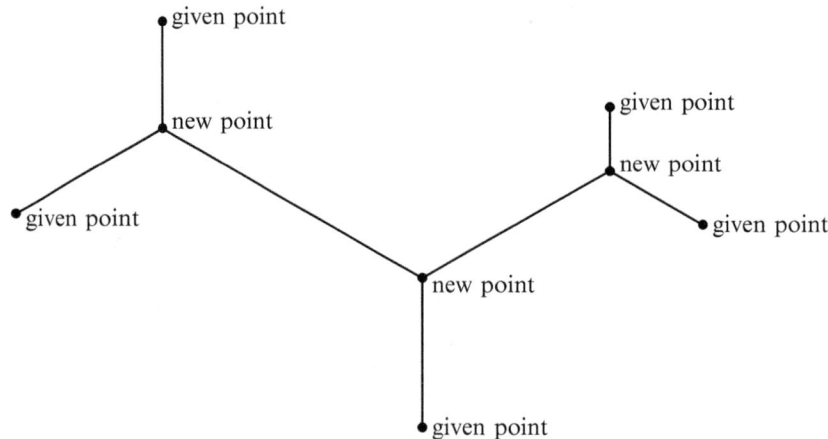

Figure 5.2 Connecting five points using the least length.

There are two lessons we should take away from the shortest path problem. The first is that allowing more complicated structures can lead to interesting solutions. That's what we did when we allowed a fork in the path. The second lesson is that, once we know solutions to a problem exist, we should

5.2 Surfaces That Minimize Area

A generalization of (5.1) is the following:

Problem: Given a closed curve (i.e., a curve begining and ending at the same place), find the surface of least area spanning the curve. (5.2)

If the closed curve lies in a plane, then the portion of the plane enclosed by the curve (Fig. 5.3) is the surface of least area spanning the curve. In particular, no area can be saved by deviating out of the plane. This fact is analogous to the fact that no length can be saved by deviating from the straight line path from one point to another.

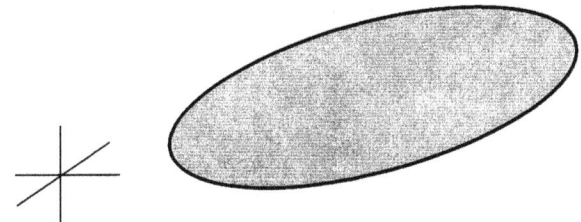

Figure 5.3 A piece of a plane is a minimal surface.

If the closed curve does not lie in a plane, then the mathematical solution to (5.2) can be very difficult indeed. In contrast to the difficulty of finding a mathematical solution to (5.2), a practical solution to the problem lies readily at hand. If one simply makes a model of the closed curve using thin wire, dips the wire model into a soap solution and removes it from the solution, then the soap film that remains on the wire model provides an excellent approximation to the surface of least area spanning the curve (Fig. 5.4). Of course, the size of the model must be well chosen and sometimes a few extraneous bubbles may need to be burst, but good results can usually be obtained. The observation that a soap film assumes the form of a minimal surface was made by the Belgian physicist Joseph Plateau. More will be said later about Plateau and his work.

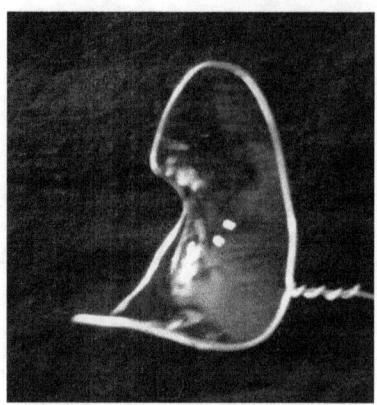

Figure 5.4 A soap film approximates a minimal surface.

The reason for the success of the soap-film method for solving (5.2) is that the surface energy of a soap film is proportional to the area of the film, so the soap film naturally assumes the area-minimizing shape. Note that a soap film is area minimizing compared to nearby configurations, not necessarily area minimizing compared to all configurations. Surfaces that are area minimizing in comparison to nearby surfaces (having the same boundary curve) are called *minimal surfaces*. Thus every area-minimizing surface is a minimal surface, but not every minimal surface is area minimizing. In Fig. 5.5 we see two copies of the same closed curve and two distinct soap films that can span the curve. Both soap films represent minimal surfaces, but only the soap film in Fig. 5.5b is area minimizing.

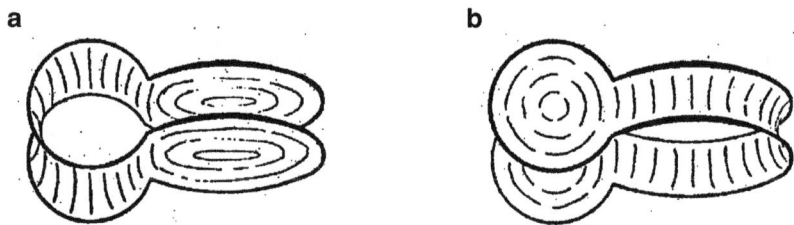

Figure 5.5 Two minimal surfaces with the same boundary. (a) A minimal surface that is not area minimizing. (b) A minimal surface that is area minimizing.

The earliest formula found for a minimal surface (other than the trivial formula for a plane) was found under the additional assumption that

5.2. Surfaces That Minimize Area

the surface be rotationally symmetric. This surface is called the *catenoid* (Fig. 5.6); it was discovered by Leonhard Euler in 1744. The discovery of the catenoid illustrates that it is sometimes easier to solve a problem with more conditions rather than with fewer conditions. Thus Euler considered area minimization *and* rotational symmetry and was able to find a solution.

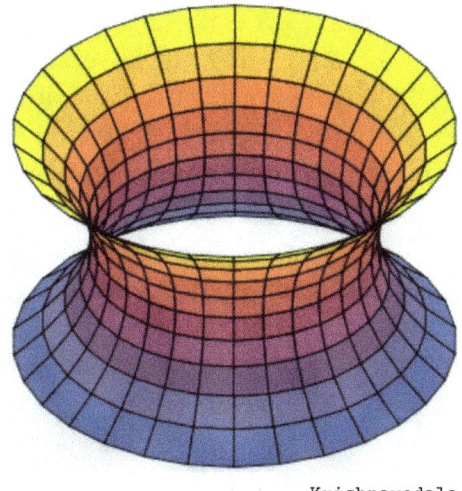

Figure 5.6 The catenoid.

Leonhard Euler

Leonhard Euler

Leonhard Euler (1707–1783) was born and educated in Switzerland, but to find academic employment it was necessary for him to leave his homeland. He spent the period 1727–1741 in St. Petersburg, Russia, the period 1741–1766 in Berlin, and then he returned to St. Petersburg where he spent the remainder of his life. The mathematical world pronounces Euler's last name as if it were "Oiler"; doing so will mark you as one of the cognoscenti. Of course, there are Americans with the last name Euler, and they use an anglicized pronunciation.

Euler was the most prolific mathematician ever. The quantity and scope of his work is monumental. It is known that he wrote about 380 articles during the 25 years he spent in Berlin, and the St. Petersburg Academy continued the process of publishing his previously unpublished work for 50 years after his death.

Euler's productivity is yet more amazing when we consider that he was burdened with vision problems from about 1740 and went blind in 1771. Euler also had a full family life. His wife gave birth to 13 children, 5 of whom survived infancy. Euler claimed to have made some of his greatest discoveries while holding a baby with other children playing round his feet. While Euler worked with talented assistants (including one of his own sons), his output was nonetheless phenomenal.

About 20 years after Euler's work, Lagrange developed a generalization of Euler's methods. Lagrange's generalization allows one to investigate area-minimizing surfaces more general than surfaces of revolution. As a particular application of his method, Lagrange derived the partial differential equation that must be satisfied by a function whose graph is an area-minimizing surface.[2] Quite appropriately, that equation is called the *minimal surface equation*. Lagrange did not find any new examples of area-minimizing surfaces, but deriving the equation that they must satisfy was a fundamental contribution to their study.

---

[2] A partial differential equation is an equation for an unknown function that depends on that function and on its rates of change in two or more directions. Most partial differential equations cannot be solved by simple formulas, and even the theoretical results concerning their solution can be disappointingly weak.

5.2. Surfaces That Minimize Area

Joseph-Louis Lagrange

J.-L. Lagrange

We usually think of Joseph-Louis Lagrange (1736–1813) as having been a French mathematician. It is true that most of Lagrange's work was written in French, and that he spent the last 26 years of his life living in Paris. On the other hand, he was born and educated in Turin (now in Italy) and was baptized Giuseppe Lodovico Lagrangia. In 1755, at age 19, Lagrange was appointed Professor of Mathematics at the Royal Artillery School in Turin. It was not until 1766 that Lagrange was induced to accept an academic position away from Turin. That position was in Berlin, where he remained for 20 years. Finally in 1787, at age 51, Lagrange moved to Paris to accept a position with the Academy of Science. Part of the attraction of the Paris position was that it entailed no teaching. The French Revolution put an end to that no-teaching arrangement, and as a result of the Revolution, Lagrange was obliged to become the first professor of analysis at the École Normale.

Lagrange began his collegiate studies with the aim of a career in law, but fortunately for mathematics he was diverted from that path. In fact, Lagrange became one of the prominent mathematicians of the late 1700s. Much of his fame came from his work on celestial mechanics. An early triumph for Lagrange was winning the Paris Academy of Science's 1764 prize for his work on the libration of the moon. We all know that the same side of the moon is always facing the Earth, but there is some irregularity of the moon's motion—called the libration—that over time allows 59% of the moon's surface to be visible from Earth. Lagrange's work on the libration of the moon and his work that produced the minimal surface equation was done while Lagrange was still in Turin.

5.3 Curvature of a Plane Curve

Consider a smooth curve in the plane, for instance the curve shown in Fig. 5.7a. At a point p on the curve, there will be many circles that are tangent to the curve at the point—in fact, there are infinitely many tangent circles. Usually among all those tangent circles there will be one circle that, when compared to any other circle, gives the best possible approximation to the curve near the point. We call that best-fitting circle the osculating circle ("osculate" is a synonym for "kiss"). For example, the plane curve in Fig. 5.7a is shown in Fig. 5.7b along with its osculating circle at p. The curvature of the curve at a point is defined to be the reciprocal of the radius of the osculating circle at that point. So a circle of small radius has large curvature and a circle of large radius has small curvature.

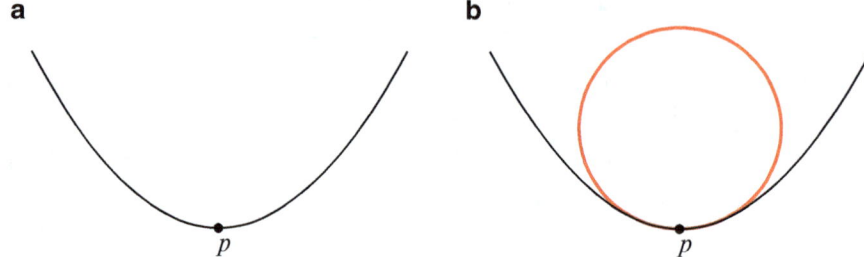

Figure 5.7 A curve and an osculating circle. (**a**) The curve. (**b**) The osculating circle in red.

Sometimes the tangent line to a curve at a point gives an even better fit to the curve than does any tangent circle at that point. The point p on the curve in Fig. 5.8a illustrates this phenomenon. The tangent line at p shown in Fig. 5.8b fits the curve better at p than does any tangent circle at p. In this case the curvature of the curve at p is defined to equal 0. If it doesn't bother you too much, you could think of the tangent line as being the tangent circle with infinite radius and then it would be consistent for the curvature to be 1 over infinity which, if it is to equal anything, ought to equal 0.

5.4 Curvature of a Surface

Next we consider a smooth two-dimensional surface in three-dimensional space, such as the surface shown in Fig. 5.9a. We will be interested in the

5.4. Curvature of a Surface

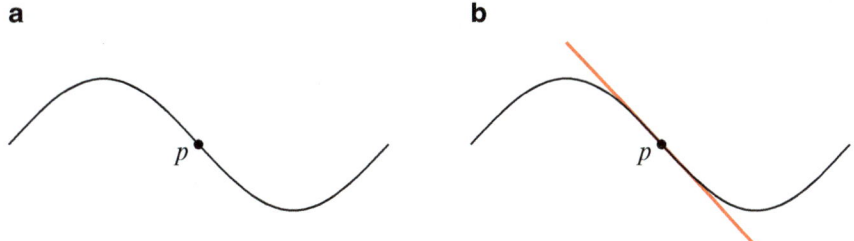

Figure 5.8 A point at which a straight line is a better fit than any circle. (a) The curve. (b) The best fit is the line in red.

behavior of the surface at the point p at the top where the tangent plane is horizontal. The curvature of the surface at that point p is analyzed by studying the curves formed when the surface is cut by a plane perpendicular to the surface at the point, that is, when the surface is cut by a vertical plane through the point. The resulting curve is called a *section* of the surface. Figure 5.9b, c show the sections formed when the surface is cut by two such planes.

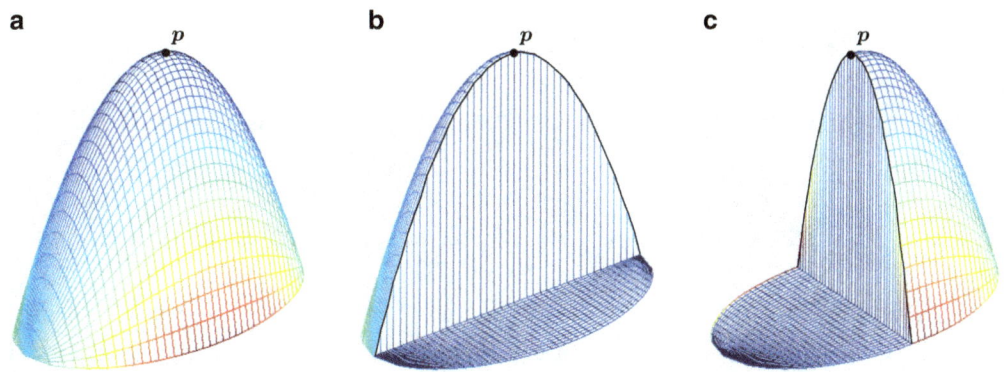

Figure 5.9 Sections of a surface. (a) The surface. (b) Cross section by a vertical plane. (c) Cross section by a second vertical plane.

Each perpendicular plane determines a section. Each section is of course a curve in the plane that determined it and thus has curvature at the point p. Figure 5.10a shows several of the curves that result when the surface in Fig. 5.9a is cut by a plane. In Fig. 5.10b the curves in Fig. 5.10a have been rotated so that they all lie in one plane and can be compared. Notice that one of the curves has the largest curvature that can be produced by cutting the surface with a vertical plane through the point at the top. Similarly, one of the curves has the smallest curvature that can be produced by cutting the surface with a vertical plane through the point. The corresponding osculating circles are shown in Fig. 5.10c. Those largest and smallest values for the curvature are called the *principal curvatures* of the surface at the point. The average of the principal curvatures is called the *mean curvature* of the surface at the point.

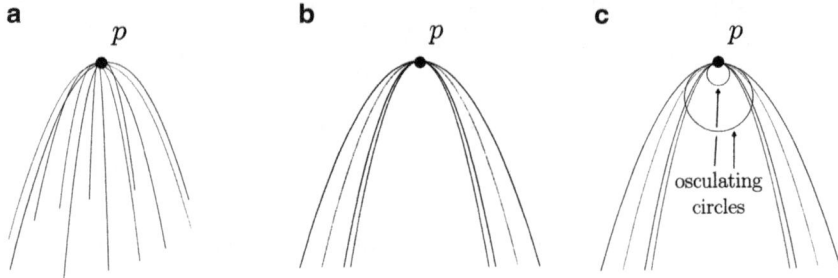

Figure 5.10 The principal curvatures. (**a**) Several cross-sectional curves. (**b**) Comparing the curves. (**c**) The smallest and largest osculating circles.

The surface in Fig. 5.9a is not special. Whenever a surface is convex in shape, as is the surface in Fig. 5.9a, there will be a maximum curvature and a minimum curvature for the curves formed by intersecting a perpendicular plane with the surface.

Of course, not every surface is convex in shape, as is the surface in Fig. 5.9a. A surface might be saddle shaped, as is the surface shown in Fig. 5.11a. We will study the surface in Fig. 5.11a near the point p in the center. The horizontal plane through p is considered to be the tangent plane to the surface at p, but here "tangent plane" means that, among planes, the horizontal plane most closely approximates the surface near p. The tangent plane will certainly intersect the surface at many points, not just at p, so the mathematical usage disagrees with ordinary parlance.

When a saddle-shaped surface such as that shown in Fig. 5.11a is cut by a vertical plane through p, the resulting curve can curve either upward[3] as in Fig. 5.11b or downward as in Fig. 5.11c, depending on the choice of the plane. Figure 5.11d shows several of the curves that result when the surface

---

[3]An *upward curve* is so called because a point on the curve goes up when it moves away from p. The analogous metaphor inspires the term *downward curve*.

5.4. Curvature of a Surface

in Fig. 5.11a is cut by a plane. All the curves in Fig. 5.11d have been rotated about the vertical axis so that they lie in one plane and can be compared. Note that one of the curves has the largest curvature among upward curves and one of the curves has the largest curvature among downward curves. The osculating circles corresponding to those largest curvatures are also shown in Fig. 5.11d.

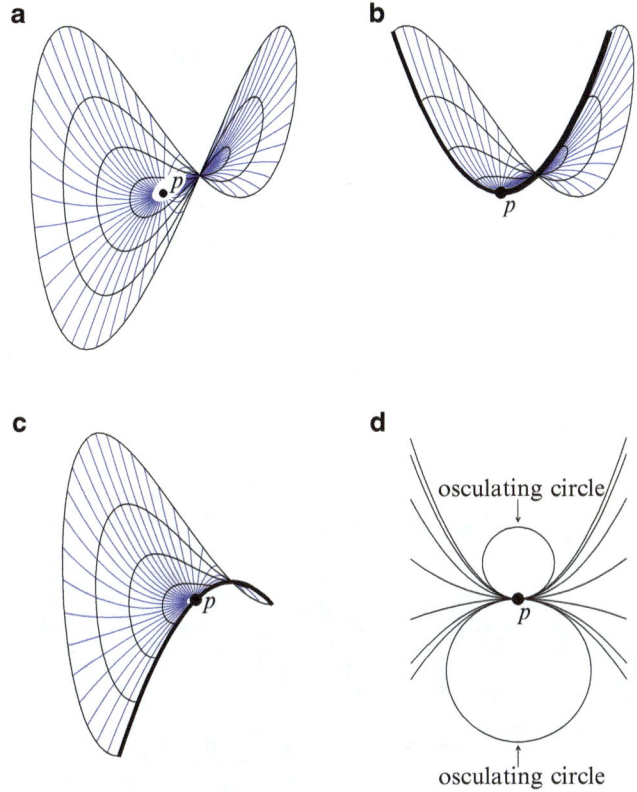

Figure 5.11 The principal curvatures in a saddle surface. (**a**) The surface. (**b**) The cross-sectional curve with the greatest upward curvature. (**c**) The cross-sectional curve with the greatest downward curvature. (**d**) Comparing the curves.

The osculating circles for a saddle surface are used to define the principal curvatures, but because the corresponding curves go in opposite directions (i.e., upward and downward), we assign one principal curvature a plus sign and the other principal curvature a minus sign. The average of the principal curvatures is again called the mean curvature of the surface at the point.

The mean curvature of a saddle-shaped surface can be positive, negative or zero. The assignment of the plus and minus signs to the curvatures is your free choice, so a positive versus a negative value for the mean curvature is not significant. On the other hand, zero mean curvature is significant.

5.5 Curvature of Minimal Surfaces

Observation of soap films will quickly convince you that they are saddle shaped. It turns out that soap films and, likewise, minimal surfaces are very special saddle surfaces because they have mean curvature 0 at each point of the surface. This remarkable fact was discovered by Meusnier in the 1770s and was one of the results in Meusnier's only mathematical paper. Meusnier also added the helicoid (Fig. 5.12) to what was still a very short list of minimal surfaces. As of 1800, the only known minimal surfaces were the plane, the catenoid, and the helicoid. It was not until the 1830s that Heinrich Ferdinand Scherk (1798–1885) added five more examples to the list of known minimal surfaces.

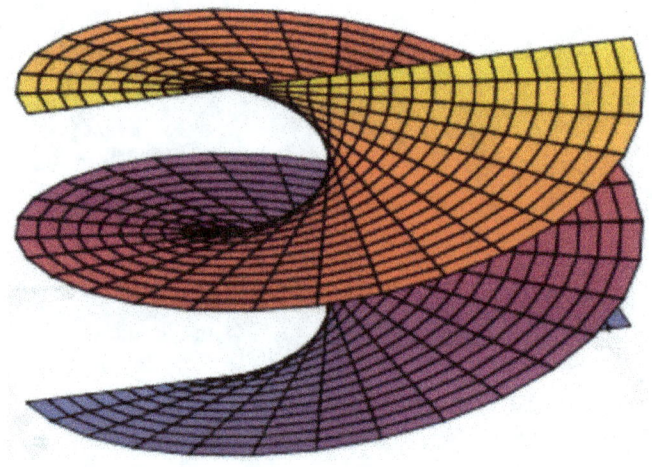

Figure 5.12 A helicoid.

J.-B.-M. Meusnier

Jean-Baptiste-Marie Charles Meusnier de la Place (1754–1793) was born in Tours, France. He was first educated at home by his father, then he received

5.5. Curvature of Minimal Surfaces

some tutoring in Paris, as preparation for the École Royale du Génie Militaire at Mézières. At the École Royale du Génie, Meusnier had the opportunity to study under Gaspard Monge. It was Monge who encouraged Meusnier to study Euler's work on the curvature of surfaces. The fruit of those studies was Meusnier's paper mentioned above, a paper which led to his being made a corresponding member of the Paris Academy at the age of 21.

Much of Meusnier's career was devoted to military engineering. He also, in 1783, presented to the Academy designs for dirigible balloons (Fig. 5.13). The more ambitious of his two proposals was for an airship 260 feet long that would carry a crew of 30. To indicate the scale of this proposal, we note that the famous rigid dirigible the Hindenberg, that flew during the mid-1930s (and burned spectacularly on May 6, 1937), was slightly longer than 800 feet and carried about 100 people.

Meusnier rose to the rank of field marshal in 1792. In June 1793, he was mortally wounded in the siege of Mainz.

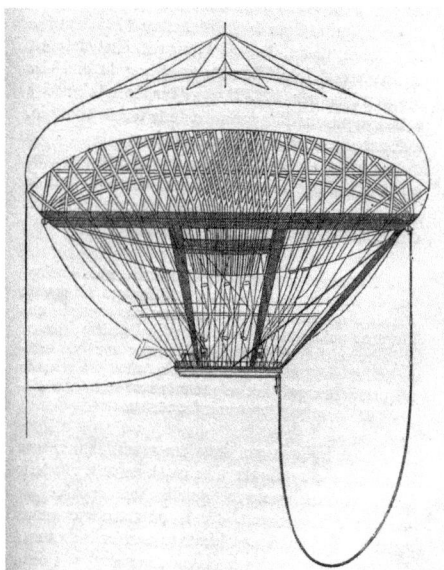

Figure 5.13 Meusnier's airship design (1784).

5.6 Plateau's Observations

Recall that one way to produce a minimal surface that spans a given curve is to make a wire model of the curve and dip it in soapy water to make a soap film. The general behavior of soap films and soap bubbles was investigated by the physicist Joseph Plateau in the mid-nineteenth century. Plateau made the following four general observations about soap films and soap bubbles:

(P1) Soap films and soap bubbles are made of collections of smooth surfaces.

(P2) The mean curvature is constant on each smooth portion of soap film or soap bubble—that constant is 0 in the case of a soap film.

(P3) Smooth portions of a soap films and soap bubbles can meet in threes along a smooth curve. When they do meet, they meet at an angle of 120°. Such a curve where three smooth pieces meet is called a Plateau border (Fig. 5.14a).

(P4) When Plateau borders meet, they do so in groups of four coming together at a point in such a way that the six angles formed are all equal (Fig. 5.14b).

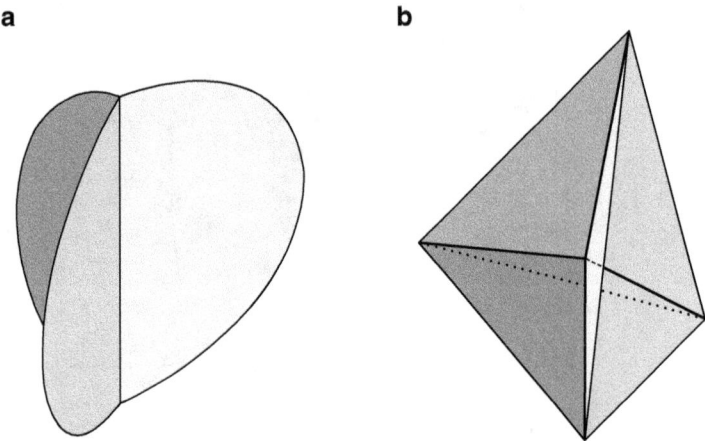

Figure 5.14 Types of singularities in soap films and soap bubbles. **(a)** Plateau border. **(b)** Four Plateau borders meeting at a point.

Plateau's experiments also led him to state that any curve can be spanned by some minimal surface. It was this last assertion that gave us *Plateau's*

5.7. Types of Spanning Surfaces

problem which asks one to prove that, given a simple closed curve in three-dimensional space, there exists a minimal surface spanning that curve.

Plateau's Problem:
Given a closed curve, find a minimal surface that spans the curve. (5.3)

Joseph Plateau

Pelizzaro (1843)
Joseph Plateau

Joseph Antoine Ferdinand Plateau (1801–1883) was a man of many talents. His artistic father sent him to the Academy of Fine Arts, but died when Joseph was 14. His mother having died a year earlier, Joseph's care fell to his uncle, a lawyer. Joseph earned a bachelor's degree in philosophy and literature, and following that a bachelor's degree in law. But science was his love, so he then registered to study mathematics and physics.

Plateau's study of surface phenomena was serendipitous. In 1840, a servant spilled oil into a mixture of water and alcohol. Plateau noticed that the drops formed perfect spheres. From that observation he went on to investigate many similar phenomena. The results of these studies were published in a number of papers and finally consolidated in 1873 in the book *Statique expérimentale et théorique des liquides soumis aux seules forces moléculaires*.

5.7 Types of Spanning Surfaces

Before we can discuss the solution of Plateau's problem in any detail, we must settle on a definition of what it means for a surface to "span" a given boundary. For example, Fig. 5.15 shows four surfaces that might or might not be considered to span the curve. Everyone should agree immediately that the surface in Fig. 5.15a *does* span the curve. Everyone should also agree immediately that the surface in Fig. 5.15b does *not* span the curve, because there is part of the curve that the surface does not even touch. The surface in Fig. 5.15c does touch the entire curve, but there is a hole in the center of the surface. If the surface in Fig. 5.15c were a soap film, the hole would grow and the surface would shrink to the bounding curve. So it is reasonable to agree that the surface in Fig. 5.15c does not span the curve. The surface in

Fig. 5.15d is like the surface in Fig. 5.15a, but with a handle added to the surface. It is as if two holes have been cut in the surface and a tube attached to connect the holes (Fig. 5.16). One might, or might not, allow a surface like that in Fig. 5.15d.

The surface in Fig. 5.15a is a disc-type surface spanning the circle because a disc can be stretched, without tearing, into the shape of the surface. Furthermore, the stretching of the disc can be done so that the boundary of the disc is simultaneously stretched into the shape of the curve and the curve is covered exactly once by the deformed boundary of the disc. Stretching without tearing is what mathematicians call a *continuous map*. The disc-type surface in Fig. 5.15a is considered to span the boundary because the continuous map that sends the disc to the surface also sends the boundary of the disc onto the curve and the map from the boundary of the disc to the

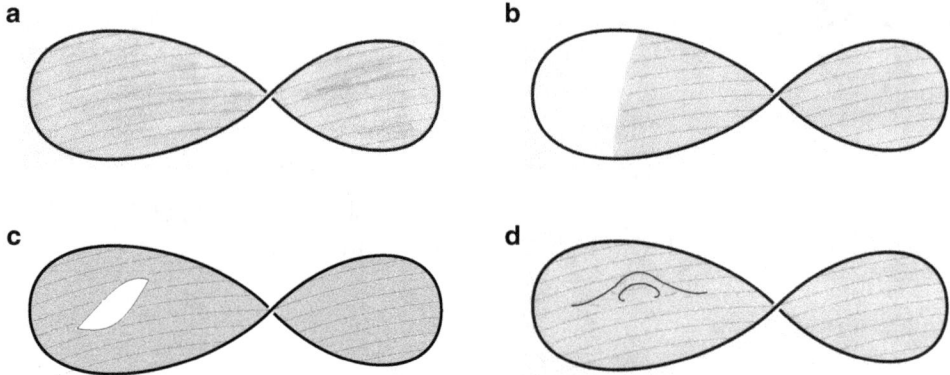

Figure 5.15 Spanning a boundary. (**a**) A disc-type surface spanning the boundary. (**b**) The surface does not touch all of the boundary. (**c**) The surface has a hole torn out of it. (**d**) The surface spans the boundary, but is not disc-type.

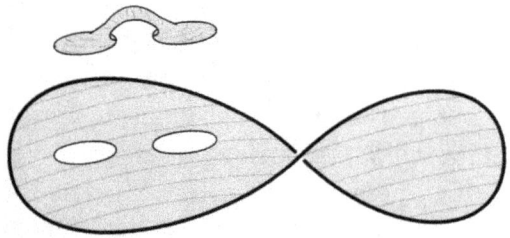

Figure 5.16 Adding a handle to a disc.

curve can be reversed to a continuous map from the curve to the boundary of the disc. The surface in Fig. 5.15d is not a disc-type surface, because a continuous map cannot create the tunnel that goes through the handle.

5.8 The Enneper–Weierstrass Formula

Spanning a simple closed curve with a disc-type surface is the conceptually simplest definition of "spanning," and it puts Plateau's problem in the context of finding a particular set of three coordinate functions defined on the disc. Not surprisingly, then, the first attacks on Plateau's problem were in the language of disc-type surfaces.

The most satisfying way to solve Plateau's problem would seem to be by giving a formula for the solution. The idea is that you would provide input data, namely, the simple closed curve, and the formula would then produce a minimal surface spanning the curve. Indeed, there does exist a formula, the *Enneper–Weierstrass formula* [discovered independently by Karl Weierstrass (1815–1897) and Alfred Enneper (1830–1885)] that takes as its input a pair of complex-analytic functions satisfying some modest compatibility conditions and produces as output a minimal surface. If one is given a polygonal path, then it is often possible to find the appropriate input functions for the Enneper–Weierstrass formula such that the output minimal surface will span the boundary. Being given a polygonal path to span is important because, as you travel along the polygon, at each point on the polygon, the perpendicular direction to any surface spanning the polygon must also be perpendicular to the polygonal line segment containing the point.

In a remarkable 1928 paper René Garnier pursued the line of attack using the Enneper–Weierstrass formula to its logical conclusion. Garnier developed a limiting argument that could be applied to a smooth curve when it is approximated more and more closely by a sequence of polygons. The result was a solution of Plateau's problem for any simple closed curve consisting of a finite number of unknotted arcs, provided the curve also has bounded total curvature. Garnier's work completed a program begun by Weierstrass some 60 years earlier.

5.8.1 Costa's Surface

In the 1980s there were exciting developments in the theory of minimal surfaces. These were new results on applying the venerable Enneper–Weierstrass formula. Now, a surface created using the Enneper–Weierstrass formula can have the property that it cuts right through itself. While such a surface certainly can be interesting, it is not the sort of surface a soap film would form. It was an open problem to create a new surface via the Enneper–Weierstrass formula that does not self-intersect and that can be extended arbitrarily far in every direction.

Important contributions to solving this problem were made by Luquésio Jorge, William Meeks, Celso Costa, and David Hoffman. Hoffman and Meeks finalized the result, and crucial to their effort was using computer graphics to visualize and dissect the surface. That part of the work needed the assistance of James Hoffman—no relation—a software engineer and inventor who was then a graduate student at the University of Massachusetts at Amherst. The surface in question is now known as the Costa–Hoffman–Meeks surface (Fig. 5.17). The full story of that famous and award-winning work appears in [Hof 87].

Anders Sandberg

Figure 5.17 The minimal surface of Costa, Hoffman, and Meeks.

5.9 Solutions by Douglas and Radó

Despite what was said earlier about a solution of the Plateau problem by a formula being the most satisfactory solution to the problem, this turned out not to be the case. Garnier's paper was *not* hailed as *the* solution of

Plateau's problem. Instead, another more abstract approach used by Jesse Douglas (1897–1965) attracted acclaim as the solution of Plateau's problem. So celebrated was that work that in 1936 Douglas was awarded one of the first two Fields Medals for the solution of Plateau's problem.[4]

Ironically, a third independent solution to Plateau's problem was provided by Tibor Radó (1895–1965) at around the same time as the results of Garnier and Douglas appeared. Whereas Douglas's work was published in 1931, Radó's was published in 1930, and Garnier's work predated both Douglas's and Radó's, so one might well wonder, "Why was it Douglas who received the Fields Medal?" It is unlikely that we will ever know the full answer to that question. Certainly both Douglas and Radó achieved greater generality than Garnier, and Douglas's solution applies in greater generality than Radó's. It also may be important that the work of Douglas and Radó was fully understood by the mathematical community, while it is quite possible that no one other than Garnier himself fully understood his 1928 paper.

5.10 Surfaces Beyond Disc Type

The subject of algebraic topology emerged in the mid-twentieth century and gave mathematicians an alternative way to describe when a surface spans a particular curve. The idea is that you can piece together a surface while maintaining a coherent and useful definition of the boundary. Thus, one can piece together the triangles shown in Fig. 5.18a. When the triangles are pushed together, as shown in Fig. 5.18b, the segments running in opposite directions "cancel out." In Fig. 5.18b the line segments without directional arrows are the "ghosts" of the oppositely oriented paths that cancel. The orientations in the solid triangular regions in Fig. 5.18b, indicated by the directed swirl, are all in the same direction—exactly because the edges had opposite orientation. When we erase the line segments that canceled each other, we are left with the solid oriented rectangle in Fig. 5.18c. The boundary of the rectangle consists of the four edges shown in Fig. 5.18c, and that boundary has the counterclockwise orientation indicated by the arrowheads drawn on it.

---

[4]The Fields Medal is generally regarded to be the "Nobel Prize" of mathematics, which is notable because there is no Nobel Prize in mathematics. One significant difference is that a Fields Medal Awardee must be under 40 years old, while Nobel Laureates are usually older.

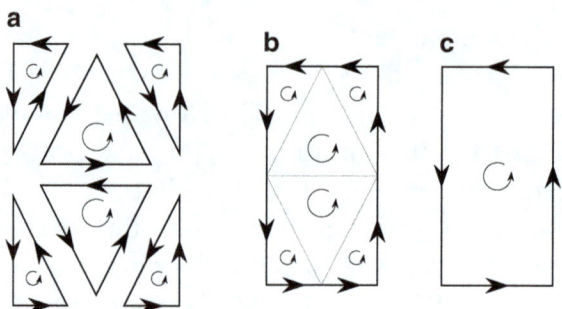

Figure 5.18 Piecing together an oriented surface. (**a**) Six oriented traingles. (**b**) The traingles have been pushed together. (**c**) Oppositely oriented edges have been removed.

Reversing the process of Fig. 5.18, Fig. 5.19 shows how to split a surface with a handle into 12 curvy triangular patches. Figure 5.19a shows the beginning surface. In Fig. 5.19b, curves and points have been drawn on the surface to divide it into four curvy pentagons. In Fig. 5.19c the four pentagons have been pulled apart, and in Fig. 5.19d, we see how a representative curvy

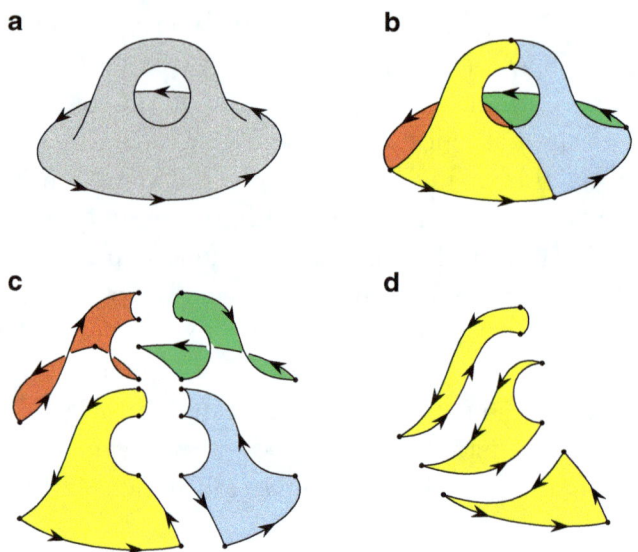

Figure 5.19 Disassembling a surface with a handle. (**a**) The beginning surface. (**b**) Curves marked to cut along. (**c**) The surface cut into four curvy pentagons. (**d**) One of the curvy pentagons cut into three curvy triangles.

5.11 Currents

pentagon can be cut into three curvy triangles. Thus the original surface with a handle can be split into 12 curvy triangles.

5.11 Currents

A standard trick of mathematicians is to enlarge the space of allowable objects so that a solution to a problem will exist. For example, enlarging the space of "numbers" from the real numbers to the complex numbers allowed mathematicians to find solutions to otherwise unsolvable polynomial equations, such as $x^2 = -1$. Similarly, at the start of this chapter, when we were considering shortest paths, we enlarged the space of paths by allowing paths in which several segments come together.

In the late 1950s a number of mathematicians expanded the collection of mathematical objects that they were willing to consider to be surfaces. One type of generalized surface that proved to be useful is called a *current* (the name "current" is used only for historical reasons). Currents are allowed to be jagged and to have handles. Currents are also allowed to occur with *multiplicity*, meaning several sheets may coincide in parts. A current has an orientation and a current has a boundary. The boundary of a current is formed in a general and flexible way, analogous to the method illustrated in Fig. 5.18. A one-dimensional current is shown in Fig. 5.20. The boundary of the current in Fig. 5.20 consists of the points on the left and right sides. The boundary of a current is also a current, so the points are oriented and given a multiplicity by assigning an integer to each point—in this case -1 and $+1$.

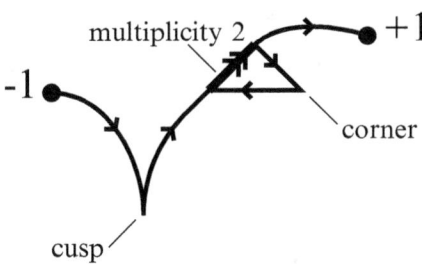

Figure 5.20 A one-dimensional current.

The use of more general surfaces, such as currents, in studying Plateau's problem allowed the solution of the problem for higher-dimensional surfaces in higher-dimensional spaces. These more general methods allowed mathematicians to prove solutions existed without any prior knowledge of what the solution surfaces might look like. Fundamental contributions were made by Ennio De Giorgi (1928–1996), Ernst Reifenberg (1928–1964), Herbert Federer (1920–2010), and Wendell H. Fleming (b. 1928).

5.12 Regularity Theory

One drawback to obtaining a solution to Plateau's problem by an abstract argument applied in a family of generalized surfaces is that the solution thus obtained might be as irregular as the least well-behaved surfaces in the family. While empirical evidence indicates that solutions of Plateau's problem should be made up of pieces of smooth surface and should obey Plateau's rules (P1)–(P4), that empirical evidence is not a proof and also is not applicable to higher-dimensional cases of Plateau's problem. The goal of regularity theory then is to close that potential gap between the generalized surface solutions and well-behaved surfaces. Unfortunately, regularity theory is quite technical, so we can give only the most general indications of how it works.

One fundamental insight is that, as a surface gets closer and closer to being horizontal, the problems of minimizing area and of minimizing elastic energy become closer and closer to being equivalent. The importance of this observation is that the problem of minimizing elastic energy is much more tractable. For elastic energy, knowledge of the boundary data can be used to obtain estimates on the interior behavior of the surface. Such estimates for the interior behavior based on the boundary, without specifically computing the solution, are called *a priori estimates*. "A priori" means "coming before" and here the significance is that one obtains the estimates before computing the solution. By making careful choices of comparison surfaces, one can use the a priori estimates for minimizing elastic energy to obtain limiting estimates for area minimization. The breakthrough work in regularity theory was done by Ennio De Giorgi in the late 1950s. Independent work that applied in more generality than De Giorgi's was done by Ernst Reifenberg in the early 1960s.

5.13 Plateau's Rules

A further extension of regularity theory was made by Jean E. Taylor. In her 1973 Princeton Ph.D. thesis she gave a mathematical proof of Plateau's Rule (P3) that when sheets of an area-minimizing surface meet, they do so in threes along a smooth curve where they meet at 120° angles. One should note the date of Taylor's thesis, because it came 100 years after the appearance of Plateau's treatise on surface phenomena. In her thesis, Taylor worked in a special algebraic-topological context designed so that a Plateau border could exist in her mathematical model of an area-minimizing surface. In her later work [Tay 76], Taylor was able to complete the program of showing—in a general setting—that Plateau's rules are mathematical facts, not solely empirical observations.

A Look Back

Plateau's problem, as treated in this chapter, was first raised by Joseph-Louis Lagrange in 1760. However it was Joseph Plateau who later experimented with soap films in an effort to understand the problem. Thus the close ties between this problem of theoretical mathematics and its basic underpinnings in mechanics and the physics of surface tension date back to the mid-nineteenth century.

Various specialized forms of the problem were solved over the years, but it was not until 1930 that Jesse Douglas and Tibor Radó independently solved the problem in some generality. Radó did it for a broad but somewhat specialized class of boundary curves, while Douglas used an entirely new method—that of graphs of harmonic functions—to find a solution for all curves. Jesse Douglas in fact was one of the winners of the first Fields Medal in 1936 for his work. Unfortunately Douglas had severe mental problems, and soon thereafter became incapacitated. It was only many years later that he returned to mathematics, and then only as a teacher of elementary courses.

The Plateau problem turns out to be a cornucopia of riches. There has been great interest in studying higher-dimensional versions of the problem. This rather general and abstract situation was finally tamed by De Giorgi and Federer/Fleming in the 1960s. Again, powerful new techniques had to be developed for this advance.

The original Plateau problem—a fairly simple question about minimal surfaces spanning a given closed curve in space—has seen many generalizations and extensions over the years. One of the most fascinating of these is the so-called double bubble conjecture. This problem asks, "what is the most efficient surface in space that will enclose two volumes of given equal size?" The problem was solved by Frank Morgan and a group of his students at Williams College in 2002. The answer, as was conjectured, is that the most efficient shape is two spherical shells which interface at a flat join. That is, it is a double bubble.

In 1976, Jean Taylor successfully studied the way in which three soap bubbles meet—the angles formed by the boundaries must be (as was anticipated from physical experiments) 120°.

There has been great interest in studying the singularities of Plateau's problem: must the minimizing surface be smooth, or will there be corners or cusps or other irregularities? This is a complicated question requiring powerful new mathematics for its solution. The prime mover behind answering the question has been Frederick Almgren (1933–1997). His 1750-page paper, published posthumously, is the capstone of many decades of effort. In that paper, Almgren proved that the set of singularities in an area-minimizing surface must be two dimensions smaller than the dimension of the surface being considered. Almgren's teacher Herbert Federer laid the foundations, together with Wendell Fleming, for studying the higher-dimensional Plateau problem. Almgren took up the gauntlet and led the charge to tame the regularity question.

Mathematicians who study Plateau's problem, and the attendant mathematical problems, have fun dipping metal frames into soap solutions and studying the resulting soap films. For soap films provide a model for minimal surfaces—this is what the surface tension equations tell us—and one can get ideas by reading nature's message. Of course, for the higher-dimensional problem, one must use one's imagination. Soap films are two dimensional surfaces—there are no three-dimensional soap films.

References and Further Reading

[**Alm 66**] Almgren, F.: Plateau's Problem: An Invitation to Varifold Geometry. W.A. Benjamin, New York (1966)

References and Further Reading

[AT 76]   Almgren, F., Taylor, J.: Geometry of soap films. Scientific American **235**, 82–93 (1976)

[Boy 20] Boys, C.V.: Soap Bubbles: Their Colours and the Forces Which Mould Them. Macmillan, New York (1920)

[Bre 30] Brewster, D.: Elements of Geometry and Trigonometry: With Notes. Oliver & Boyd, Edinburgh (1824). Translated[5] from the French of A.M. Legendre. In: Brewster, D. (ed.) With Notes and Additions, and An Introductory Chapter on Proportion

[Hof 87] Hoffman, D.: The computer-aided discovery of new embedded minimal surfaces. Mathematical Intelligencer **9**, 8–21 (1987)

[Mor 09] Morgan, F.: Geometric Measure Theory: A Beginner's Guide. Academic, Burlington, Massachusetts; Elsevier, Amsterdam (2009)

[Oss 69] Osserman, R.: A Survey of Minimal Surfaces. Van Nostrand Reinhold, New York (1969)

[Pla 73] Plateau, J.: Statique Expérimentale et Théorique des Liquides Soumis aux Seules Forces Moléculaires. Gauthier-Villars, Paris (1873)

[Tay 76] Taylor, J.E.: The structure of singularities in soap-bubble-like and soap-film-like minimal surfaces. Annals of Mathematics **103**, 489–539 (1976)

---

[5]The translation was by Thomas Carlyle. He was paid £50 for that work.

Chapter 6
Euclidean and Non-Euclidean Geometries

6.1 The Concept of Euclidean Geometry

Ancient mathematics was motivated by very practical reasoning. What we now call land management and commerce were the overriding considerations, and calculational questions grew out of those transactions. As a result, many of the ideas considered involved meshing rectangles and triangles, their areas, and their relative proportions. Basic geometry and trigonometry grew out of these largely pragmatic considerations.

It is not well understood how the concrete methodology described in the last paragraph led to the abstract subject of mathematics as we know it today. In particular, the genesis of the ideas of theorem and proof is lost in the sands of time. It was finally Euclid of Alexandria (325–265 BCE) whose writings formalized mathematics as we now conceive it. Euclid's *Elements*, a great masterpiece of ancient mathematical theory, has gone through a thousand editions and is perhaps the book of greatest duration and influence in all of intellectual history.

In the millennia preceding Euclid, questions of geometry were treated heuristically. One posited assertions about areas and ratios, and one verified them empirically. Around 600 BCE, Greek scholars transformed geometry from its practical roots into a branch of pure mathematics. Thales of Miletus is credited with initiating this transition. Most of the writings of these earliest Greek scholars have been lost, but fortunately Euclid's *Elements* has survived. From Euclid we have learned of the formalization of geometry and of the idea that a geometric assertion requires a precise statement and a precise proof.

In particular, Euclid introduced five axioms on which all of our standard, heuristic geometry is based. These axioms each have intuitive appeal, and do not demand proof because they are so "obvious." The five books of Euclid's geometry are in large part premised on an understanding of how to use these five axioms to develop and prove the basics of modern geometry. Prior to work of the Greek scholars there had never been such a formal treatment of mathematics—certainly never a formal enunciation of axioms and a methodology for proving theorems from those axioms. This is the model for the way that we do all of mathematics today.

It is worthwhile now for us to examine Euclid's axioms. We do this in part to understand what Euclid accomplished, and to appreciate the structure of his geometry. Our other motivation is to develop the ideas of *non-Euclidean* geometry and explore how they grow naturally out of the classical formalism. The version of Euclid's axioms that we use today are in fact the product of the efforts of many scholars. We formulate the famous Parallel Postulate in the language of John Playfair (1748–1819). The overall form and structure of the axioms as we present them here is due to David Hilbert (1862–1943).

What is important about Euclid's *Elements* is the paradigm it provides for the way that mathematics should be studied and recorded. Euclid begins with several definitions[1] of terminology and ideas for geometry, and then he records five important postulates (or axioms) of geometry. A version of these postulates is as follows:

P1. Through any pair of distinct points there passes a line.

P2. For each segment $\overline{AB}$ and each segment $\overline{CD}$ there is a unique point E on $\overleftrightarrow{AB}$ (the line through A and B) such that B is between A and E and the segments $\overline{CD}$ and $\overline{BE}$ are congruent,[2] written $\overline{CD} \cong \overline{BE}$ (Fig. 6.1a).

P3. For each point C and each point A distinct from C there exists a circle with center C and radius CA (Fig. 6.1b).

---

[1] We cannot take the time to describe and discuss Euclid's definitions. But they are fascinating, and indeed charming. For example, his definition of point is "that which has position but no mass."

[2] "Congruent" means they coincide exactly when superimposed.

6.1. The Concept of Euclidean Geometry

P4. All right angles are congruent.

These are the standard four axioms which give our Euclidean conception of geometry. The fifth axiom, a topic of intense study for 2000 years, is the so-called parallel postulate (stated next in Playfair's formulation):

P5. For each line ℓ and each point P that does not lie on ℓ there is a unique line m through P such that m is parallel to ℓ (Fig. 6.1c).

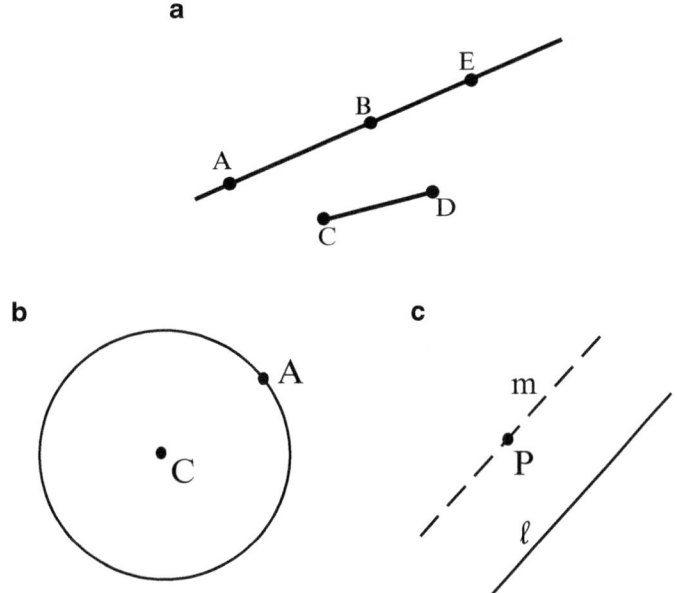

Figure 6.1 Euclid's axioms. (**a**) The second axiom: constructing a segment. (**b**) The third axiom: constructing a circle. (**c**) The fifth axiom: the Parallel Postulate.

Of course, prior to this enunciation of his celebrated five axioms, Euclid had defined "point," "congruent," "line" "between," "circle," and the other terms he used. Although Euclid borrowed freely from mathematicians both earlier and contemporaneous with himself, it is generally believed that the famous "Parallel Postulate," that is Postulate **P5**, is Euclid's own creation.

Euclid derives all of his geometry—by rigorous mathematical methods (in other words, using *definitions* and *theorems* and *proofs*)—from just these five axioms. This is a lovely illustration of the idea that would later become known as *Occam's Razor*. Named after William of Occam (1288–1348), this is a precept which advocates that any logical system should be premised on a minimal and streamlined collection of axioms or assumptions. This has been a most influential model in the development of modern logic, and in particular in the way that we set out mathematics.

Let us take a moment to describe what Euclid's axioms entail.

The Meaning of Euclid's Axioms

The First Axiom: Lines are the most basic construct in Euclidean geometry. We typically determine a line by specifying two points through which it passes. This axiom tells us that this is a valid construction.

The Second Axiom: Segments are congruent if they have the same length, so this axiom tells us that segments of a given length can be constructed.

The Third Axiom: This axiom specifies the existence of a circle with a given center and a given radius.

The Fourth Axiom: Of all angles in geometry, right angles are the most fundamental and most basic. This specifies the congruence relation on right angles.

The Fifth Axiom: This is the celebrated Parallel Postulate. It tells us of the existence of lines parallel to a given line. As important as the existence of parallel lines provided by the axiom is the uniqueness the axiom guarantees. For any point not on the given line, there is *exactly one* parallel to the given line passing through the point. This axiom is fundamental to many constructions in traditional Euclidean geometry and is what sets Euclidean geometry apart from other geometries.

6.2 A Review of the Geometry of Triangles

In this section we review some basic ideas from the Euclidean geometry of triangles. We do this in part to remind ourselves of the rigid nature of Euclidean geometry, and in part to set the stage for some of the reasoning that follows.

In fact we shall state some simple results from planar geometry and prove them in the style of Euclid. For the reader with little background in proofs,

6.2. A Review of the Geometry of Triangles

this will open up a whole world of rigorous reasoning and geometrical analysis. Let us stress that, in the present text, we are only scratching the surface.

In the ensuing discussion we shall use the fundamental notion of *congruence*. In particular, two triangles are congruent if their corresponding sides and angles are equal in magnitude. See Fig. 6.2. Thus, for two triangles to be congruent, six quantities must be equal: three side lengths and three angle measures. As a result, the traingles coincide when they are superimposed.

If two triangles can be shown to be congruent by checking five or fewer of the required six equalities, that fact is noteworthy and it must be either assumed as an axiom or proved as a theorem. In fact, there are a variety of ways to verify that two triangles are congruent without checking all six equalities[3]:

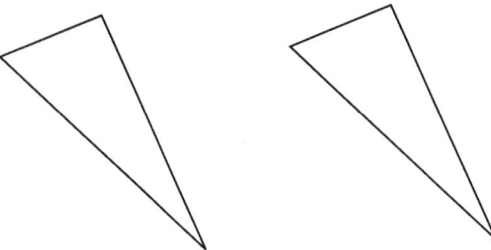

Figure 6.2 Two congruent triangles.

- If the two sets of sides may be put in one-to-one correspondence so that corresponding pairs are of equal length, then the two triangles are congruent. We call this device "Side-Side-Side" or SSS. See Fig. 6.3.

---

[3]In this discussion we use corresponding markings to indicate sides or angles that are equal. Thus if two sides are each marked with a single hash mark (or double or triple hash marks), then they are understood to be equal in length. If two angles are each marked with a single hash mark (or double or triple hash marks), then they are understood to have equal measure.

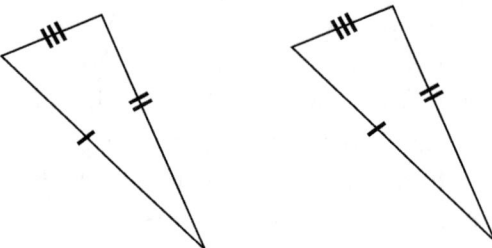

Figure 6.3 Side-Side-Side axiom.

- If just one side and its two adjacent angles correspond in each of the two triangles, so that the two pairs of angles are of equal measure and each of the corresponding sides is of equal length, then the two triangles are congruent. We call this device "Angle-Side-Angle" or ASA. See Fig. 6.4.

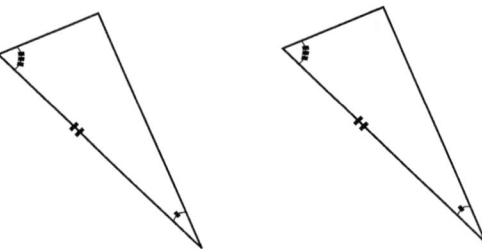

Figure 6.4 Angle-Side-Angle axiom.

- If two sides and the included angle correspond in each of the two triangles, so that the two pairs of sides are of equal length, and the included angles are of equal measure, then the two triangles are congruent. We call this device "Side-Angle-Side" or SAS. See Fig. 6.5.

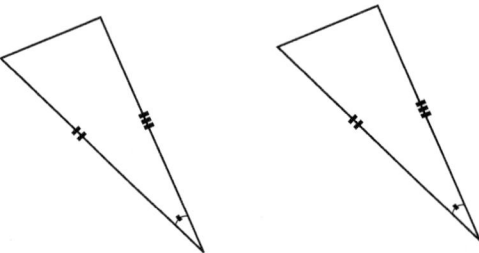

Figure 6.5 Side-Angle-Side axiom.

6.2. A Review of the Geometry of Triangles

We will take these three paradigms for congruence as intuitively obvious and thus effectively as axioms (Euclid actually provides arguments for these three paradigms).

At this point you might wonder if any set of three equalities is sufficient to guarantee congruence. This is not the case. In Euclidean geometry, similar triangles have the same shape, but they exist in infinitely many different sizes (Fig. 6.6), so there is no Angle-Angle-Angle axiom for congruence. For another example, observe that in Fig. 6.7 the two triangles $\triangle ABC$ and $\triangle ABD$ share the angle at A and the side $\overline{AB}$ while the sides $\overline{BC}$ and $\overline{BD}$ are the same length because C and D are two points on a circle centered at B. Thus Fig. 6.7 shows us that Angle-Side-Side is not sufficient to imply congruence of two triangles.

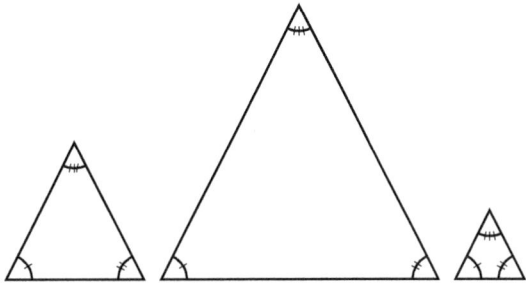

Figure 6.6 Angle-Angle-Angle does not imply congruence.

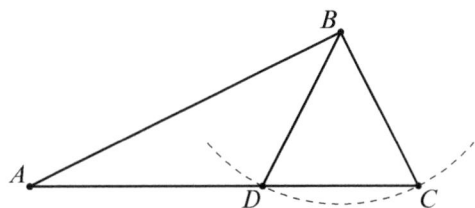

Figure 6.7 Angle-Side-Side does not imply congruence.

Of course we cannot describe the full development of Euclidean geometry here. Perhaps a sample theorem and corollary will give the gist of the subject:

Theorem 6.2.1 *Let $\triangle ABC$ be an isosceles triangle with equal sides $\overline{AB}$ and $\overline{AC}$. See Fig. 6.8. Then the angles $\angle B$ and $\angle C$ are equal.*

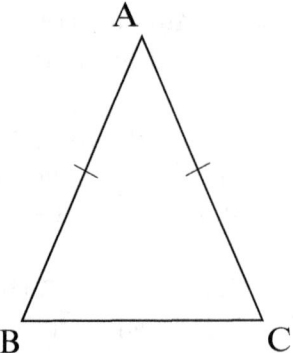

Figure 6.8 An isosceles triangle.

Proof Draw the *median* from the vertex A to the opposite side BC (here the definition of the median is that it bisects the opposite side). See Fig. 6.9. Thus we have created two subtriangles $\triangle ABD$ and $\triangle ACD$. Notice that these two smaller triangles have all corresponding sides of equal length (Fig. 6.10): side $\overline{AB}$ in the first triangle is congruent to the side $\overline{AC}$ in the second triangle; side $\overline{AD}$ in the first triangle is congruent to side $\overline{AD}$ in the second triangle; and side $\overline{BD}$ in the first triangle is congruent to side $\overline{CD}$ in the second triangle (because the median bisects side $\overline{BC}$). As a result (by SSS), the two subtriangles are congruent. All the corresponding artifacts of the two triangles are the same. We may conclude, therefore, that $\angle B = \angle C$.
□

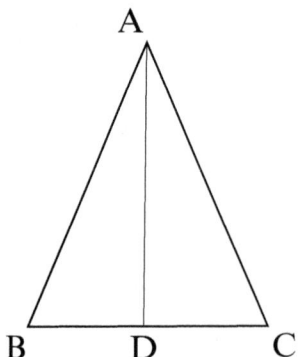

Figure 6.9 The median.

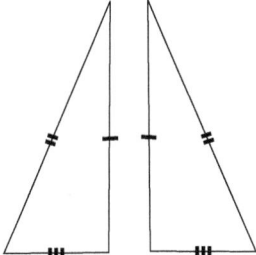

Figure 6.10 Analysis of an isosceles triangle.

Corollary 6.2.2 *Let $\triangle ABC$ be an isosceles triangle as in the preceding theorem (Fig. 6.8). Then the median from A to the opposite side BC is also perpendicular to BC.*

Proof We have already observed that the triangles $\triangle ABD$ and $\triangle ADC$ are congruent. In particular, the angles $\angle ADB$ and $\angle ADC$ are equal. But those two angles also must sum up to $180°$ or π radians. The only possible conclusion is that each angle is $90°$ or a right angle. $\square$

6.3 Some Essential Properties of Euclidean Geometry

One of the characteristic properties of Euclidean geometry is the fact that, when two parallel lines are crossed by a third line, called a *transversal*, the alternate interior angles formed are of equal measure. In Fig. 6.11, l_1 and l_2 are parallel and m is the transversal. In the figure, one pair of alternate interior angles consists of the angles labeled α and δ and the other pair of alternate interior angles consists of the angles labeled β and γ.

Theorem 6.3.1 *In Euclidean geometry, the alternate interior angles formed by a transversal intersecting two parallel lines have equal measures.*

Proof To prove that the alternate interior angles have equal measure, we will argue by contradiction. So referring to Fig. 6.11, we suppose that $\angle \alpha$ and $\angle \delta$ have unequal measure. In fact, we may suppose that $\angle \alpha$ has larger measure than $\angle \delta$, because if it were the other way around, we could simply rename the lines and angles.

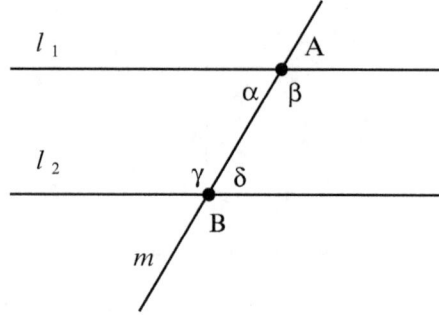

Figure 6.11 Two parallel lines and a transversal.

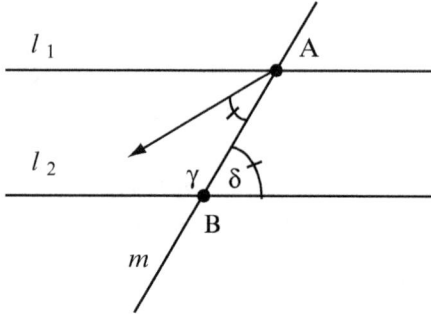

Figure 6.12 Constructing a ray.

Construct a ray originating at A that makes an angle with the ray $\overrightarrow{AB}$ that has the same measure as $\angle \delta$ (Fig. 6.12). The tic marks on the arcs tell us that those angle measures are equal. Since we are arguing by contradiction, we must accept the tic marks, even though the angles *look* unequal.

The ray just constructed must intersect l_2 in a point that we label C (Fig. 6.13). This is where we use the Euclidean parallel axiom that there can be but one line through A that is parallel to l_2.

The distance from A to C, $d(A, C)$, is now a determined quantity, so we can find a point D on l_2 that is on the opposite side of B from C and that is at the same distance from B as C is from A, that is, $d(B, D) = d(A, C)$ (Fig. 6.14).

6.3. Some Essential Properties of Euclidean Geometry

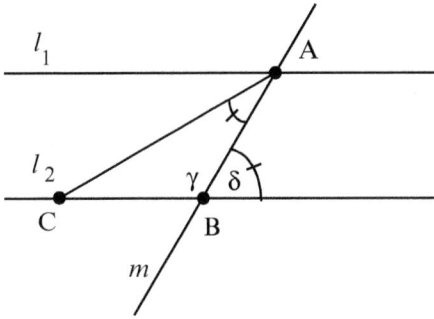

Figure 6.13 Intersecting the ray and the line.

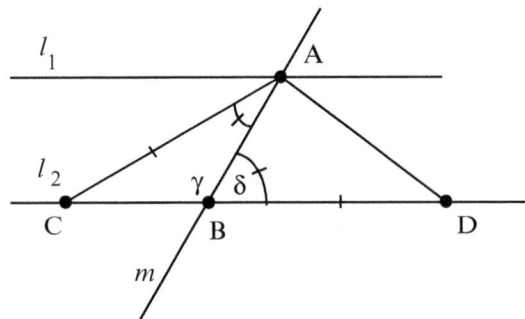

Figure 6.14 Triangles △BAC and △ABD.

Now we consider the two triangles △BAC and △ABD as in Fig. 6.14. These two triangles share the side $\overline{AB}$. The angles ∠BAC and ∠ABD are of equal measure by construction and the sides $\overline{AC}$ and $\overline{BD}$ are of equal length, also by construction. Thus the triangles are congruent by the SAS axiom.

The congruence of the triangles tells us that ∠ABC and ∠BAD must have equal measures as shown by the tick marks in Fig. 6.15. Since the angles ∠γ and ∠δ are supplementary,[4] the angles ∠BAC and ∠BAD are also supplementary, that is, C, A, and D are colinear (of course, they cannot *look* colinear in our figures).

Because C, A, and D are colinear, the line segment $\overline{CD}$ contains the point A. On the other hand, C and D were defined to be points on l_2, so the line segment $\overline{CD}$ is a subset of l_2. We conclude that l_1 and l_2 have the point A in common, contradicting the fact that they are distinct (nonintersecting) parallel lines. □

---

[4] Angles are *supplementary* if their measures add to 180°.

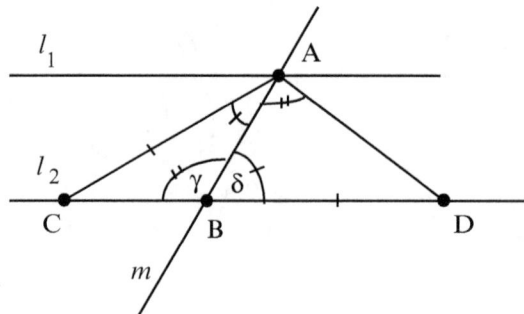

Figure 6.15 Congruent angles.

Another characteristic property of Euclidean geometry is the fact that the measures of the interior angles of a triangle add to 180°. We prove this fact next using the preceding result about alternate interior angles.

Theorem 6.3.2 *In Euclidean geometry the sum of the measures of the interior angles in a triangle is 180°.*

Proof Let the triangle $\triangle ABC$ be given. Let l be the line through C parallel to the line $\overleftrightarrow{AB}$, and name two points D and E on that parallel line on opposite sides of C (Fig. 6.16).

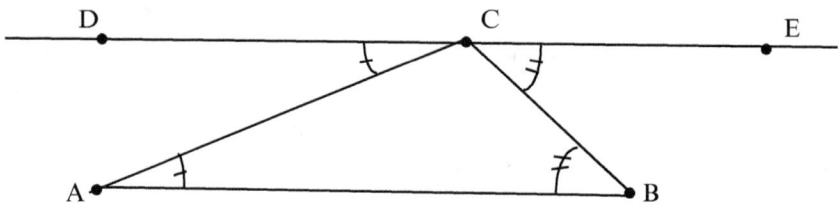

Figure 6.16 Sum of the angles in a triangle.

Next we make the crucial observation that the angles $\angle ABC$ and $\angle ECB$ have the same measure and the angles $\angle BAC$ and $\angle DCA$ also have the same measure (by Theorem 6.3.1). Thus the three angles $\angle DCA$, $\angle ACB$, and $\angle EBC$ have the same measures as the three interior angles in the triangle $\triangle ABC$. Since the measures of the three angles $\angle DCA$, $\angle ACB$, and $\angle EBC$ clearly add to 180°, we see that the measures of the three interior angles in the triangle $\triangle ABC$ also add to 180°. □

6.4 What is Non-Euclidean Geometry?

For more than 2000 years people wondered whether the Parallel Postulate (number 5 in our enumeration) is redundant, that is, does it actually follow from the other axioms? Does this axiom in fact follow from the other four? It is an intuitively appealing idea that this should be the case, and it is what people believed for a good many years—indeed millennia.

Euclidean geometry gives a very accurate, though idealized, representation of lines, angles, triangles, and other figures on a flat surface such as a sheet of paper or a blackboard. In the outside world of fields and farms and city streets, Euclidean geometry still serves well, but compromises with the idealization are evident. The shortest path between two points is a straight line segment, but knowing that is little help when mountains get in the way or, in town, if the streets don't align with that shortest path. Certainly the shortest distance—the best flight path, for instance—on the surface of the earth is not a straight line. In fact it is given by what we call a "great circle."

So it is that we can imagine that there may be geometries that are not identical to that of an infinitely large blackboard. Perhaps the role of lines should be played by other paths. Then we can ask whether, in every such geometry for which the first four postulates are true, it is also the case that the parallel postulate is true.

If we do find a geometry in which the first four postulates are true, but in which the parallel postulate is false, then we will know that the parallel postulate is *not* redundant—that the parallel postulate is, in fact, *independent* of the other postulates.

6.5 Spherical Geometry

Carrying on with the train of thought from the preceding section, we consider geometry on the unit sphere in three-dimensional space (Fig. 6.17). The role of straight lines in this new geometry is played by great circles. Recall that a *great circle* is a circle obtained by intersecting the sphere with a plane through the center of the sphere. See Fig. 6.18. On the surface of the sphere the shortest path between two points is an arc of a great circle, so it is appropriate that great circles should play the same role in the geometry of the sphere that straight lines play in the geometry of the plane. One may verify directly that the first four postulates of Euclid are satisfied, or at least nearly satisfied (see the discussion below), in this new geometry.

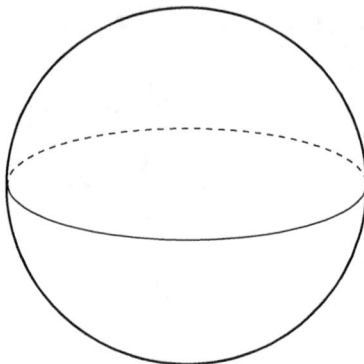

Figure 6.17 The sphere as a domain for geometry.

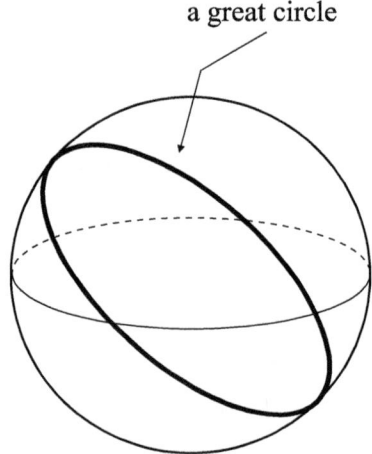

Figure 6.18 A great circle playing the role of a line.

- The first Euclidean postulate states that "Through any pair of distinct points there passes a line." On the surface of the sphere, it is true that through any pair of distinct points on the sphere there passes a great circle.

In most instances the two given points and the center of the sphere determine a plane, and the intersection of that plane and the sphere is the great circle containing the two points. But if the two points on the sphere are antipodal points (that is to say, opposite points), then the construction just given does not work because the origin and the two given points are collinear. In case the two given points on the sphere are antipodal, we can pick any of the

6.5. Spherical Geometry

infinitely many planes containing the two points, and the intersection of that plane and the sphere is a great circle containing the two points. Thus we see that, unlike the situation for points and lines in the plane (or for points and lines in space) for which there is exactly one line through two distinct points, there may be more than one great circle passing through a pair of points.

- The second Euclidean postulate states that "For each segment $\overline{AB}$ and each segment $\overline{CD}$ there is a unique point E on $\overleftrightarrow{AB}$ such that B is between A and E and $\overline{CD} \cong \overline{BE}$." The interpretation of this postulate on the sphere is hampered by the requirement that B be between A and E. The fundamental difficulty here is that a great circle, while unbounded (i.e., it has no end point(s)), is of only finite extent. If we take $\overline{AB}$ to be an arc of a great circle, then we can start at B and append another arc of a great circle and call its other endpoint E, but the location of E might be such that we cannot reasonably say that B is between A and E.

- The third Euclidean postulate states that "For each point C and each point A distinct from C there exists a circle with center C and radius CA." The distance between two points on the sphere should be measured on the surface of the sphere. If A is the antipodal point from C, then the set of points on the sphere at the same distance from C as A is from C will contain only A. If A is not antipodal from C, then the set of points on the sphere at the same distance from C as A is from C will be an ordinary circle that lies on the surface of the sphere. If we allow a single point to be called a circle, then the third postulate holds.

- The fourth Euclidean postulate states that "All right angles are congruent." The angle formed by the intersection of a pair of great circles is defined to be the angle between their tangent lines at the point of intersection. With this definition, the fourth postulate holds unequivocally.

Our new spherical geometry is a geometry similar to Euclid's geometry. But we shall now see that the fifth axiom—the Parallel Postulate—fails dramatically. The reason is that if l is a given line and P is a point not on that line—see Fig. 6.19—then there will not exist a line through P that is parallel to l. As we see, any "line" through P—that is, any great circle through P—will in fact intersect l. The geometry on the surface of the sphere is an example of a non-Euclidean geometry.

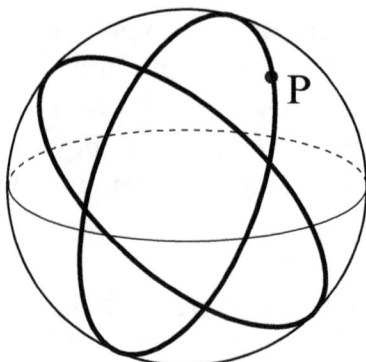

Figure 6.19 Failure of the parallel postulate.

One might feel that we have not truly shown the independence of the Parallel Postulate. For one thing, we needed a certain amount of equivocation to say that we had satisfied the first three postulates. Additionally, if Euclid were to comment on the sphere as a model of his geometry, he would probably object to the following features of spherical geometry:

- Points cannot be arbitrarily far apart on a sphere.
- A single point does not divide a great circle into two separate parts.

6.6 Neutral Geometry

To be completely convincing in asserting that the Parallel Postulate is independent of the other postulates, we should have a model that satisfies the other postulates with no equivocation. We would also want to satisfy any hidden assumptions that Euclid might have made.

In addition to the postulates, Euclid also noted some facts that he called "Common Notions."

- Things which are equal to the same thing are also equal to one another.
- If equals are added to equals, the wholes are equal.
- If equals are subtracted from equals, the remainders are equal.
- Things which coincide with one another are equal to one another.
- The whole is greater than the parts.

6.6. Neutral Geometry

Of course, these common notions are obviously true and we use them without thinking about it. But close study of Euclid reveals that he made assumptions beyond those that he codified in his postulates and common notions. As one trivial example, the actual existence of points is not a postulate or common notion, nor is it included in a definition!

Among the many who sought to prove the parallel postulate from the others was the Jesuit priest Giovanni Girolamo Saccheri (1667–1733). Saccheri's attempted proof of the parallel postulate centered on detailed consideration of a quadrilateral with two sides of equal length that are perpendicular to a third side. These figures are now known as *Saccheri quadrilaterals*. That third side perpendicular to the two equal sides is called the *base* of the Saccheri quadrilateral.

In studying a Saccheri quadrilateral, attention is focused on the remaining two angles, the *summit angles* of the Saccheri quadrilateral. Those angles are equal and apparently they might be acute angles, right angles, or obtuse angles.[5] Figure 6.20 illustrates the shapes a Saccheri quadrilateral might take in a geometry in which the lines are not necessarily the familiar straight lines of Euclidean geometry. Figure 6.20b is what a Saccheri quadrilateral must look like in Euclidean geometry—lines are straight and the figure is a rectangle.

Figure 6.20 Saccheri quadrilaterals. (**a**) Acute summit angles. (**b**) Right summit angles. (**c**) Obtuse summit angles.

Saccheri's planned path to proving the parallel postulate was to show that the summit angles could neither be acute nor obtuse, so the only possibility would be that they are always right angles. Saccheri gave a correct proof that the summit angles cannot be obtuse, i.e., Fig. 6.20c does not actually occur in any geometry that is a truly faithful model of Euclid's first four postulates and his other assumptions. On the other hand, we now know with

---

[5] Angles measuring less than 90° are *acute*, while angles measuring more than 90°, but less than 180°, are *obtuse*.

certainty that it *is* possible for the summit angles in a Saccheri quadrilateral to be acute. So Saccheri was destined to fail, and Fig. 6.20a is a schematic representation of a valid possibility.

Over the centuries many incorrect proofs purported to show that the Fifth Postulate was a consequence of the first four. Saccheri's attempt, which we described above, was recorded in his 1733 book *Euclides ab omni nævo vindicatus* (the title translates roughly to "Euclid freed of all defects"). Saccheri's book was rediscovered after 150 years and recognized as a signal contribution to "neutral geometry," i.e., geometry that omits the parallel postulate, but is otherwise that of Euclid.

6.7 Hyperbolic Geometry

In the early nineteenth century, a few mathematicians finally were willing to contemplate the possibility that the Parallel Postulate *does not* follow from the other four. The two most prominent people associated with this new point of view are János Bolyai (1802–1860) and Nikoali Ivanovich Lobachevsky (1792–1856). They proposed what we now call "hyperbolic geometry." Ironically, Bolyai's father, Farkas, was friends with the noted mathematician Carl Friedrich Gauss (1777–1855). Farkas Bolyai wrote to Gauss and told him of János's wonderful discovery. Gauss replied as follows (translated from the original German):

> If I begin with the statement that I dare not praise such a work, you will of course be startled for a moment: but I cannot do otherwise; to praise it would amount to praising myself; for the entire content of the work, the path which your son has taken, the results to which he is led, coincide almost exactly with my own meditations which have occupied my mind for from 30 to 35 years. On this account I find myself surprised to the extreme.
>
> My intention was, in regard to my own work, of which very little up to the present has been published, not to allow it to become known during my lifetime. Most people have not the insight to understand our conclusions and I have encountered only a few who received with any particular interest what I communicated to them. In order to understand these things, one must first have a keen perception of what is needed, and upon this point the majority are quite confused. On the other hand, it was my plan to

6.7. Hyperbolic Geometry

> put all down on paper eventually, so that at least it would not finally perish with me.
>
> So I am greatly surprised to be spared this effort, and am overjoyed that it happens to be the son of my old friend who outstrips me in such a remarkable way.

This seemingly well-intentioned statement from Gauss had a devastating effect on young Bolyai, and the novice was tormented by it for the rest of his life. Nonetheless, János Bolyai and Nikoali Lobachevsky are credited today with the creation of non-Euclidean geometry. They developed a geometry that is true if the Euclidean parallel postulate is replaced by the following postulate.

Hyperbolic Parallel Postulate. Through a given point not on a given line there are at least two distinct lines parallel to the given line.

Many of the results in hyperbolic geometry are counterintuitive to anyone who believes Euclidean geometry is the one true geometry. For example, in hyperbolic geometry (1) any two similar triangles must be congruent, (2) there are no rectangles, and (3) the sum of the interior angles in a triangle is *always strictly less* than 180°.

6.7.1 The Question of Consistency

Now it is one thing to posit a set of axioms for a putative discipline. It is quite another to show that those proposed axioms are not mutually contradictory. A set of axioms that is not mutually contradictory is also called a *consistent* axiom system. Of course, a set of mutually contradictory axioms might be such that one can easily see a contradiction or maybe a simple argument could reveal a contradiction. More worrisome is the possibility that an argument that is very clever or very long or both is needed to reveal a contradiction. Given a mathematical theory that *seems* to be free of internal contradictions, the way mathematicians show it is, *in fact*, free of contradictions is by constructing what is called a *model* for the theory. This involves using another mathematical theory, say set theory (Sect. 12.6), to produce a concrete mathematical object that satisfies the axioms of the theory being investigated. Once a model has been constructed, then we know that if set theory itself is free of internal contradictions, then the same is true of the theory being investigated. In practice, one does not often go all the way back to set theory to construct a model—mathematicians work with higher level constructions—but, in principle, they could start with set theory and work up from there.

In the case of classical Euclidean geometry, there is a standard "model" of lines and points and circles that indeed fits the description provided by Euclid's original axioms. For non-Euclidean geometry—a geometry that does *not* satisfy the Euclidean Parallel Postulate—a new model is needed with the new, unexpected properties Bolyai and Lobachevsky (as anticipated by Gauss) discovered. This geometry must satisfy Euclid's other four postulates, but not the Parallel Postulate. Neither Bolyai nor Lobachevsky provided a model, so the theory remained suspect, until 1868 when Eugenio Beltrami filled the gap with not just one model, but with three.

6.7.2 Models of Hyperbolic Geometry

The first model of hyperbolic geometry was the pseudosphere constructed by Beltrami in the 1860s.

Simpler models developed later proceed by embedding the hyperbolic plane into a subset of the usual Euclidean plane. To accomplish such an embedding, distances must be distorted. In some models both distances and angles are distorted. One of the nicest models is the upper half-plane model. In the upper half-plane model, we fix a horizontal Euclidean line L and represent the hyperbolic plane by the points above, but not on, that line. The hyperbolic lines are of the following two types:

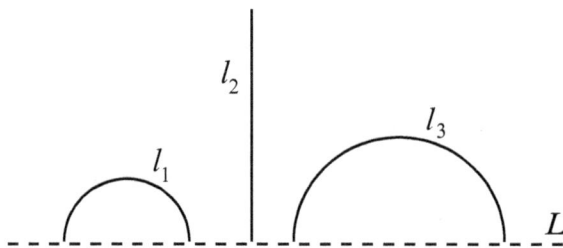

Figure 6.21 Three hyperbolic lines in the upper half-plane model.

6.7. Hyperbolic Geometry

- vertical Euclidean half-lines starting at L (for example, l_2 in Fig. 6.21),
- Euclidean half-circles that intersect L perpendicularly (for example, l_1 and l_3 in Fig. 6.21).

The hyperbolic lines shown in Fig. 6.22 do not intersect, but for a pair of hyperbolic lines that do intersect, the angle between them is the angle their tangent lines make. Thus Fig. 6.22 shows two pairs of intersecting hyperbolic lines and each intersection forms a right angle.

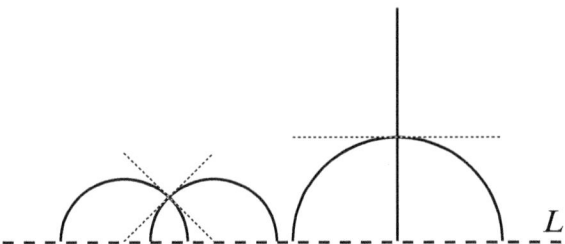

Figure 6.22 Two pairs of perpendicular hyperbolic lines.

Because the angle of intersection of a pair of hyperbolic lines is defined to equal the Euclidean angle of intersection of the Euclidean lines that are tangent to the hyperbolic lines at the point of intersection, the upper half-plane model does not distort angles. Distances are distorted, and the distance between two points is given by a complicated formula. The crucial feature of the distance is that the hyperbolic distance from any hyperbolic point to the line L is infinite. Therefore, for any point P on a hyperbolic line l, P divides l into two parts, each of which extends infinitely far away from P as measured using hyperbolic distance.

We can now verify the first four Euclidean postulates:

- The first Euclidean postulate states that "Through any pair of distinct points there passes a line." To find the hyperbolic line containing two given hyperbolic points, we do one of two things:

(1) If the points are on a vertical Euclidean line, then they are on the part of that line that is above L.

(2) If the points are not arranged vertically, then the Euclidean perpendicular bisector of the Euclidean segment they define intersects L, and that intersection point is the center of the Euclidean circle that contains both points—the top half of that Euclidean circle is the hyperbolic line containing the two points. See Fig. 6.23.

In every instance, there is a unique hyperbolic line that contains the two distinct points.

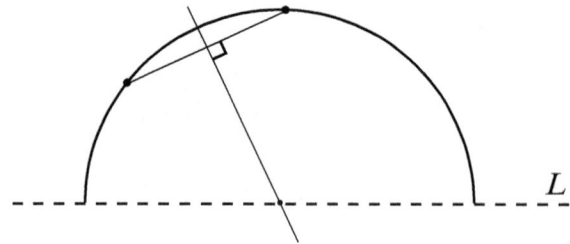

Figure 6.23 The hyperbolic line determined by two points.

- The second Euclidean postulate states that "For each segment $\overline{AB}$ and each segment $\overline{CD}$ there is a unique point E on $\overleftrightarrow{AB}$ such that B is between A and E and $\overline{CD} \cong \overline{BE}$." This postulate requires that we be able to extend a segment in either direction by any length. This is possible because each hyperbolic line extends an infinite hyperbolic distance on either side of a point or a segment.

- The third Euclidean postulate states that "For each point C and each point A distinct from C there exists a circle with center C and radius CA." A hyperbolic circle with center C is of course the set of all points at a given hyperbolic distance from C. Thus the hyperbolic circle with center C and containing A will consist of all the hyperbolic points that are at the same distance from C as is A.

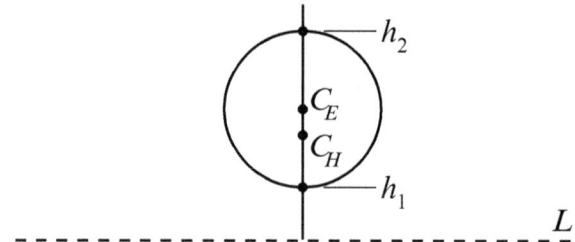

Figure 6.24 The Euclidean center and the hyperbolic center of a circle.

It is not obvious, but each hyperbolic circle is also a Euclidean circle and each Euclidean circle that lies entirely above L is a hyperbolic circle. So one and the same set of points in the upper half-plane can be considered both a hyperbolic circle and a Euclidean circle, but the hyperbolic center

6.7. Hyperbolic Geometry

and the Euclidean center are never the same point. Figure 6.24 illustrates this phenomenon and shows how to find the hyperbolic center. Construct the vertical line through the Euclidean center of the circle. That line intersects the circle at two points with heights h_1 and h_2 above L. The hyperbolic center is on the line at height $\sqrt{h_1 h_2}$ (the point labeled C_H), the geometric mean of the two heights. Of course, the Euclidean center is at height $(h_1 + h_2)/2$ (the point labeled C_E), the arithmetic mean of the two heights.

- The fourth Euclidean postulate states that "All right angles are congruent." Because the angle formed by the intersection of a pair of hyperbolic lines is defined to be the angle between their tangent lines at the point of intersection, this postulate is satisfied.

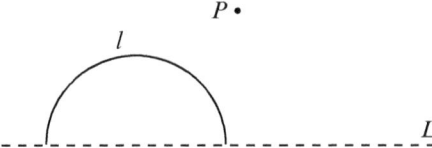

Figure 6.25 A point P and a line l in hyperbolic space.

Let us now verify that the upper half-plane model satisfies the Hyperbolic Parallel Postulate. Let a hyperbolic line l and a point P be given, say as in Fig. 6.25. Specifically, it is convenient (but not essential) to consider a point P that is not vertically above the hyperbolic line l. That simplifying assumption guarantees that the vertical half-line through P is parallel to l.

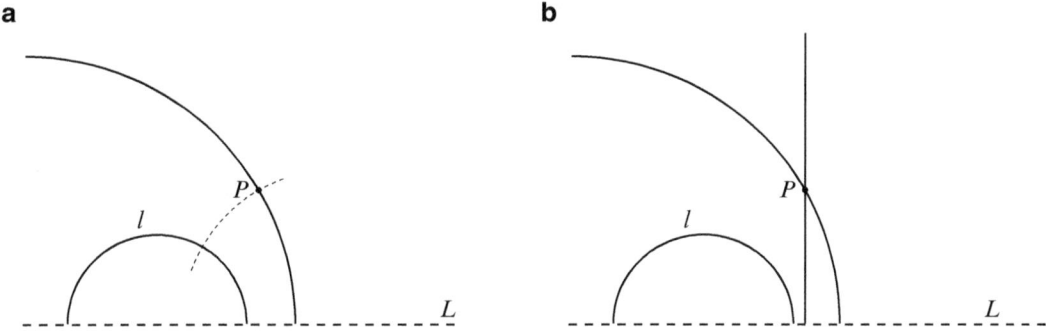

Figure 6.26 Satisfying the hyperbolic parallel postulate. **(a)** Constructing a line through P parallel to l. **(b)** Two lines through P parallel to l.

To construct a second parallel line through P we proceed as follows: drop a hyperbolic line segment through P that is perpendicular to l (the dashed arc in Fig. 6.26a), and then at P construct a hyperbolic line that is perpendicular to the hyperbolic line segment just constructed. Figure 6.26b shows two hyperbolic lines parallel to l passing through P, namely, the hyperbolic line constructed in Fig. 6.26a and the vertical half-line through P which is also a hyperbolic line.

Finally, Fig. 6.27 illustrates a Saccheri quadrilateral in the upper half-plane model of hyperbolic space. Notice that the summit angles are acute.

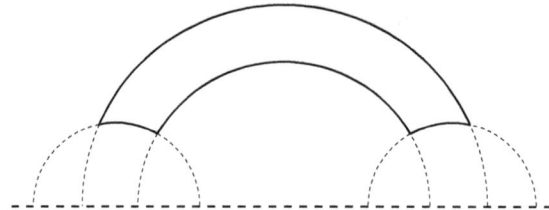

Figure 6.27 A Saccheri quadrilateral in hyperbolic space.

A Look Back

Early mathematics was based, in part, on land management issues. Basic geometry and trigonometry grew out of questions about rectangular and triangular plots of land. As a result the ancient Greeks and others had a very tactile, concrete understanding of geometry. They could not have conceived of non-Euclidean geometry.

It is a tribute to Bolyai and Lobachevsky (and also to Gauss) that they had the imagination to conceive of a geometry in which the parallel postulate fails. But the person who really put non-Euclidean geometry into a broad and profound context was Bernhard Riemann (1826–1866). Although he only lived to 40 years of age (he died tragically as a consequence of poverty and ill health), Riemann's ideas have revolutionized mathematics.

When Riemann was a student of Gauss, the master assigned young Bernhard to give a series of lectures on geometry. It was in these lectures that Riemann introduced the foundations of what is now known as Riemannian geometry. Riemann's basic idea was that we can vary the way that we measure the length of a curve from point to point in space. In this way we can put a geometry on a bounded domain, or on a surface in space, which reflects physical or mechanical parameters.

Albert Einstein did not know much mathematics (by his own admission), but he had the insight to realize that Riemann's geometry was the correct language for his new general relativity (the version of relativity that takes into account the bending of space by gravitational force). This development put both relativity theory and Riemannian geometry on the map.

In the twentieth century, a number of outstanding mathematicians have brought Riemann's geometry to full fruition. Among them were Élie Cartan (1869–1951) of France, who developed the theory of integration on manifolds. Shing-Shen Chern of China developed the higher-dimensional theory, and made profound contributions to the geometric theory of complex analysis. Shing-Tung Yau (1911–2004) of China has had a major impact on all aspects of geometry. Yau's work is mainly in differential geometry—especially in geometric analysis. His contributions have had an influence on both physics and mathematics and he has been active at the interface between geometry and theoretical physics. Yau's proof of the positive mass conjecture in general relativity demonstrated 60 years after its discovery that Einstein's theory is consistent and stable. His proof of the Calabi conjecture allowed physicists using Calabi–Yau compactification to show that string theory is a viable candidate for a unified theory of nature. Calabi-Yau manifolds are part of the standard toolkit for string theorists today. One of the important and startling consequences of string theory, and especially of Yau's contributions, is that space is not three-dimensional (as Isaac Newton thought) or four-dimensional (as Albert Einstein thought) but 26 dimensional.

REFERENCES AND FURTHER READING

[**Bel 68**]   Beltrami, E.: Saggio di interpretazione della geometria non-euclidea. Giornale di Mathematiche **6**, 285–315 (1868)

[**Bel 69**]   Beltrami, E.: Teoria fondamentale degli spazii di curvatura costante. Annali di Matematica Pura ed Applicata **2**(2), 232–255 (1868–1869)

[**Gre 07**]   Greenberg, M.J.: Euclidean and Non-Euclidean Geometries, 4th edn. W.H. Freeman, New York (2007)

[**Sta 93**]   Stahl, S.: The Poincaré Half-Plane: A Gateway to Modern Geometry. Jones & Bartlett Learning, Burlington (1993)

[**WW 92**]   Wallace, E.C., West, S.F.: Roads to Geometry. Prentice-Hall, New York (1992)

Chapter 7
Special Relativity

7.1 Introduction

The theories of special and general relativity are two triumphs of twentieth century thought. It is remarkable that both are attributable to Albert Einstein (1879–1955). Popular culture considers an understanding of relativity to be beyond the ability of most people. Certainly gaining a deep enough understanding of relativity to perform calculations and make predictions is a very difficult undertaking. On the other hand, the principles underlying relativity (which arise from physics, but are explained in terms of mathematics) are straightforward, and one *can* hope to understand those principles and to appreciate their consequences. In this chapter, we concentrate our attention on special relativity.

7.2 Principles Underlying Special Relativity

Our life experiences and our senses tell us that physical phenomena happen in space, while time marches on unaffected by whatever events are happening in space. In actuality, that model of the physical world—with time and space clearly and cleanly separated—is only an approximation to reality, but a very good approximation indeed when the speeds considered are small compared to the speed of light.[1]

---

[1] The speed of light is roughly 600 million miles per hour. Physicists denote the speed of light with the letter c.

In our everyday lives we constantly use electronic devices, so it is nearly impossible to imagine that 300 years ago mankind's understanding of electricity and magnetism, (collectively called electromagnetism) was almost nil. Not until the nineteenth century did physicists understand electromagnetism well enough to realize that electromagnetic radiation, i.e., radio waves, must exist.

The nineteenth century understanding of electromagnetism is encapsulated in the equations known as Maxwell's laws.[2] One curious feature of Maxwell's laws is that they require a strict distinction between the following two situations

(1) a moving magnet and a wire at rest,

(2) a moving wire and a magnet at rest.

This distinction is disturbing: the equations should depend on the motion of the magnet and the wire *relative* to each other. In his 1905 paper "On the electrodynamics of moving bodies,"[3] Einstein showed how to eliminate this defect in Maxwell's equations. In that paper Einstein referred to the resulting theory as the "principle of relativity." Einstein's later work further extended the principle of relativity, thus the principle he described in his 1905 paper is called "special relativity."

Einstein's theory of special relativity is based on two postulates. The first postulate is that the laws of physics, in particular those governing electrodynamics (i.e., electromagnetism and motion), should be the same in any uniformly moving frame of reference; (1) and (2) above tell us that this postulate is *not* satisfied by Maxwell's laws. Einstein's second postulate is that light always travels with the same speed in a vacuum. Of course, measuring the speed of light is difficult—in fact, it was not until the 1600s that the speed of light could be shown to be finite—but by 1905 the speed of light could be measured with sufficient precision that Einstein's postulate of a fixed light speed could be believed.

7.3 Some Consequences of Special Relativity

To see how special relativity works, we will consider some thought experiments involving a rapidly moving vehicle, say a spaceship. We will assume that the

---

[2] Maxwell's laws are named in honor of James Clerk Maxwell (1831–1879).

[3] "Zur Elektrodynamik bewegter Körper," *Annalen der Physik* **17** (1905), 891–921.

7.3. Some Consequences of Special Relativity

spaceship is far from any star, so gravity does not have to be considered—the extension of relativity to include gravity is "general relativity." To aid us, we will use coordinate systems, because it is a coordinate system that constitutes a frame of reference. We begin with a coordinate system for which time is plotted along the vertical axis and for which there is only one spatial direction with spatial positions plotted along the horizontal axis (Fig. 7.1). In this coordinate system each horizontal line represents all the points of space at one time and each vertical line indicates one point of space at all times. Such a coordinate system is called a *space-time diagram*.

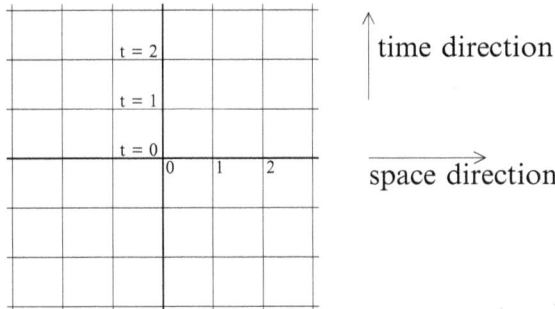

Figure 7.1 Simplified space-time coordinates with one spatial direction.

In the coordinate system of Fig. 7.1, a spaceship will be idealized as a point traveling along a curve. Physicists call such a path a *worldline*. If the spaceship is not moving in relation to the coordinate system of Fig. 7.1, then the worldline that represents it is simply a vertical line. If the spaceship is moving with constant velocity in relation to the coordinate system of Fig. 7.1, then the worldline that represents it is a straight line that is tipped away from vertical; the greater the speed of the spaceship the further away from vertical is the line representing it.

The unit of measurement for speed is determined by the choices of a unit of measurement for distance and a unit of measurement for time. In the everyday context of highway travel, we usually measure distance in miles and time in hours, so that miles per hour is the natural unit of speed (in the United States). Since the speed of light is over 600 million miles per hour, it is inconvenient to use miles and hours as units of distance and time when speeds comparable to the speed of light are considered. Instead, it is customary to let the unit of time be the second and to choose the unit of distance to make the speed of light, c, turn out to be 1. The unit of distance that accomplishes this is the "light-second"; those are the units used in Fig. 7.1 and subsequent figures.

Figure 7.2 uses two pink wedges to show all the points that can reach the origin, i.e., the point with coordinates $(0,0)$, or can be reached from the origin, by traveling at or below the speed of light. The boundary edges of the wedges (the red lines) are paths at exactly the speed of light, thus those edges represent paths that light would travel. Because of our choice of units (i.e., the second for time and the light-second for distance), each edge of a pink wedge, and in fact any path followed by light, must have slope ± 1. The totality of paths that light through the origin can take is called the *light cone* at $(0,0)$. Because we have simplified space to be one-dimensional, the light cone consists of only the two red lines. If we had two spatial dimensions in our coordinate system, then the light cone would look like a familiar cone. If we were faithful to reality and had three spatial dimensions in our coordinate system, then the light cone would be a three-dimensional object in a four-dimensional space; drawing useful figures would be problematic.

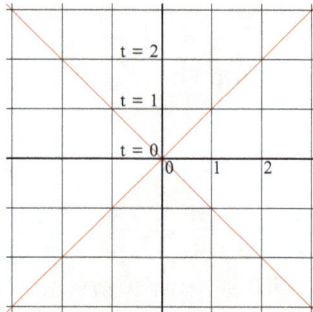

Figure 7.2 Space-time coordinate system showing a *light cone.*

Now consider the worldline of the spaceship as graphed in the space-time diagram of Fig. 7.2. We will assume that the spaceship is moving at one-half the speed of light. This speed is fast enough for us to demonstrate the effects of relativity. The worldline of the spaceship is as shown in Fig. 7.3. To help us keep track of things, we also imagine that there are "signposts" set up in space that are one light-second apart. The dots on the spaceship's worldline in Fig. 7.3 each indicate the event of passing one of the signposts. Since the spaceship is traveling at half the speed of light, two seconds elapse between each pair of signposts.

7.3. *Some Consequences of Special Relativity* 167

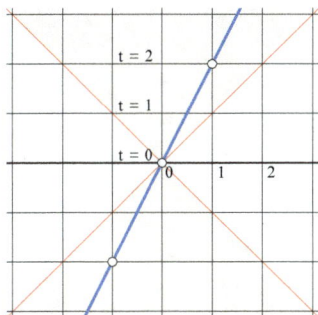

Figure 7.3 Worldline of a spaceship traveling at half the speed of light.

If the spaceship carries a clock and a ruler, then a coordinate system moving with the spaceship can be constructed. Any specific point in space-time will have a pair of coordinates (t, x) as measured with respect to the original coordinate system of Fig. 7.3 and it will also have a pair of coordinates (t', x') as measured using the clock and ruler on the spaceship, that is, t' and x' are the time and position with respect to the coordinate system moving with the spaceship.

The coordinates (t, x) and (t', x') will be related to each other via a set of formulas. Such a set of formulas is a *change of coordinates*. Because the vehicle is moving at constant speed, the formulas relating (t, x) and (t', x') are of the simple type known as a *linear change of coordinates*. A linear change of coordinates is a change of coordinates for which the formulas for t' and x' involve constants and first powers of t and x, but no other powers of t and x. Only the linear changes of coordinates of the special type called *Lorentzian* will satisfy Einstein's two postulates: namely, (1) the laws of physics are the same in any uniformly moving coordinate system and (2) the speed of light is the same in any uniformly moving coordinate system.

Figure 7.4 shows a coordinate system (t', x') in the same part of space-time as in Fig. 7.2. These new coordinates have been obtained using a Lorentzian change of coordinates. Of course, we could present the formula for the change of coordinates, but instead we will rely on the diagrams. Notice that the speed of light is still equal to 1 because the red lines that represent the boundary of the light cone at $(0, 0)$ pass through the points $(t', x') = (1, 1)$ and $(t', x') = (1, -1)$. Also notice that the line $x' = 0$ is tipped from vertical by exactly the same amount as the line in Fig. 7.3 that represents the worldline of the spaceship traveling at one-half the speed of light. This last

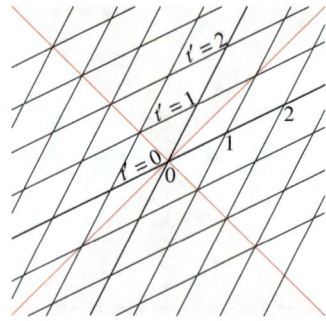

Figure 7.4 Coordinate system moving with a spaceship traveling at half the speed of light.

fact tells us that t' and x' are the coordinates as measured in the coordinate system moving with the spaceship.

Now look at the events of passing the signposts as marked by the dots along the worldline of the spaceship. In the original coordinate system, two of those signposts were passed at times $t = 0$ sec and $t = 2$ sec. Figure 7.5 shows the worldline of the spaceship in the coordinate system that moves with it. Figure 7.5 also shows the events of passing the signposts. The first signpost is passed at $t' = 0$ sec, but the next signpost is passed when t' is *strictly smaller* than 2 sec; in fact, the formulas for the Lorentzian transformation tell us that the next signpost is passed when $t' = \sqrt{3} \approx 1.73$ sec.

What we see here is the remarkable fact that as a person on the spaceship travels for a year—as timed aboard the spaceship—a person not moving relative to the coordinate system in Fig. 7.3 sees 421 days go by. That is, time, as measured inside the spaceship, is 13.5 % slower than time as measured in the initial coordinate system. This consequence of special relativity is called *time dilation*.

The magnitude of the time dilation effect increases with increasing speed. Of course, the time dilation can be computed using a formula, but it is easier and more striking to see this effect illustrated using graphical calculations based on Fig. 7.6. Figure 7.6 shows coordinate axes for t and x. The red line represents the path that a light ray would follow. If a line through the origin is used to represent the worldline of a spaceship moving with constant speed, then the intersection of that line and the green curve shows the location of $t' = 1$ and $x' = 0$ in a coordinate system moving with the spaceship. The t coordinate of that intersection point shows how much time has passed in the reference coordinate system when 1 second has passed in the spaceship.

7.3. Some Consequences of Special Relativity

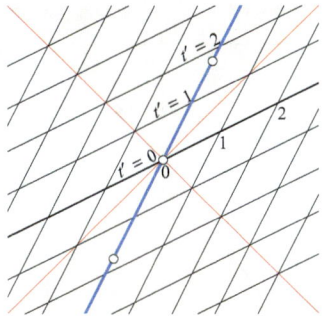

Figure 7.5 Worldline of a spaceship traveling at half the speed of light in the coordinate system that moves with it.

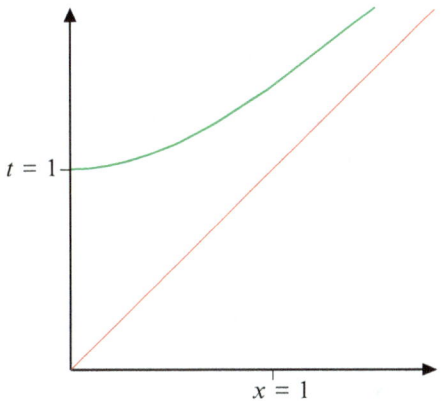

Figure 7.6 Time dilation diagram.

To illustrate this calculation, in Fig. 7.7 we show the line representing the worldline of a spaceship traveling at 3/4 the speed of light. The slope of the worldline of the spaceship is 4/3. At the point where $t' = 1$ and $x' = 0$, we see that t is about 1.51.

The magnitude of the time dilation increases without limit as the speed of the spaceship approaches the speed of light. In Fig. 7.8 we show the line representing the worldline of a spaceship traveling at 95 % of the speed of light (the line has slope 20/19). At the point where $t' = 1$, we see that t is about 3.20. If the spaceship were to reverse course and travel back to $x = 0$ at the same speed, i.e., 95 % of the speed of light, then when the spaceship arrived at $x = 0$ the time t would equal about 6.4, while the time t' as

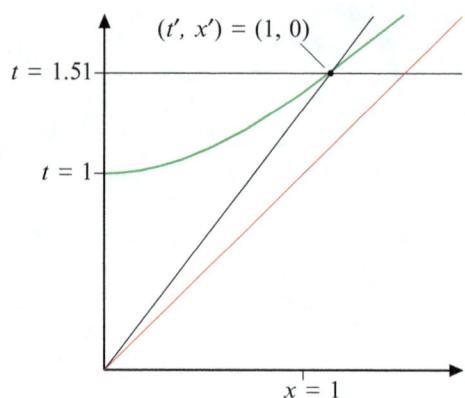

Figure 7.7 Time dilation at 75 % of the speed of light.

measured in the spaceship would equal 2. A person waiting at $x = 0$ would have aged three times as much as a person traveling in the spaceship. This is known as the *twin paradox*, because, as a thought experiment, we imagine one twin traveling on the spaceship, while the other waits on Earth. When the traveling twin returns home, he/she finds that the earthbound twin has aged more. It is crucial for the twin paradox that the paths traveled by the twins not be symmetric. If the two twins both traveled in opposite directions at equal speeds, then turned around and returned at equal speeds, they would then both age by the same amount. The twin paradox is a real phenomenon that has been confirmed experimentally.

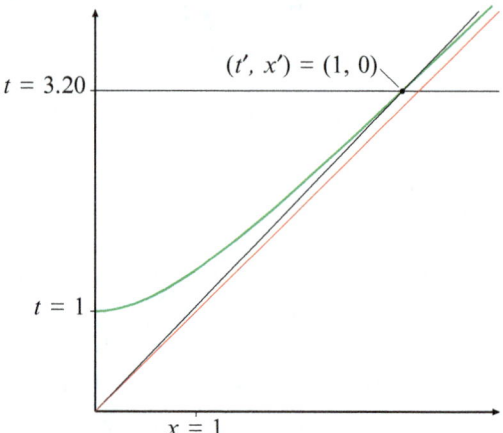

Figure 7.8 Time dilation at 95 % of the speed of light.

7.4. Momentum and Energy

Time dilation is one of the surprising consequences of relativity, but it is not the only surprise. Another surprising phenomenon is *length contraction*. To understand length contraction, we need to put the x' coordinates on our space-time diagram. The formulas tell us that the x'-axis is simply the reflection of the t'-axis through the light cone as shown in Fig. 7.9. The location at which $t' = 1$ and $x' = 0$ reflects to the location at which $t' = 0$ and $x' = 1$. Since $x \approx 1.51$ at the point where $(t', x') = (0, 1)$, we see that a length of 1.51 light-second in the (t, x)-coordinate system measures only 1 light-second in the (t', x')-coordinate system: this is the length contraction phenomenon. While length contraction is surely a real phenomenon, carrying out sufficiently accurate laboratory experiments to demonstrate it remains a challenge.

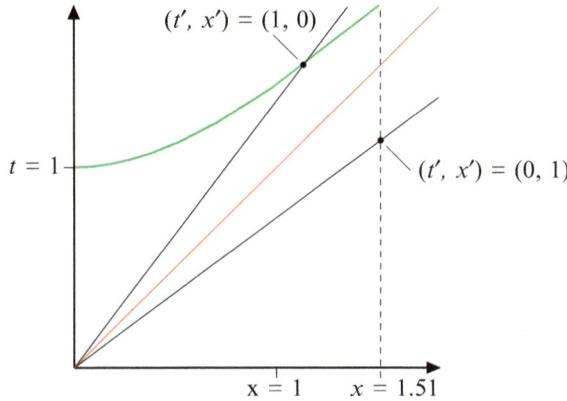

Figure 7.9 Length contraction at 75% of the speed of light.

7.4 Momentum and Energy

Before Einstein published his theory of relativity, scientists and engineers used *Newtonian mechanics*, based on Newton's laws of motion, to explain and predict the behavior of mechanical systems. A fundamental part of Newtonian mechanics is the *momentum* of an object. Momentum is the velocity of the object multiplied by the mass of the object. The object considered is often thought of as being small, but with significant mass, and is idealized as a *particle* or *point mass*.

7.4.1 Vector Quantities

It is important to remember that velocity, and hence momentum, are vector quantities; that is, they include both direction and magnitude. Because a vector quantity gives both direction and magnitude, a vector is more complicated than a number. To deal with this complexity, it is common to represent a vector by breaking it into parts that point in important or convenient directions. This process is analogous to describing a route on land as going 13 miles north and then 5 miles west. Those parts of a vector (13 miles north and 5 miles west in our example) are called the *components* of the vector.

This process of breaking a vector into components is illustrated in Fig. 7.10. The vector that goes 13 miles north and 5 miles west is the black arrow in Fig. 7.10. The red vertical arrow in Fig. 7.10 represents the *vector component* along a north/south axis, and the blue horizontal arrow represents the vector component along an east/west axis. If we assume that going north is the positive direction of motion on the north/south axis, then all north/south motions can be compared to a vector 1 mile long and pointing north; we call this vector a *basis vector*. The red vertical arrow can be obtained from the preceding north/south basis vector by scaling up the basis vector by a factor of 13. Similarly, if going east is the positive direction of motion on the east/west axis, then all east/west motions can be compared to a basis vector 1 mile long and pointing east. The blue horizontal vector is obtained from the east/west basis vector by reversing its direction and then scaling up by a factor of 5. By agreeing to consider "reversing direction" to be scaling by a factor of -1, we can describe the blue horizontal vector as being obtained by scaling the east/west basis vector by a factor of -5. Because of their relationship with scaling, the numbers $+13$ and -5 are called the *scalar components* of the vector.

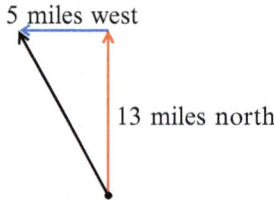

Figure 7.10 Components of a vector.

7.4. Momentum and Energy

The components of a vector depend on the choice of coordinate system. When the coordinate system is changed, the components change, but the vector stays the same. Figure 7.11 shows how the components of a vector change when the coordinate system is rotated. The vector itself is shown in black and the component in each coordinate direction is shown in the same color as the corresponding coordinate axis. Notice that the vector stays the same, but the components change significantly.

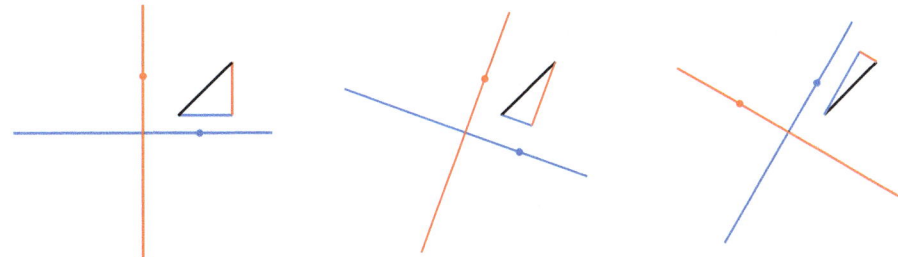

Figure 7.11 How components change when the coordinate system rotates.

Another way to represent how the components change is to compare the length of the component in a given direction with the unit length in the coordinate axis in the same direction. This is the representation of the vector in terms of its scalar components. For the example of rotating the coordinate system, the scalar components of the vector can be plotted against the angle by which the coordinate system has been rotated. The scalar components are illustrated in Fig. 7.12 by the graphs above the first coordinate system (which serves as the reference against which the angle of rotation is measured). The values corresponding to the three specific coordinate systems are marked by dots.

7.4.2 Vectors in Relativity

In relativity theory, we need to consider how the components of vectors in space-time change when the coordinate system changes, but the type of coordinate change that is of interest is one in which the coordinate systems are moving with respect to each other. Figure 7.13 shows how the components of a vector change when one coordinate system for space-time moves relative to another coordinate system. The reference coordinate system is shown on the left. The red line represents the time axis and the blue line

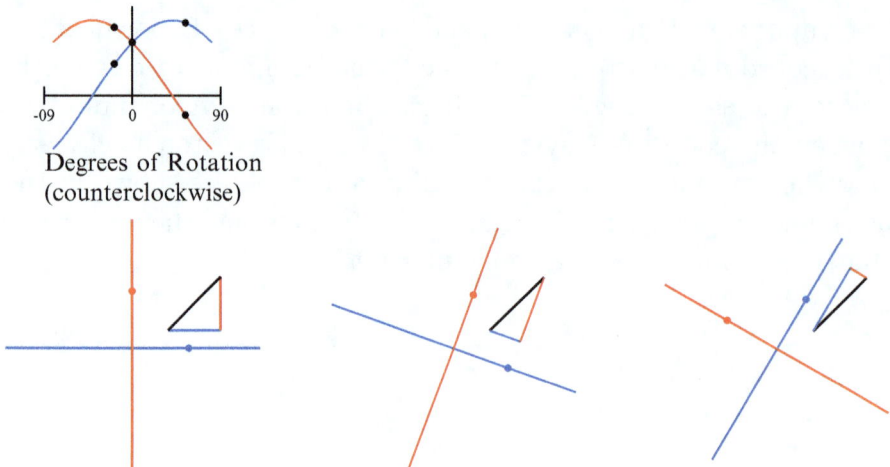

Figure 7.12 How vector and scalar components change when the coordinate system rotates.

represents the space axis—with space simplified to be one-dimensional. The vector itself is chosen to be in the direction light travels. Thus it is called a *lightlike* vector. The vector is shown in black and the vector component in each coordinate direction is shown in the same color as the corresponding coordinate axis. Again, notice that the vector stays the same, but the components change significantly.

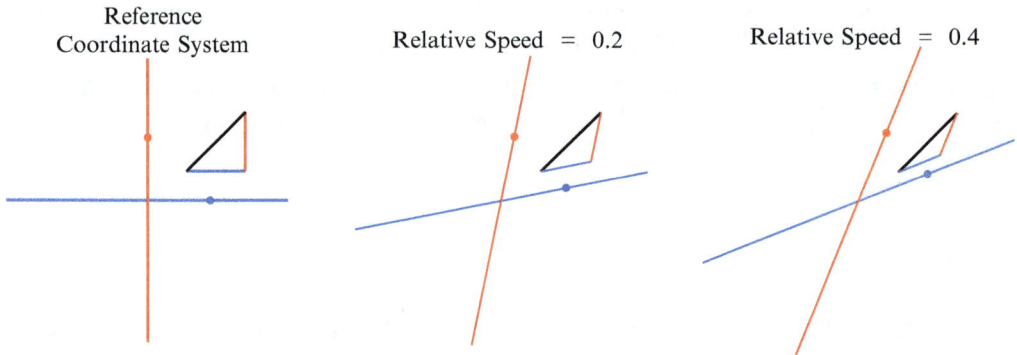

Figure 7.13 How vector components in space-time change when the coordinate system changes.

The behavior of the scalar components is also of interest when the coordinate system changes. In Fig. 7.14 the behavior of both the scalar and vector

7.4. Momentum and Energy

components is shown. For a lightlike vector in our simplified space-time, the absolute value of the scalar component in the time direction will be equal to the absolute value of the scalar component in the space direction in all coordinate systems. For the vector in Fig. 7.14 those components are positive and hence equal.

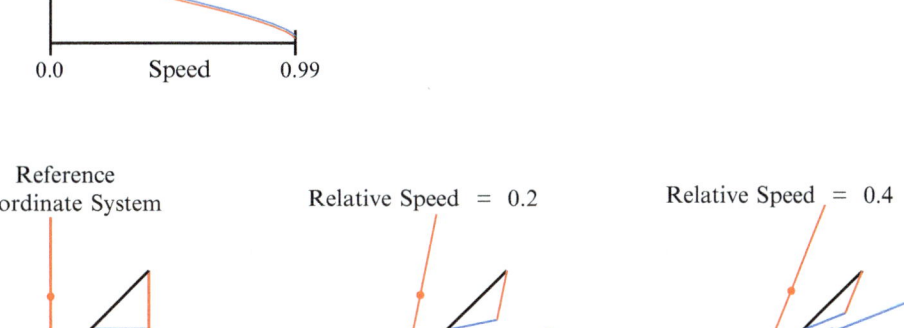

Figure 7.14 How scalar and vector components in space-time change when the coordinate system changes.

7.4.3 Relativistic Momentum

As in Newtonian physics, relativity has a notion of momentum: the *relativistic momentum*. The relativistic momentum is a vector in space-time just as the Newtonian momentum is a vector in space. For speeds that are small compared to the speed of light, the spatial vector component of the relativistic momentum is very nearly equal to the Newtonian momentum. When the speed is zero, then the spatial vector component of the relativistic momentum and the Newtonian momentum are equal—both are the zero vector.

Since the relativistic momentum is a vector in space-time, it must be unaffected by any Lorentzian change of coordinate system. Under such a change of coordinates the components are allowed, even required, to change, while the space-time vector itself remains an unchanging physical quantity. This sounds harmless enough, but remember that a Lorentzian change of coordinates allows a moving coordinate system. In particular, if the coordinate system is moving with the particle, then the Newtonian momentum as measured in that coordinate system is the zero vector in space but, in any coordinate system moving relative to the particle, the Newtonian momentum is a non-zero spatial vector. We will exploit this idea to determine the relativistic momentum.

Consider a reference coordinate system moving with the particle. We know the following facts:

• In the reference coordinate system moving with the particle, the relativistic momentum must be entirely in the time direction, since its spatial component, i.e., the Newtonian momentum, is zero.

• In the reference coordinate system moving with the particle, the relativistic momentum must equal the vector in the time direction whose spatial component matches the Newtonian momentum when we change coordinates to a coordinate system moving slowly relative to the particle ("slowly" means as compared to the speed of light).

• In any coordinate system moving slowly relative to the particle, the Newtonian momentum nearly equals the spatial component of the relativistic momentum.

Figure 7.15 illustrates how the components change for a vector that points in the time direction in the reference coordinate system. The red curve shows the component in the time direction, and the blue curve shows the component in the spatial direction. Notice that, for low speeds, the spatial component changes nearly linearly and the component in the time direction changes little.

Figure 7.16 examines this phenomenon more closely. Figure 7.16a–c show what happens for vectors that point in the time direction in the reference coordinate system, but which have differing magnitudes. It turns out that the *slope* of the line approximating the spatial component for low speeds equals the magnitude of the time component in the reference coordinate system moving with the particle. That is, for a coordinate system moving slowly relative to the particle, the spatial component is a constant multiple of the velocity. The Newtonian momentum of the particle is also a constant

7.4. Momentum and Energy

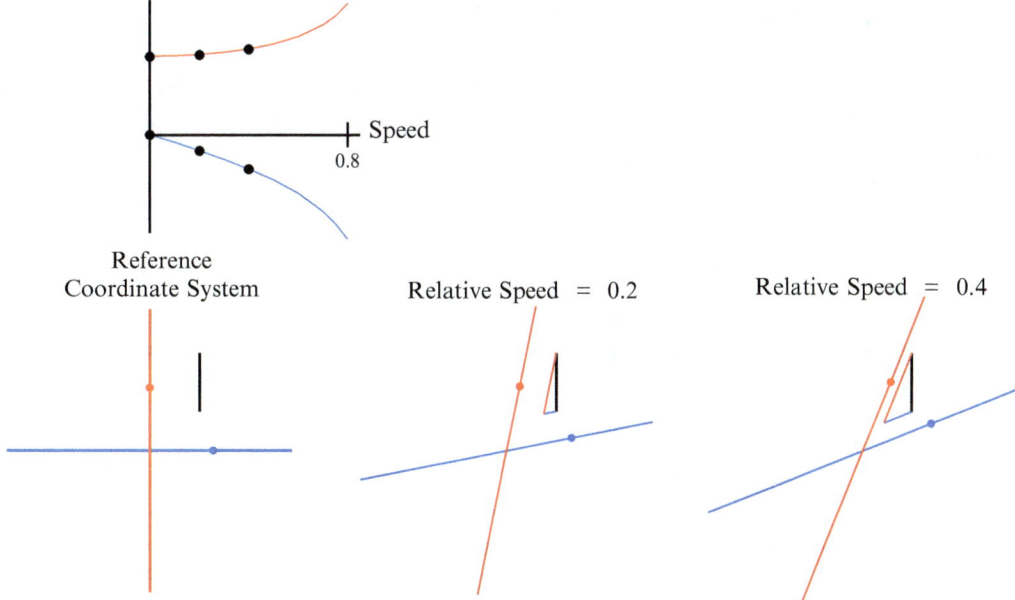

Figure 7.15 How the components of a vector in the time direction in the reference coordinate system change when the coordinate system moves.

multiple of the velocity and the constant multiplier is the mass of the particle (remember Newtonian momentum is *mass* times velocity). We conclude that

> in the reference coordinate system moving with the particle, the relativistic momentum of the particle must be the vector in the time direction with magnitude equal to the mass of the particle.

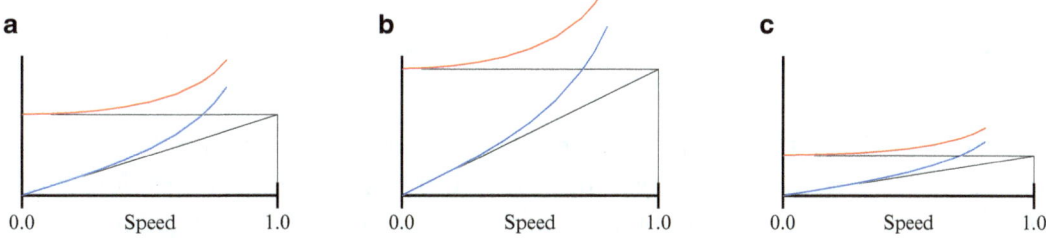

Figure 7.16 How the scalar components of three distinct timelike vectors change when the coordinate system moves. Spatial component in blue, time component in red. (**a**) Medium-size timelike vector. (**b**) Large timelike vector. (**c**) Small timelike vector.

7.4.4 Rest Mass

Next we would like to understand more completely the significance of the relativistic momentum in the coordinate system that moves with the particle. We will do so by again considering what happens when we change the coordinate system.

Because the component of the relativistic momentum in the time direction stays so nearly constant at low speeds, it is almost impossible to see how it varies. To overcome this difficulty, we can emphasize the change by subtracting its initial value and dividing by the speed; this is a process of shifting and rescaling to emphasize the detail that is of interest. Figure 7.17a–c show what happens for vectors that point in the time direction in the reference coordinate system, but that have differing magnitudes—that is, the figures correspond to particles with differing masses. Again, the red curve shows the component in the time direction, and the blue curve shows the component in the spatial direction. The red curve is shifted and rescaled, and when that is done, the time component of the relativistic momentum is seen to initially follow a line whose slope is one-half the slope of the line that the spatial component initially follows. Recall we saw above that slope is the mass of the particle. We conclude that

> for small speeds, the change in the time component of the relativistic momentum is one-half the mass of the particle multiplied by the speed of the particle squared.

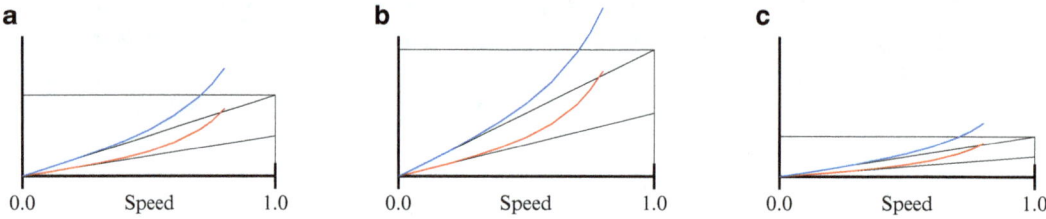

Figure 7.17 How the scalar components of three distinct timelike vectors change when the coordinate system moves and the timelike scalar component is shifted and rescaled. Spatial component in blue, shifted and rescaled time component in red. (**a**) Medium-size timelike vector. (**b**) Large timelike vector. (**c**) Small timelike vector.

In Newtonian physics, one-half the mass of the particle multiplied by the speed of the particle squared is the *kinetic energy* of the particle. So we see that, for a slowly moving particle, the change in the the time component of the relativistic momentum vector equals the Newtonian kinetic energy of the particle. Thus we have learned that the time component of the relativistic momentum is the *energy E* of the particle.

When a particle moves more and more slowly, the Newtonian kinetic energy decreases to zero, and this shows up as a decrease in the magnitude of the time component of the relativistic momentum. Now the Newtonian kinetic energy can decrease to zero, but the time component of the relativistic momentum does not. Instead the time component decreases to its minimum, and that minimum is the mass of the particle in the coordinate system moving with the particle. So we see that

the energy of a particle at rest is its mass, that is,

$$E = m.$$

We have arrived—almost—at Einstein's famous equation $E = m\,c^2$, but the c^2 seems to be missing. Now, recall we used the light-second as our unit of distance, which means $c = 1$. So we have

$$E = m \cdot 1^2 = mc^2$$

afterall. Had we used another unit for distance, then the value of c would have been explicitly visible—and it would have been squared.

Certainly the development of relativity theory is one of the grand adventures of modern science. This set of ideas has revolutionized the way that we see our world. Lovely and inviting explorations of these ideas appear in [TaW 92] and [Wil 93].

A Look Back

Although the theory of special relativity was grounded in the ideas of Hendrik Lorentz (1853–1955) and others, it was Albert Einstein (1879–1955) who fully developed the context for relativity theory in modern physics, and who came up with the broader theory of general relativity that is so important for modern cosmology.

Photo by Henri Manuel

Henri Poincaré

Interestingly, the mathematician Henri Poincaré (1854–1912)—who had a special passion for physics—developed special relativity independently of Einstein. According to Stephen Hawking, Poincaré's paper appeared just 2 weeks after Einstein's. Even less well-known is the fact that Poincaré was invited to participate in the 1904 World's Fair (a centennial celebration of the Louisiana Purchase) that was held in St. Louis—in Forest Park, just across the street from the current location of Washington University.

The heart of the campus today consists of stately buildings that were originally constructed to be administration buildings for the Fair. One of these is Cupples I, which now houses the mathematics department. Another is Holmes Lounge, where Poincaré delivered a lecture on—guess what?—special relativity. And this was in 1904, 1 year before the appearance of Einstein's celebrated paper on the same topic. In fact a translation of Poincaré's lecture was published in a journal called *The Monist*, vol. 15, January, 1905. The title is "The Principles of Mathematical Physics." While the talk is largely philosophical, it is also the case that certain pages could have been lifted from a modern freshman physics text. It treats the Michelson–Morley experiment, discusses the standard topic of a train traveling at the speed of light with two observers, and many other familiar parts of special relativity theory.

Perhaps even more surprising was that there was a major campaign, conducted over a period of at least 5 years, to obtain the Nobel Prize in Physics for Poincaré. In fact Gösta Mittag-Leffler (1846–1927) spearheaded the effort, and he had support from Paul Painlevé (1863–1933), Gaston Darboux (1842–1917), and Ivar Fredholm (1866–1927). Here is a portion of Mittag-Leffler's arguments, contained in a letter to one Professor Paul Appell (1855–1930):

> The time is come when we can hope to make Poincaré winner of the Nobel Prize. I send enclosed with the next mail a proposal written by Fredholm that he subjects to your judgment and one by Mr. Darboux. He has made considerable use of the proposal made by Darboux this year. The most important thing is first to establish the prominent part played by pure theory in physics and then to conclude with the proposition to give the prize for discoveries defined by a sufficiently simple formula. After some discussion, we have found this formula in Poincaré's discoveries

> concerning the differential equations of mathematical physics. I
> think that we will win with this program.

Mittag-Leffler went on to say that the nominators had to avoid "mathematics" and refer to "pure theory" because "like those who are only experimentalists, members of the Nobel committee for Physics are scared silly by mathematics."

Needless to say, Poincaré was never awarded the Nobel Prize. In one year, thirty-four eminent physicists and mathematicians supported the nomination. To no avail; the experimentalists were never convinced. Mittag-Leffler summarized the situation as follows:

> We have again been beaten, this time for the Nobel Prize. This crowd of naturalists who do not understand anything about the fundamentals of things has voted against us. They fear mathematics because they don't have the slightest possibility of understanding anything about it.

REFERENCES AND FURTHER READING

[Bow 86]   Bowler, M.G.: Lectures on Special Relativity. Pergamon Press, Oxford (1986)

[Car 04]   Carroll, S.M.: Space-Time and Geometry: An Introduction to General Relativity. Addison-Wesley, San Francisco (2004)

[Haw 96]   Hawking, S.W.: A Brief History of Time. Bantam Books, New York (1996)

[Mer 05]   Mermin, N.D.: It's About Time: Understanding Einstein's Relativity. Princeton University Press, Princeton (2005)

[TaW 92]   Taylor, E.F., Wheeler, J.A.: Spacetime Physics, 2nd edn. W. H. Freeman, New York (1992)

[Wil 93]   Will, C.M.: Was Einstein Right? Putting General Relativity to the Test. Basic Books, New York (1993)

Chapter 8
Wavelets in Our World

8.1 Introductory Ideas

It is part of human nature to want to break a complicated problem up into simpler components. This is a way that we have of getting our hands on the problem, and of analyzing it.

As an example, a criminal detective investigating a messy murder case will break it up into pieces. First he/she establishes the *motive*, the *means*, and the *opportunity*. Then the investigator looks at the scene of the crime, the weapon, the corpse, interviews witnesses, verifies timelines, and so on. Having all these pieces of the puzzle helps the criminologist to come to terms with all the data in the case, and is also useful at the trial in helping the jury to assess the information.

In a similar fashion a scientist studying a natural phenomenon will dissect the situation into many pieces. After all, nature is complex and baffling; we must exert all our wiles to begin to understand its many complexities. A physicist wanting to understand why water freezes at 32° Fahrenheit will want to first determine that water is made up of hydrogen and oxygen, calculate the relative proportions of these two components (two-to-one is the right answer), examine the atomic structure of the component elements, learn how the two elements interact, understand the Second Law of Thermodynamics, and so forth.

In mathematics this idea of breaking up a problem into accessible pieces has become commonplace. Every mathematician has a catalog of big, exciting problems to be solved. It is possible for a professional mathematician to dedicate a lifetime to the solution of one particular problem. But to, make any progress, to be able to publish some papers showing new insights and new inroads into the problem, or to be able to get tenure and apply for grants and get invited to conferences, the mathematician must learn how to work in increments.

One of the great paradigms in mathematics of breaking up a complicated problem into simple pieces is due to Jean Baptiste Joseph Fourier (1768–1830). Fourier was studying *functions*, which are part of the basic language of modern mathematics. A function is a rule that assigns numbers to numbers. For example, the "square function" assigns to each number its square. We write such a function as $f(x) = x^2$, where the notation tells the reader that

- the name of the function is f;
- the function is applied to a number called x;
- the function produces from x a new number x^2.

It is safe to say that much of modern mathematics consists of expressing problems in terms of functions, and then analyzing those functions. Some functions, such as the function f in the last paragraph, are rather simple. But here is a more complicated function. Look at the Dow-Jones Industrial Average over the years 1992–2006. Let the function G assign to each point in time the value of the Dow-Jones Index at that time. Figure 8.1 illustrates this function G by way of a *graph*. This graph of the Dow-Jones is quite erratic. There is no way to anticipate or describe such a function.

8.2 Fourier's Ideas

Of course nobody knows how to predict the Dow-Jones Industrial Average, nor how to calculate it in advance. It would be impossible to specify a rule or formula that defines the function G described in the last paragraph. Nonetheless it is a function, and one that we must learn to deal with. Many of the physical phenomena of nature are governed by functions, and we cannot necessarily discern what the function is or by what rules the function is formed. Our job as scientists is to come to terms with the function, to learn how to use it, and thereby understand the laws of nature.

8.2. Fourier's Ideas

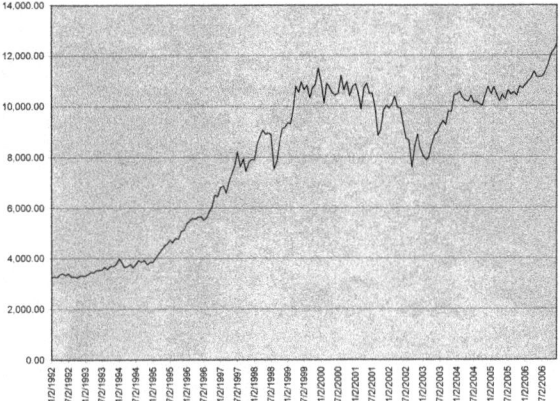

Figure 8.1 The Dow-Jones index.

Fourier's idea was to develop the technology to break up an arbitrary function up into simple component pieces. This circle of ideas is commonly referred to as "Fourier analysis." For historical reasons, and because they relate to classical physics problems in the study of heat and waves, Fourier's component pieces were sine and cosine functions. These are simple, periodic waves that are prevalent in all of nature. Refer to Figs. 8.2 and 8.3.

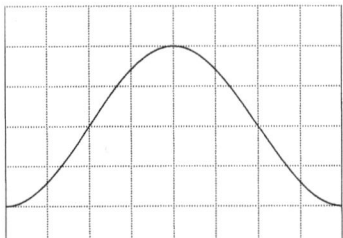 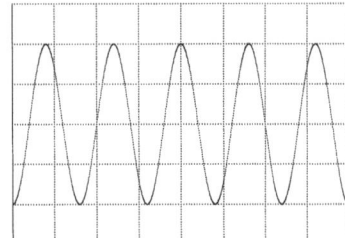

Figure 8.2 Cosine functions.

In fact, important scientists—including Leonhard Euler (1707–1783) and Daniel Bernoulli (1700–1782)—had suspected for many years that functions could be broken up into sines and cosines. But the process is counterintuitive—see Fig. 8.4. It is by no means evident how to break up the function displayed there into the very regular and smooth functions sine and cosine shown in Figs. 8.2 and 8.3. Fourier derived a set of explicit formulas that showed how to do this (today his formulas, and variants thereof, are routinely implemented on the computer—so the formulas are quite explicit and elementary).

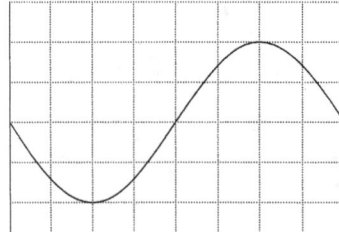 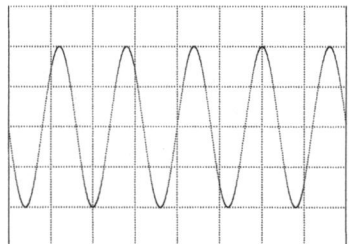

Figure 8.3 Sine functions.

And he developed applications of his ideas to the theory of heat. His book *Théorie Analytique de la Chaleur* (*The Analytic Theory of Heat*) is one of the great scientific works of all time. It has exerted an enormous influence over the development of mathematics and of analytical science. Fourier analysis is used to study and analyze differential equations, to implement image compression algorithms, and to engage in signal processing. It is essential to radio and television transmission and to the encoding of information on CDs (compact discs), DVDs, and Blu-ray discs. In short, Fourier's ideas are prevalent in all aspects of modern communication. Fourier's name lives on, and will likely do so for many centuries. In fact it would be difficult for us here, without getting rather technical, to give an explicit description of Fourier's methods, and of his formulas. But we can illustrate his ideas and give a feeling for what Fourier analysis is all about.

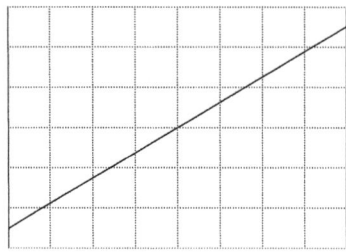

Figure 8.4 A function that we wish to approximate by sines or cosines.

Figure 8.4 shows a function, in fact the function $f(x) = x$. This function looks nothing like a cosine function or a sine function (Figs. 8.2 and 8.3). Yet Fig. 8.5 shows a good approximation given by summing (or *superimposing*, in the sense of physics) five sine functions and the even better approximation given by summing (or superimposing) ten sine functions.

8.2. Fourier's Ideas

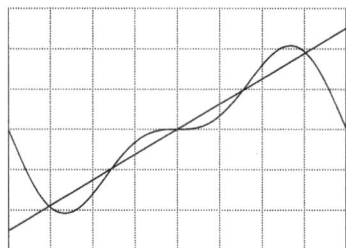 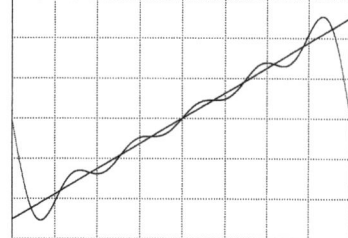

Figure 8.5 Approximation with five sine functions (*left*) and with ten sine functions (*right*).

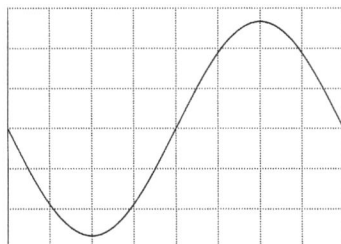 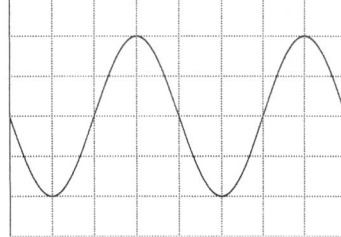

Figure 8.6 First pair of sine function approximants.

Just for fun, we exhibit in Figs. 8.6–8.8 some of the component sine functions that are superimposed to make Fig. 8.5. None of the functions shown in Figs. 8.6–8.8 looks anything like the function in Fig. 8.4, but the sum of the functions (which entails rather complicated cancellations) in Figs. 8.6–8.8 does give a reasonable approximation to the function in Fig. 8.4.

Certainly Fourier's ideas—especially because they were derived in a semi-heuristic fashion—were received at first in 1807 with some skepticism. His definitive book on the subject was not published until 1822, and in fact he himself published it in his role as Secretary of the French National Academy of Sciences.

It was not until many years later that Fourier's ideas were finally put on a rigorous and mathematically reliable footing. Certainly Johann Peter Gustav Lejeune Dirichlet (1805–1859) and Georg Friedrich Bernhard Riemann (1826–1866) played key roles in working out some of the basic ideas of Fourier series. The modern approach was not developed until the early twentieth century.

One of the important applications of Fourier analysis in the mid-twentieth century was to filter noise out of musical recordings. In the days of long-playing records, typical noise was a "pop" or a "click"—see Fig. 8.9.

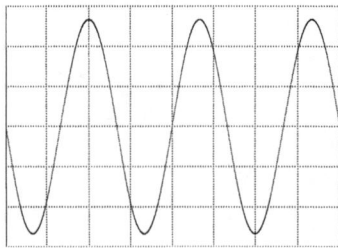 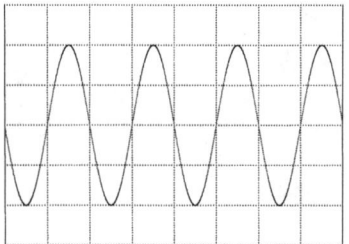

Figure 8.7 Second pair of sine function approximants.

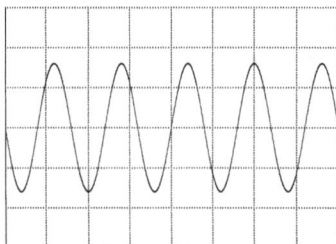 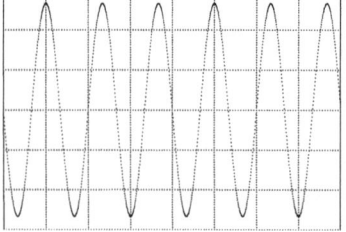

Figure 8.8 Third pair of sine function approximants.

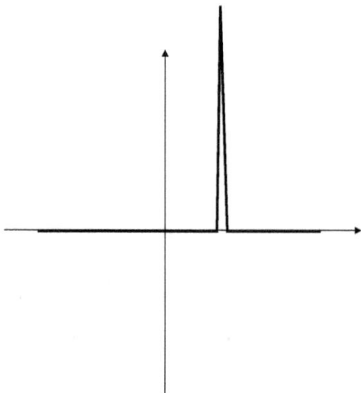

Figure 8.9 A "pop" or "click".

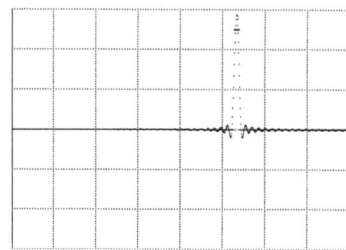

Figure 8.10 Fourier approximation of a "pop" or "click".

It stretches the imagination to try to think how to approximate such a spike with sines and cosines. Figure 8.10 shows an approximation with 100 cosine functions. You can see an "error" consisting of low level vibrations of the graph. In the music this comes across as a hiss. Thus the noise filters of the 1950s removed the pops and clicks from the music but replaced them with a low-level hiss. The Dolby noise reduction system of the 1960s helped considerably to reduce that hiss. This technology involved more sophisticated ideas, such as the Fast Fourier Transform.

8.3 Enter Wavelets

Wavelet theory is a new mathematical technology that has only been with us for about 30 years. Unlike many pure mathematical ideas, which only find their way into the technical sector (i.e., the applied world) after a few generations have passed, wavelets were being applied by engineers to very practical problems only 5 years after they were invented.

This remarkable historical event took place for a number of reasons. One is that the theory of wavelets grew out of a question that was asked of pure mathematician Yves Meyer of Paris *by an engineer*. Thus the answer was transmitted directly to an engineer (and then to the engineering community), and had immediate meaning and impact for the engineering world. The second reason is that the potential importance and utility of wavelets is immediate once one examines the components of the theory. The third reason is that many influential people—including Yves Meyer and Ronald R. Coifman of Yale—expended considerable effort to advertise and promote wavelet theory. Coifman was frequently on the train to Washington, D.C. giving testimony before Congress, before the National Science Foundation, and before military agencies.

Mathematicians tend to be a rather reticent and timid bunch, who prefer to keep to themselves. The general hoopla that surrounded wavelets for 20 years was quite out of their character. So the entire circle of events surrounding wavelets was unusual, and led to remarkable results.

The main difference between wavelet theory and Fourier theory is that Fourier theory is built up from cosines and sines (Figs. 8.2 and 8.3) while wavelet theory is built up from little localized blocks (Fig. 8.11).

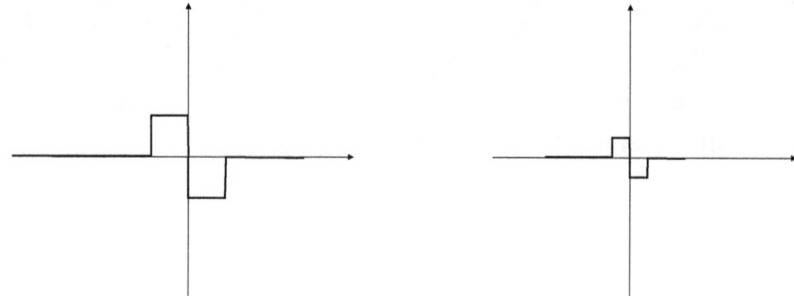

Figure 8.11 Typical wavelets.

Thus, for instance, a pop or click can be much more accurately approximated by wavelets—see Fig. 8.12—than by the Fourier cosines and sines. In Fig. 8.12 it is very easy to see the superposition of the wavelets—after all, we are just stacking boxes on top of each other—and it is easy to see that this superposition provides a good approximation to the spike that represents noise in our music.

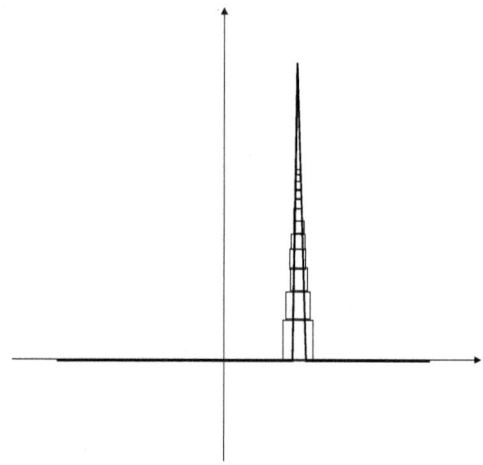

Figure 8.12 Accurate approximation of a "pop" or "click" by wavelets.

The tricky part of wavelet theory is to complete the simple-minded, box-like functions in Fig. 8.11 into a full system of approximating functions that can generate "almost any" function that will arise in practice. And to do so in such a way that the approximating expansions can be readily and accurately manipulated and calculated on a computer. The advantage of the cosines and sines in the Fourier theory was that the necessary algebraic relationships among the component functions are built into the theory of trigonometry that everyone learns in high school. For wavelets, these relationships (which are very complicated, and there are infinitely many of them) had to be constructed by hand.

8.4 What Are Wavelets Good for?

One of the remarkable facts about wavelets is that they work in roughly the same way as the human eye. What does this mean? An important part of vision—certainly something that people who work on the vision problem (teaching machines to see) think about all the time—is the so-called *edge detection problem*. If you are looking at an object, you want to detect its edge(s). Detecting the edges is tantamount to seeing the basic shape of the object. What wavelet theory does—because it deals with translated and scaled bumps as depicted in Fig. 8.11—is that it performs edge detection at many different scales (or orders of magnitude). This is what "seeing" really is: at the coarsest level, edge detection determines the basic shape of an object. At finer levels, edge detection calculates the detailed features of the object.

Wavelets perform two functions: analysis and synthesis.

For the Analysis Aspect: wavelet theory does three distinct things: **(a)** It scales the component functions (as indicated in Fig. 8.11) to adjust the frequency being analyzed; **(b)** It translates the component functions in order to position the analysis at the right place (for example, at the location of a pop or click); **(c)** It performs filtering processes.

For the Synthesis Aspect: the wavelet is used to model the signal at different scales. This is done in such a way that the synthesized or reconstructed signal matches the original input signal. The wavelet theory gives a flexible approach to the problem, and rapid and accurate results.

Today wavelets are used in

- image compression
- image denoising

- image enhancement
- image recognition
- feature detection
- texture classification
- signal processing
- signal design
- edge detection
- signal compression

In short, wavelet theory has had a profound impact on the way we understand sound and sight. The vision problem is a question of particular interest to the military. The military would like to have a machine that can "look" down the road and tell whether two jeeps or a tank (or something else) are approaching. Without accurate edge detection, all the machine sees is a blob. The problem is to analyze that blob at various scales and resolutions and see what the object or objects actually are. We are some ways yet from achieving this goal, but great strides have occurred in recent years because of wavelet theory.

The reader may refer to [Chu 77], [Dau 92], [Mey 92], [Wal 06], and [Wal 08] for further (mathematical) details about these aspects of wavelet theory.

8.5 Key Players in the Wavelet Saga

Certainly one of the important figures in the modern development of wavelet theory has been Ingrid Daubechies. Now a professor at Duke University, Daubechies began her career on the scientific staff at AT&T Bell Labs. Of course the latter was a natural venue in which she could learn the ideas of signal processing and communications technology. When wavelet theory came along she was well positioned to contribute to it. And she has been contributing decisively to the scientific development for over 20 years.

8.5. Key Players in the Wavelet Saga

Indeed, Daubechies's book *Ten Lectures on Wavelets* [Dau 92] has proved to be the canonical textbook for those who want a quick and accurate introduction to wavelets. Daubechies is in great demand on the lecture circuit, and as a collaborator on scientific projects ranging from pure mathematical analysis to applied image compression.

Daubechies is also a gifted teacher. She takes great joy in explaining her ideas about wavelets to undergraduates students—even those who are not mathematics majors. One of her favorite educational devices is to bring an ordinary music CD into class. She takes a nail and scratches the playing surface viciously. It turns out that if the scratch is inflicted radially—see Fig. 8.13a—then the result cannot be heard when the disc is played (and she of course demonstrates this on the spot in her class). But if the scratch is rotational—see Fig. 8.13b—then the damage can be heard immediately when the disc is played. Both of these phenomena can be explained immediately by way of wavelets, and the way that the signal processing algorithm works. The radial scratch can be ignored by wavelets because the localized "bumps" shown in Fig. 8.11 can dance around the resulting spikes. But the rotational scratch gives a signal that runs parallel to the wavelet bumps, and therefore cannot be ignored.

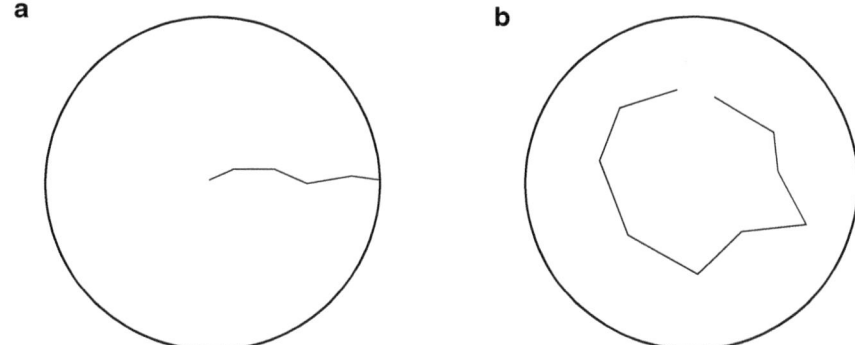

Figure 8.13 Scratches in a CD. (**a**) Radial scratch. (**b**) Rotational scratch.

Daubechies's spouse Robert Calderbank is also a distinguished scientist. A Ph.D. in pure mathematics, Calderbank worked for 20 years on engineering projects at AT&T Bell Labs. While there, he was one of the project managers in charge of developing the new cell phone technology; he relates that a good deal of mathematics went into that development. Just as an instance, the coding of signals uses *Cayley numbers*—an eight-dimensional generalization of the familiar real numbers. Every cell phone has a copy of the Cayley numbers embedded in it.

As indicated earlier Coifman has been another major player in the development and dissemination of wavelet theory. Coifman earned his Ph.D. in pure mathematics in 1965 at the Université de Genève. By luck he met the prominent harmonic analyst Guido Weiss, who was on leave in Switzerland (largely because his wife, a microbiologist, had a position there). Weiss ended up teaching Coifman all about modern harmonic analysis, and Coifman soon thereafter moved to Weiss's home institution (and the home institution of the first author of this book)—Washington University in St. Louis. There Coifman worked through the academic ranks and became one of the most prominent and successful harmonic analysts of our time. He ended up moving to Yale University in 1981 and soon became involved in the development of wavelet theory. He has had a number of important students in the field, notably Victor Wickerhauser (now a professor at Washington University). Coifman was awarded the National Medal of Science for his work on wavelets. Guido Weiss continues to run an active wavelet seminar at Washington University.

Yves Meyer was a student of the prominent French analyst Jean-Pierre Kahane. Meyer himself has had many successful students, including Guy David and Stéphane Jaffard. Although the roots of wavelet theory go back to work of Alfred Haar (1885–1933), J. E. Littlewood (1885–1977) and R. E. A. C. Paley (1907–1933) in the 1930s, Yves Meyer may be said to be the modern father of wavelet theory. Meyer has taught at a number of Parisian universities, including the École Normale Supérieure. He is a member of the French National Academy of Sciences.

A Look Back

Wavelets give a new way of decomposing a function into fundamental units. What is powerful about wavelets is that one can customize the units to specific problems and applications. Wavelet theory has revolutionized applied mathematics and engineering. The number of scientists actively engaged in wavelet research numbers in the many thousands. There are now many dozens of books on the subject, and some of the triumphs of wavelet theory are truly startling:

- Jamie Howarth and Steve Rosenthal received the 2008 Grammy Award for producing an acoustically nearly perfect re-recording of Woody

A Look Back

Guthrie's only live album *The Live Wire: Woody Guthrie in Performance 1949*. Of course they used wavelets to take the original wire-recording signal and turn it into a modern digital signal of good quality.

- In 1993, Victor Wickerhauser of Washington University used wavelets to help the FBI develop a technology for compressing computer files of fingerprints. It turns out that the FBI acquires hundreds of thousands of new prints every day, and the images are of course digital and take a lot of disc space. Wickerhauser's image compression technology reduces the space needed by an order of magnitude.

- Steve LaVietes received a Sci-Tech Oscar for work on Katana, software used in *The Amazing Spider-Man* and *Paranorman* that allows artists to make changes efficiently to large computer graphics scenes.

- The current standard for image compression is JPEG200, and that is a technology that was developed using wavelet theory.

- A typical movie is 2 hours long, which is 7,200 sec. There are 30 images per second in a movie. Each image is about $1,000 \times 1,000$ in a rectangular array, or 1 million pixels (see Fig. 8.14), and each pixel has at least 10 color and brightness attributes. A quick calculation shows that this all gives rise to 2.16 trillion bytes. Far too large for a DVD disc, or even for a Blu-ray disc. Wavelet technology makes it possible for a movie to fit on a single disc that you can slip into your DVD or Blu-ray player.

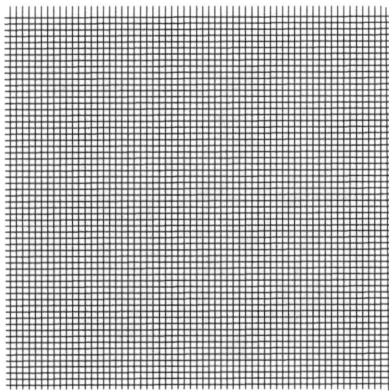

Figure 8.14 The pixel grid for a movie image.

- In medical diagnostics, important information can be gleaned from the analysis of data obtained from radiological, histological, and chemical tests, and this is important for arriving at an early detection of potentially dangerous tumors and other pathologies.

What is surprising is that the analysis of these very different types of data can be performed efficiently using the type of mathematics based on ideas coming from wavelet theory. The practical impact of these ideas in applications is remarkable. For example, several hospitals have adopted medical diagnostics methods that were developed by Coifman and his group using wavelet-based methods.

Wavelet theory continues to grow and develop. The higher-dimensional theory has particularly been the subject of intense study for the past 10 years or more. Many of the best mathematical analysts in the world focus their research programs on wavelet theory. Part of the reward is that they get to interact with engineers, physicists, and other applied scientists to create a productive scientific symbiosis.

REFERENCES AND FURTHER READING

[**Chu 77**] Chui, C.K.: Wavelets: A Mathematical Tool for Signal Analysis. Society for Industrial and Applied Mathematics, Philadelphia (1997)

[**Dau 92**] Daubechies, I.C.: Ten Lectures on Wavelets. Society for Industrial and Applied Mathematics, Philadelphia (1992)

[**Mey 92**] Meyer, Y.F.: Wavelets and Operators. Cambridge University Press, Cambridge (1992)

[**Wal 06**] Walker, J.S.: Wavelet-based image processing. Applicable Analysis **85**, 439–458 (2006)

[**Wal 08**] Walker, J.S.: A Primer on Wavelets and Their Scientific Applications, 2nd edn. Chapman & Hall, Boca Raton (2008)

Chapter 9
RSA Encryption

9.1 Basics and Background

Modern security considerations make it desirable for us to have new types of encryption schemes. It is no longer enough to render a message so that only the intended recipient can read it (and outsiders cannot). In today's complex world, and with the advent of high-speed digital computers, there are new demands on the technology of cryptography. We would now like to have secure messages that *anyone* can encode, but only select people can read. We would like cryptography systems that are protected against various types of eavesdropping and fraud. The important system that we shall describe here, developed during the 1980s by Ron Rivest, Adi Shamir, and Leonard Adleman, enables many such technologies.

In the old days (beginning even with Julius Caesar), it was enough to have a method for disguising the message that we were sending. For example, imagine that the alphabet is turned into numeric symbols by way of the scheme

$$A \longmapsto 0$$
$$B \longmapsto 1$$
$$C \longmapsto 2$$
$$\vdots$$
$$Z \longmapsto 25$$

S.G. Krantz and H.R. Parks, *A Mathematical Odyssey: Journey from the Real to the Complex*, DOI 10.1007/978-1-4614-8939-9_9,
© Springer Science+Business Media New York 2014

Then use the encryption

$$
\begin{aligned}
0 &\longmapsto 3 \\
1 &\longmapsto 4 \\
2 &\longmapsto 5 \\
&\vdots \\
22 &\longmapsto 25 \\
23 &\longmapsto 0 \\
24 &\longmapsto 1 \\
25 &\longmapsto 2,
\end{aligned}
$$

which we will see later can be written

$$n \longmapsto n + 3 \bmod 26. \tag{9.1}$$

[Here "mod" denotes clock or modular arithmetic—discussed in greater detail in Subsection 9.2.2. We render any integer n modulo 26 by subtracting off all multiples of 26. So 29 modulo 26 is 3, 59 modulo 26 is 7, and 10 modulo 26 is 10.] And now convert these numbers back to roman letters. As a simple example, the phrase

WHAT ME WORRY?

translates to the string of integers

22 7 0 19 12 4 22 14 17 17 24

Notice that it is common, in elementary cryptography, to ignore punctuation and spaces.

The encryption (9.1) turns this string of integers into

25 10 3 22 15 7 25 17 20 20 2

and this, in turn, transliterates to

ZKDWPHZRUUC (9.2)

It is clear that anyone receiving the message (9.2) would have no idea what it means—nor even what the message was about, or what context it fits into. On the other hand, such an encoded message is pretty easy to decrypt. Especially if the decryptor knows that we have used a simple "shift

9.1. Basics and Background

algorithm" to encode the message, and if in addition he or she knows that the most commonly used letters in the alphabet are E and then T, it would then be a fairly simple matter to reverse-engineer this encryption and recover the original message.

Today life is more complex. You can imagine a scenario in which

(1) You wish to have a means that a minimum-wage security guard (whom you don't necessarily trust) can check that people entering a facility know a password—but you don't want him to know the password,

(2) You wish to have a technology that allows anyone to encrypt a message—using a standard, published methodology—but only someone with special additional information can decrypt it,

(3) You wish to have a method to be able to convince someone else that you can perform a procedure, or solve a problem, or prove a theorem, without actually revealing the details of the process.

This may all sound rather dreamy, but in fact—thanks to the efforts and ideas of Rivest, Shamir, and Adleman—it is now possible. The so-called RSA encryption scheme is now widely used; it is named by the initials of Rivest, Shamir, and Adleman, the coauthors of the seminal paper [RSA 78]. For example, the e-mail messages received on a cell phone are encrypted using RSA. Banks, secure industrial sites, high-tech government agencies (for example, the National Security Agency), and many other parts of our society routinely use RSA to send messages securely.

Leonard Adleman

2010, Adleman

Len Adleman

Leonard Max Adleman (b. 1945) grew up in San Francisco, majored in mathematics at the University of California at Berkeley, and received his Ph.D. in computer science from Berkeley in 1976. He is renowned for his part in developing RSA encryption. Surprisingly, in addition to being a professor of computer science at the University of Southern California, he is also a professor in USC's Department of Biological Sciences. His 1994 paper "Molecular computation of solutions to combinatorial problems" in *Science* reported on the first known successful use of DNA to carry out a computational algorithm.

In this discussion we describe how RSA encryption works, and then encrypt a message using the methodology. We will describe all the mathematics behind RSA encryption, and prove the results necessary to flesh out the theory behind RSA. We will also describe how to convince someone that you can prove any particular theorem (if you really can)—without revealing any details of the proof. This is a fascinating idea—something like convincing your mother that you have cleaned your room without letting her have a look at the room. But in fact the idea has profound and far-reaching applications.

9.2 Preparation for RSA

9.2.1 Background Ideas

We now sketch the background ideas for RSA. These are all elementary ideas from basic mathematics. Many new mathematical ideas are so profoundly complicated, and require so much background to understand, that they are beyond the reach even of mathematicians who are not specialists. It is remarkable that, with RSA, all that is needed are very elementary and classical ideas. We can present them here in just a few pages. These are new ideas that have decisively affected modern cryptography. The National Security Agency (NSA) in Washington, D.C. employs more Ph.D. mathematicians than any other institution or agency in the world. And the primary focus of NSA work is cryptography. This cryptography is based on algebra and number theory, of which RSA encryption is a fundamental example.

9.2.2 Modular Arithmetic

Modular arithmetic is an easy idea that many of us first encounter in grade school. But it is important for many parts of mathematics, and is used extensively in number theory.

When we write $k \bmod n$ we mean simply the remainder when k is divided by n. Thus
$$25 = 1 \bmod 3,$$
because
$$25 = 8 \cdot 3 + 1,$$
$$15 = 3 \bmod 4,$$

9.2. Preparation for RSA

because
$$15 = 3 \cdot 4 + 3,$$
$$-13 = -3 \bmod 5 = 2 \bmod 5$$

because
$$-13 = -2 \cdot 5 - 3 = -3 \cdot 5 + 2.$$

It is an important, but easily checked, fact that modular arithmetic respects sums and products. That is,

$$(a+b) \bmod n = (a \bmod n) + (b \bmod n)$$

and

$$(a \cdot b) \bmod n = (a \bmod n) \cdot (b \bmod n).$$

We shall use these facts in a decisive manner below.

9.2.3 Euler's Theorem

Let a and b be two positive integers. We say that a and b are *relatively prime* if they have no common prime factors. For example,

$$72 = 2^3 \cdot 3^2,$$
$$175 = 5^2 \cdot 7,$$

hence 72 and 175 are relatively prime.

If n is an integer, let $\mathcal{P}(n)$ be the set of positive integers less than n that are relatively prime to n. Let $\varphi(n)$ be the number of elements in $\mathcal{P}(n)$. For example, if $n = 15$, then

$$\mathcal{P}(15) = \{\, 1,\, 2,\, 4,\, 7,\, 8,\, 11,\, 13,\, 14 \,\} \qquad \text{and} \qquad \varphi(15) = 8.$$

In fact, we could have predicted that $\varphi(15) = 8$ without actually looking at the list of integers that are relatively prime to 15. That is because 15 equals the product of the two prime numbers 3 and 5, and whenever p and q are prime numbers, then

$$\varphi(p \cdot q) = (p-1) \cdot (q-1). \tag{9.3}$$

The reason for this last equation is that the only numbers less than or equal to n that are not relatively prime to n are $p, 2p, 3p, \ldots, q \cdot p$ and $q, 2q, 3q, \cdots,$ $(p-1)q$. There are q numbers in the first list and $p-1$ numbers in the second list. The set $\mathcal{P}(n)$ of numbers relatively prime to n is the complement of these two lists, and it therefore has

$$pq - q - (p-1) = pq - q - p + 1 = (p-1) \cdot (q-1) \equiv \varphi(p \cdot q)$$

elements.

The function $\varphi(n)$ was introduced by Leonhard Euler (1707–1783), so it is called *Euler's phi function* or simply the *phi function*. Euler proved the following theorem about the phi function:

Theorem *If n is a positive integer and k is relatively prime to n, then*

$$k^{\varphi(n)} = 1 \bmod n.$$

Euler's theorem is true for the following reason. First, notice that $\mathcal{P}(n)$ is closed under multiplication modulo n—just because the product of two integers that are relatively prime to n will also be relatively prime to n. Thus, for k in $\mathcal{P}(n)$, the various powers of k,

$$k, \; k^2, \; k^3, \; \ldots$$

cannot all be distinct modulo n (since $\mathcal{P}(n)$ is finite), so there is a smallest power s such that $k^s = 1 \bmod n$. That s is called the *order of k*. Another theorem—that we shall explain below—tells us that the order of any k in $\mathcal{P}(n)$ must divide the number of elements in $\mathcal{P}(n)$. Thus we have $\varphi(n) = s \cdot t$, for some integer t, and hence

$$k^{\varphi(n)} = (k^s)^t = 1^t = 1 \bmod n.$$

To illustrate these phenomena in the case $n = 15$ and $k = 2$, observe that $2^2 = 2 \cdot 2 = 4$, $2^3 = 4 \cdot 2 = 8$, and $2^4 = 8 \cdot 2 = 1 \bmod 15$. Of course, $1 \cdot 2 = 2$, so we see that multiplication by 2 mod 15 cycles through the powers of 2 mod 15 as shown in Fig. 9.1. The number of elements in that cycle, 4, is the order of 2 in $\mathcal{P}(15)$. Now, the order of 2 in $\mathcal{P}(15)$ divides $8 = \varphi(15)$.

9.2. Preparation for RSA

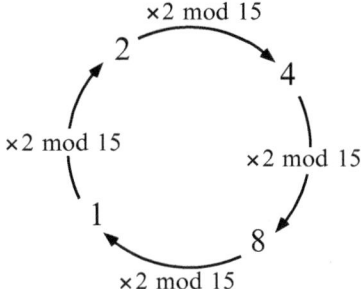

Figure 9.1 Powers of 2 mod 15.

To see why the order of 2 *must* divide $\varphi(15)$, pick any of the numbers in $\mathcal{P}(15)$ that *did not appear* in the cycle of numbers in Fig. 9.1, say 7. Observe that $7 \cdot 2 = 14$, $14 \cdot 2 = 13 \bmod 15$, $13 \cdot 2 = 11 \bmod 15$, and $11 \cdot 2 = 7 \bmod 15$. Thus multiplication by 2 mod 15 cycles through the numbers 7, 14, 13, and 11 as shown in Fig. 9.2.

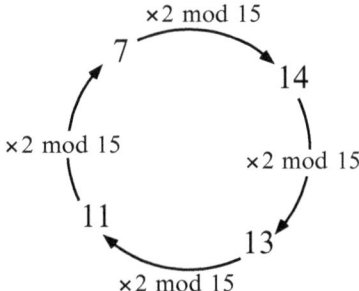

Figure 9.2 Number 7 multiplied by the powers of 2 mod 15.

The number of elements in the new cycle is automatically the same as the number of elements in the first cycle. In this case our two cycles have captured all the elements of $\mathcal{P}(15)$. For another choice of k and n, if we had not exhausted $\mathcal{P}(n)$ after forming two (or more) cycles, we would pick any number in $\mathcal{P}(n)$ that still had not appeared and form yet another cycle. Eventually, by this process, we would break all of $\mathcal{P}(n)$ into disjoint cycles each containing k elements. Therefore k divides $\varphi(n)$.

9.2.4 Relatively Prime Integers

Two integers a and b are *relatively prime* if they have no prime factors in common. As noted above, for example, 72 and 175 are relatively prime.

A fundamental fact of elementary number theory is that, if a and b are relatively prime, then we can find other integers x and y such that

$$xa + yb = 1. \tag{9.4}$$

For example, we have noted that $a = 72$ and $b = 175$ are relatively prime. The corresponding integers x, y are $x = -17$ and $y = 7$. Thus

$$(-17) \cdot 72 + 7 \cdot 175 = 1.$$

One can prove this result using Euler's theorem above. For, since b is relatively prime to a, we see that

$$b^{\varphi(a)} = 1 \bmod a.$$

But this just says that

$$b^{\varphi(a)} - 1 = k \cdot a$$

for some integer k. Unraveling this equation gives (9.4).

In practice, one finds x and y using the Euclidean algorithm (otherwise known as long division). In the example of 72, 175, one calculates:

$$\begin{aligned} 175 &= 2 \cdot 72 + 31 \\ 72 &= 2 \cdot 31 + 10 \\ 31 &= 3 \cdot 10 + 1. \end{aligned}$$

You know you are finished when the remainder is 1. For now we have

$$\begin{aligned} 1 &= 31 - 3 \cdot 10 \\ &= 31 - 3 \cdot (72 - 2 \cdot 31) \\ &= 7 \cdot 31 - 3 \cdot 72 \\ &= 7 \cdot (175 - 2 \cdot 72) - 3 \cdot 72 \\ &= 7 \cdot 175 - 17 \cdot 72. \end{aligned}$$

That is the decomposition we seek.

9.3 The RSA System Enunciated

Now we can quickly and efficiently describe how to implement the RSA encryption system, and we can explain how it works.

9.3. The RSA System Enunciated

Imagine that the Chief Executive Officer (CEO) of Enteric Corporation has an important message to send to the company's Chief Financial Officer (CFO). Of course the CFO is a highly-placed person with many responsibilities, and you can imagine that the Chair's message is quite secret. So the desire is to encode the message:

Your time is up. Hasta la vista, baby.

Thus the Chair goes to the company library and finds the RSA encryption book. This is a readily available book that anyone can access. It is not secret. A typical page in the book reads like this:

Title	Value of n	Value of c
Chair of the board	4431...7765	8894...4453
Chief executive officer	6668...2345	1234...9876
Chief financial officer	7586...2390	4637...4389
⋮	⋮	⋮

Table 9.1 The RSA encryption book.

What does this information mean? Of course we know, thanks to Euclid, that there are infinitely many primes.[1] So we can find prime numbers with as many digits as we wish. Each number n in the RSA encryption book is the product of two 75-digit primes p and q: Thus $n = p \cdot q$. Each number c is chosen to be a number with at least 100 digits that is relatively prime to $\varphi(n) = (p-1) \cdot (q-1)$. Of course we do not publish the prime factorization of the number n; we also do not publish $\varphi(n)$. All that we publish is n and c for each individual. On the other hand, each listed individual *does know*

---

[1] Why is that? Suppose that there were only finitely many primes $p_1, p_2, \ldots, p_k$. Define $P = (p_1 \cdot p_2 \cdots p_k) + 1$. If we divide P by any p_j we get the remainder 1. So P is not composite. It must be prime. But P is greater than all the primes in our exhaustive list $p_1, p_2, \ldots, p_k$. That is a contradiction. Hence there are infinitely many primes.

the prime factorization of his or her own value of n and thus he or she knows $\varphi(n)$. It is this knowledge that allows the individual to decode messages received.

Now an important point to understand is that the CEO does not need to understand any mathematics or any of the theory of RSA encryption in order to encode the message. All that is required is this:

(1) First, break the message into units of five letters. We call these "words," even though they may not be English language words. For the message from the Chair to the CFO, the "words" would be

$$\text{YOURT} \quad \text{IMEIS} \quad \text{UPHAS} \quad \text{TALAV} \quad \text{ISTAB} \quad \text{ABY}$$

(2) Each "word" is translated into a sequence of numerical digits, using our usual scheme of translation (that is, A corresponds to 0, B corresponds to 1, and so forth).

(3) Then each transliterated word w is encoded with the rule

$$w \longmapsto w^c \bmod n.$$

The Chair will send to the CFO this sequence of encrypted words. That's all there is to it.

The real question now is:

What does it take to decrypt the encoded message? How can the CFO read the message?

This is where some mathematics comes into the picture. We have to use Euler's theorem and our ideas about relatively prime integers. But the short answer to the question is the following: If $\widetilde{w}$ is a word encrypted according to the simple scheme described above, then we decrypt it with this algorithm:

(1) We find integers x and y so that $xc + y\varphi(n) = 1$.

(2) Then we calculate

$$\widetilde{w}^x \bmod n.$$

That will give the decrypted word w with which we began. (We shall provide the mathematical details of this assertion in the next section.) Since w has only five characters, and n has 150 characters, we know that $w \bmod n = w$, so there is no ambiguity arising from modular arithmetic. We can translate w back into roman characters, and we recover our message.

Now here is the most important point in our development thus far:

> In order to encrypt a message, we need only look up n and c in the public record RSA encryption book. But, in order to decrypt the message, we must know x. Calculating x necessitates knowing $\varphi(n)$, and finding $\varphi(n)$ when given n necessitates knowing the prime factorization of n.

There is no known algorithm for calculating the prime factorization of an integer with k digits for which the time required to run the algorithm is bounded by a power of k. For an integer with 150 digits, using a reasonably fast computer, it would take several years to find the prime factorization.

You might note that if each piece of the message is just five letters long, then there are only 26^5 possible encryptions. This is a bit less than 12 million possibilities, so someone could decrypt the message by having a computer compare those 12 million possible encryptions to the encryption that has been sent. Because of considerations like this, we often find it convenient to append a 50-digit random integer to each piece.

9.4 The RSA Encryption System Explicated

In fact, with all the preliminary setup we have in place, it is a simple matter to explain the RSA encryption system.

Now suppose that one has selected an $n = p \cdot q$ and a c relatively prime to $\varphi(n) = (p-1) \cdot (q-1)$ corresponding to a particular person listed in the RSA encryption book. If one is that individual or a certified decryptor, then one knows the prime factorization of n—that is, one knows that $n = p \cdot q$ for p and q prime.

One therefore knows that $\varphi(n) = (p-1) \cdot (q-1)$ and so one can calculate the x and the y in the identity $xc + y\varphi(n) = 1$. Once one knows x, then one knows everything. For

$$
\begin{aligned}
\widetilde{w}^x \bmod n &= \left(w^c\right)^x \bmod n \\
&= w^{cx} \bmod n \\
&= w^{1-y\varphi(n)} \bmod n \\
&= w \cdot \left(w^{\varphi(n)}\right)^{-y} \bmod n \\
&= w \cdot 1^{-y} \bmod n \\
&= w \bmod n \\
&= w,
\end{aligned}
$$

since w is certainly relatively prime to n.

This shows how one recovers the original word w from the encrypted word $\widetilde{w} = w^c \bmod n$.

9.5 Zero Knowledge Proofs: How to Keep a Secret

We shall now give a quick and dirty description of how to convince someone that you can prove **Theorem A** without revealing any details of the proof of **Theorem A**. The idea comes across most clearly if we deal with an example from graph theory.

So suppose that we are given a graph. See Fig. 9.3a. A *Hamiltonian circuit* in a graph is a path through the graph that comes back to where it started and passes through every other vertex exactly once. For the graph in Fig. 9.3,

$$A, B, C, E, G, F, D, A \tag{9.5}$$

describes a Hamiltonian circuit. The circuit is also illustrated in Fig. 9.3b using thicker lines to indicate the edges that make up the Hamiltonian circuit.

There are some graphs that have no Hamiltonian circuit. The graph in Fig. 9.4 is one such graph. (The difference between the graphs in Figs. 9.3 and 9.4 is that there is an edge connecting B to C in Fig. 9.3, but none in Fig. 9.4.) In general, there is no known algorithm to find Hamiltonian circuits in graphs in polynomial time, so finding such circuits is a difficult problem.

9.5. Zero Knowledge Proofs: How to Keep a Secret

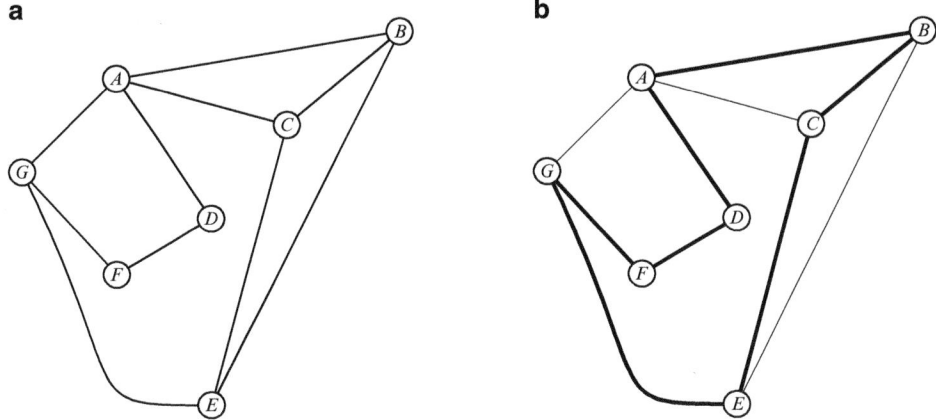

Figure 9.3 (a) A graph with a Hamiltonian circuit, (b) The Hamiltonian circuit emphasized.

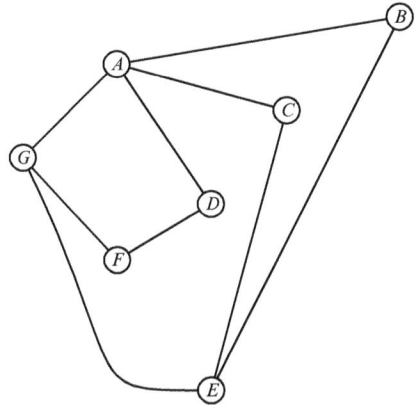

Figure 9.4 A graph containing no Hamiltonian circuit.

Let us suppose that we wish to convince someone, the *verifier*, that we, the *prover*, know a Hamiltonian circuit for the graph in Fig. 9.3. The verifier knows the graph, but not the Hamiltonian circuit. Of course, one way to prove that we know a Hamiltonian circuit is simply to provide the list (9.5) of vertices. But suppose we want to convince the person without revealing the circuit itself. Then we must be much more subtle.

A complete description of the graph can be provided in the form of an *adjacency matrix* which is a table listing all possible pairs of vertices and whether any two are connected by an edge. This data is exhibited in Table 9.2. A "1" in a particular position indicates that the two vertices labeling the row and column are connected by an edge. A "0" indicates that the corresponding vertices are not connected.

	A	B	C	D	E	F	G
A	0	1	1	1	0	0	1
B	1	0	1	0	1	0	0
C	1	1	0	0	1	0	0
D	1	0	0	0	0	1	0
E	0	1	1	0	0	0	1
F	0	0	0	1	0	0	1
G	1	0	0	0	1	1	0

Table 9.2 The adjacency matrix.

Our first step as prover is to randomly assign the numbers 1 through 7 to the vertices. For example,

$$\begin{aligned} 1 &\leftrightarrow G \\ 2 &\leftrightarrow A \\ 3 &\leftrightarrow B \\ 4 &\leftrightarrow C \\ 5 &\leftrightarrow D \\ 6 &\leftrightarrow E \\ 7 &\leftrightarrow F \end{aligned} \qquad (9.6)$$

If there are n vertices in the graph, then there are $n!$ ways of numbering the vertices. To prevent the verifier from being able to gain any information about the Hamiltonian circuit, the prover should see to it that all $n!$ possibilities for the numbering should be equally likely to be used.

The adjacency matrix will depend on the numbering of the vertices. With the numbering above, the new adjacency matrix is

	1	2	3	4	5	6	7
1	0	1	0	0	0	1	1
2	1	0	1	1	1	0	0
3	0	1	0	1	0	1	0
4	0	1	1	0	0	1	0
5	0	1	0	0	0	0	1
6	1	0	1	1	0	0	0
7	1	0	0	0	1	0	0

(9.7)

9.5. Zero Knowledge Proofs: How to Keep a Secret

The information in (9.6) and (9.7) can be succinctly presented in the following form:

$$
\begin{array}{c|ccccccc}
G & 0 & 1 & 0 & 0 & 0 & 1 & 1 \\
A & 1 & 0 & 1 & 1 & 1 & 0 & 0 \\
B & 0 & 1 & 0 & 1 & 0 & 1 & 0 \\
C & 0 & 1 & 1 & 0 & 0 & 1 & 0 \\
D & 0 & 1 & 0 & 0 & 0 & 0 & 1 \\
E & 1 & 0 & 1 & 1 & 0 & 0 & 0 \\
F & 1 & 0 & 0 & 0 & 1 & 0 & 0 \\
\end{array}
\tag{9.8}
$$

As prover, we encode the 56 pieces of information in (9.8) as 56 messages with 56 distinct keys. We can do this using RSA encryption. Metaphorically, we are locking the entries of (9.8) in 56 separate boxes, so that we will be able to reveal as much or as little of (9.8) as desired.

The verifier receives the 56 encoded facts in a form looking as meaningless as the following (but each encryption will involve many more than 5 digits):

$$
\begin{array}{cccccccc}
28918 & 52162 & 81647 & 53060 & 70997 & 49626 & 88974 & 48237 \\
63553 & 48663 & 30995 & 77233 & 13363 & 54407 & 20044 & 50001 \\
19429 & 54164 & 76393 & 77452 & 58763 & 26766 & 84610 & 76797 \\
10365 & 32639 & 27756 & 89643 & 18731 & 42206 & 59193 & 86645 \\
27119 & 81525 & 98872 & 31573 & 35878 & 86324 & 58151 & 98947 \\
51085 & 29676 & 18876 & 23261 & 46901 & 67917 & 35806 & 45766 \\
32368 & 91921 & 17453 & 47070 & 84673 & 30883 & 46557 & 71500 \\
\end{array}
\tag{9.9}
$$

The verifier now asks to see either

(1) The full graph

or

(2) The Hamiltonian circuit.

To make it as difficult as possible for the prover to get away with being deceptive, the verifier should make this choice randomly with equal likelihood for either option.

After the verifier has made his or her choice, the prover responds as follows:

(1) In case the full graph has been requested by the verifier, the prover provides all 56 keys, the verifier converts (9.9) back to (9.8), and then the verifier checks that the data in (9.8) really does correspond to the original graph in Fig. 9.3.

(2) In case the Hamiltonian circuit has been requested, the prover provides only seven keys for the edges involved in the circuit. Of course the prover tells the verifier which location each of the seven keys will decode. After using the seven keys, the verifier will have the following:

$$\begin{matrix} 28918 & 52162 & 81647 & 53060 & 70997 & 49626 & 88974 & 1 \\ 63553 & 48663 & 30995 & 1 & 13363 & 54407 & 20044 & 50001 \\ 19429 & 54164 & 76393 & 77452 & 1 & 26766 & 84610 & 76797 \\ 10365 & 32639 & 27756 & 89643 & 18731 & 42206 & 1 & 86645 \\ 27119 & 81525 & 1 & 31573 & 35878 & 86324 & 58151 & 98947 \\ 51085 & 1 & 18876 & 23261 & 46901 & 67917 & 35806 & 45766 \\ 32368 & 91921 & 17453 & 47070 & 84673 & 1 & 46557 & 71500 \end{matrix} \quad (9.10)$$

From (9.10), the verifier can tell that there are seven edges as promised and that the Hamiltonian circuit is

$$1, 7, 5, 2, 3, 4, 6, 1$$

using the numbering of the vertices by row and column in (9.10). Of course the verifier does not know how the vertices were numbered because he or she does not have the keys to decode that part of (9.9). Thus the verifier does not receive any information about which edges of the graph in Fig. 9.3 form the Hamiltonian circuit.

Now let us consider why the verifier should believe that the prover really does know a Hamiltonian circuit for the graph in Fig. 9.3. Certainly if the prover does not know a Hamiltonian circuit, then that failure will be apparent if the verifier makes choice (2) and asks to see the circuit. But the prover could cheat by forming (9.9), using the adjacency matrix for a different graph—a graph for which the prover does know a Hamiltonian circuit. In this case, the prover would be caught in a lie if the verifier makes choice (1) and asks to see the whole graph. Thus the verifier knows that if the prover is being deceptive, then the verifier has probability 1/2 of catching the deception.

To increase the verifier's confidence that the prover is not being deceptive, the whole process must be repeated a second time. To protect the secrecy of the Hamiltonian circuit, the prover must make another independent random choice for the numbering of the vertices, and the prover must use a new encryption of the data so that the old keys no longer work. Similarly, to guard against the possibility of the prover changing the mode of deception, the verifier must make the choice of (1) seeing the whole graph or (2) seeing the Hamiltonian circuit randomly and independently of the choice that was made the first time. After the second repetition of the process, the probability that the prover can get away with deception will be reduced to $1/2 \cdot 1/2 = 1/4$.

While the verifier can never be absolutely certain that the prover is not being deceptive, after k repetitions the probability that the prover can successfully deceive the verifier will be reduced to $1/2^k$.

Of course the example we presented is for Hamiltonian circuits in a graph, just because it is simple to describe. For a more complex situation, such as a logical deduction, a more elaborate arrangement of codes—the metaphorical locked boxes containing the information—will be needed. Nonetheless, the principle will be the same. The verifier can check that the prover is working with the given inputs—in our example, using the graph given in Fig. 9.3—or the verifier can check that the prover is doing allowable steps—in our example, forming a Hamiltonian circuit. Meanwhile, nothing is revealed that allows the verifier to reconstruct the deduction itself.

A Look Back

Cryptography is an old idea with a long history. Textbooks tell us that Julius Caeser encrypted messages by shifting each letter of a message three to the right in the alphabet. That is one of the very simplest encryption methodologies.

Today, with the advent of high-speed digital computers, we have sophisticated means of encrypting messages. And, given the nature of international politics and corporate intrigue, this technology is really necessary. The largest employer in the world of Ph.D. mathematicians is the National Security Agency in Washington, D.C. (over 2,000 employees), and their primary activities are cryptological.

The cell phone of one of these authors has a message on the screen saying the calls are encoded using RSA. One of the basic ideas behind RSA is that it would take even a very fast computer a long time (a year or more) to factor a large integer into prime factors. The Robert Redford film *Sneakers* is about a mathematician who finds the means, and produces the machine to implement the idea, to factor large integers quickly. This obviously blows RSA encryption out of the water. Such an eventuality is not likely to happen any time soon, but quantum computing could change the picture.

Rivest, Shamir, and Adleman created RSA encryption in the early 1980s. They gave a talk on their work at the International Congress of Mathematicians in Berkeley in 1986. Around the same time, they applied for a U.S. patent for some of their ideas.

Now one of the reasons that the Patent Office exists is to give the government the opportunity to review new ideas, and to coopt those that it thinks it needs. Well, RSA was a prime candidate. And, indeed, the government stepped in and said, "We want this technology for national security purposes. We order you to stop disseminating this idea. You can no longer give talks on this idea. And you should write to those to whom you have mailed your papers and ask for those papers back."

Well, this caused quite a scandal in academic circles. Nothing quite like this had ever happened before. While Rivest, Shamir, and Adleman could certainly stop lecturing about their work, it was virtually impossible for them to recall the many hundreds of preprints that they had distributed (this was in the days before the Internet, so at least nothing was posted on the Web). And obviously they were very upset to be so hemmed in by the government.

Interestingly, the entity that saved the day was none other than the National Security Agency—headquarters of cryptography in this country. They said that the ideas behind RSA were in the public domain anyway, and nobody could squelch these ideas. In addition, they argued, the actual value of RSA in practice was in no way compromised by a great many people knowing how it worked.

So, in the end the shackles were taken off the creators of RSA. And today there are many interesting variants of the idea. One can read papers about applications of RSA with dishonest encryptors, corrupt decryptors, with evesdroppers, and many other variants.

Another exciting innovation of the past few years is that there are now applications of algebraic geometry and elliptic curve theory—parts of mathematics that go back to Abel and Gauss in the 1800s—to developing new encryption schemes. It is heartening to see old mathematical ideas brought back to life in new contexts.

REFERENCES AND FURTHER READING

[Blu 86]   Blum, M.: How to prove a theorem so no one else can claim it. Proceedings of the International Congress of Mathematicians (Berkeley, CA, 1986), pp. 1444–1451. American Mathematical Society, Providence (1987)

[BG 84]   Blum, M., Goldwasser, S.: An efficient probabilistic public-key encryption scheme which hides all partial information. Advances in Cryptology (Santa Barbara, 1984). Lecture Notes in Computer Science, vol. 196, pp. 289–299, Springer, Berlin (1985)

[BDMP 91] Blum, M., De Santis, A., Micali, S., Persiano, G.: Noninteractive zero-knowledge. SIAM Journal on Computing **20**, 1084–1118 (1991)

[FFS 88]   Feige, U., Fiat, A., Shamir, A.: Zero-knowledge proofs of identity. Journal of Cryptology **1**, 77–94 (1988)

[RSA 78]   Rivest, R.L., Shamir, A., Adleman, L.M.: A method for obtaining digital signatures and public-key cryptosystems. Communications of the Association for Computing Machinery **21**, 120–126 (1978)

[DMP 88]   De Santis, A., Micali, S., Persiano, G.: Noninteractive zero-knowledge proof systems. Advances in Cryptology—CRYPTO '87 (Santa Barbara, 1987). Lecture Notes in Computer Science, vol. 293, pp. 52–72, Springer, Berlin (1988)

Chapter 10
The P/NP Problem

10.1 Introduction

Mathematicians are always interested in problems and their solutions. In this chapter we will be interested specifically in problems that can be solved using an *algorithm*, that is, by a step-by-step procedure.

Every child in school learns algorithms for solving math problems. As a student's education progresses, the algorithms tend to get more complicated and sophisticated. A student also learns that sometimes there is both an easy way and a hard way to solve a particular type of problem. It is natural then to want to make precise the intuitive feeling that some algorithms are harder than others, and to investigate whether some problems inherently require more difficult algorithms than others. Various formalisms can be used for this study, but it is most natural in our modern world to think in terms of using a computer to carry out an algorithm.

The "P" in the chapter title stands for "polynomial" as in *polynomial time*. Precisely, it refers to computations for which the time required increases no faster than the size of the problem raised to a fixed power. For example, consider the problem of adding two whole numbers. Without doing any of the actual calculation, you can be sure that adding two 18-digit numbers, such as

$$
\begin{array}{r}
289,186,957,888,231,332 \\
+\ 767,099,779,936,568,650 \\
\hline
1,056,286,737,824,799,982
\end{array}
$$

will take about twice as long as adding two nine-digit numbers, such as

$$\begin{array}{r}585,901,063\\+159,501,547\\\hline 745,402,610\end{array}$$

You can make that estimate because you know from experience that the process of addition takes approximately the same amount of time for each pair of digits to be added (i.e., for each column). That is, the difficulty of adding whole numbers with n digits is proportional to n.

As another example, you might ask yourself about an algorithm for sorting. Suppose a group of n children is required to line up by height. Each of the n children will need to compare his or her height to the heights of the other children; each child might even need to compare his or her height to all $n-1$ of the other children. So lining up by height may require as many as, but certainly no more than, $n(n-1) = n^2 - n$ height comparisons.

A particular type of problem is said to be of Class **P** if there is an algorithm for solving problems of that type *and* if there is a polynomial p such that, when the input size of the problem is n, the algorithm will produce the solution after at most $p(n)$ computational steps. Problems of Class **P** are considered to be *tractable*.

Despite what you might reasonably guess, the "NP" in the chapter title *does not* stand for "Not Polynomial." In fact, "NP" stands for "Nondeterministic Polynomial." A particular type of problem is said to be of Class **NP** if there is an algorithm for verifying the validity of possible problem solutions *and* there is a polynomial p such that, when the input size of the problem is n and a possible solution is proposed, the verification algorithm will *decide its validity* after at most $p(n)$ computational steps. You can think of the word *nondeterministic* as referring to the fact that the computer does not need to determine the solution, but only to check that a possible solution is correct.

An example of a problem for which verification of a solution is easier than finding a solution is factoring whole numbers. For instance, by simply calculating the product of 8707 and 7759, it is easy to verify that 67,557,613 factors into

$$8707 \times 7759.$$

But starting from scratch to find nontrivial factors of 67,557,613 by hand would be tedious work indeed. (Of course, the feeling we have about the

10.2. Complexity Theory

difficulty of factoring does not prove that there is no fast algorithm to do it. It is just that such an algorithm is not known at this time.) To summarize the preceding discussion:

- For a problem in Class **P**, the solution can be *found* in polynomial time.

- For a problem in Class **NP**, a possible solution can be *verified* in polynomial time.

The P/NP Problem referred to in the chapter title is the meta-problem of determining whether the Class **NP** and the Class **P** are the same. For the record, we note that the Class **P** is automatically a subclass of the Class **NP** for the following simple reason: if you have in hand a polynomial time algorithm for solving a problem, then you can certainly verify any proposed solution by simply comparing it with the actual solution. That actual solution can be obtained in polynomial time and the comparison is proportional to the size of the solution, so the verification can be accomplished in polynomial time. We also note here that, while there is no simple way to see it, not every algorithmically solvable problem is in Class **P**. Thus, it very well could be that the Class **NP** is strictly larger (indeed, much much larger) than the Class **P**. Indeed, most experts believe that the Class **NP** is in fact strictly larger than the Class **P**, but as yet nobody has been able to prove this assertion.

Many people consider the question described in the last paragraph to be the most important problem in the mathematical sciences.

10.2 Complexity Theory

Complexity theory is a conceptual framework for measuring the difficulty of computational problems. The aim of the theory is to estimate how much computer time it will take to solve a particular type of problem. There are three crucial factors involved in developing such an estimate: one factor is the speed at which the computer itself can perform the computational steps required by the calculation; this factor is the *hardware speed*. The second factor is the procedure that is to be carried out to solve the type of problem; this factor is the *software*. The third factor is the size of the specific instance of the problem; this factor is the size of the data file needed to input the problem. Note that memory requirements are not our worry—we assume we have all the memory we need. (Of course, one could ask about memory requirements, but that is a different question.)

The speed at which a computer can perform the steps of a calculation—the hardware speed—is dependent on the specific computer employed. Until roughly one hundred years ago a computer was a human being—"computer" was a job title, not a piece of equipment.[1] The human computers of old gained speed and accuracy when teamed with mechanical devices: one hundred years ago, such a mechanical device would have been an adding machine (see Fig. 10.1). Those mechanical devices were steadily improved, incorporated electronics, and finally became what we now call computers.

Figure 10.1 A mechanical adding machine.

Since next year's computers will no doubt be faster than this year's computers, it makes little sense to specify a hardware speed that is based on actual computers. Instead complexity theory is constructed on the basis of a standardized model of a computer called a *Turing machine* (about which

---

[1] Often the invention of the computer is attributed to Charles Babbage (1791–1871), who designed several mechanical computers in the 1800s. While Babbage's machines were partially built during his lifetime, none was completed. A functioning model of Babbage's *Difference Engine No. 2* was completed in 1991. It was built using machining tolerances achievable in the nineteenth century, providing final validation of Babbage's work. A fairly comprehensive history of computing machines can be found in [Kra 11, Chap. 6].

we will say more later). Using that standardized model, one constructs algorithms to solve problems. If you devise an algorithm to solve a type of problem and if that method runs in polynomial time on a Turing machine (as measured by the number of Turing machine operations required), then by definition you have proved that the problem is in the Class **P**.

The Turing machine metaphor initially envisions building a custom machine to carry out each algorithm. But, even in Turing's first paper on the topic, he described the *universal Turing machine* which would be capable of emulating the behavior of any specific Turing machine. Effectively, the universal Turing machine functions as a programmable computer, so the distinction between Turing machine hardware and Turing machine software is not important.

Any algorithm designed for a Turing machine can be translated into your favorite high-level programming language (e.g., `Python`, `JAVA`, `C++`) and run on an actual computer. Any programming language that is not adequate to allow one to do everything a Turing machine can do would be rejected immediately because it is too weak. Consequently, any problem in the Class **P** can be solved by an algorithm that runs in polynomial time on an actual computer.

Translating high-level software into Turing machine operations is a different proposition. Once you know more about Turing machines, you might amuse yourself by figuring out how you would translate the commands available in your own favorite high-level programming language into Turing machine operations. That can surely be done, but you may tire of the exercise before it is complete. It is conceivable that there is some yet-to-be-invented language used on some yet-to-be-invented computer that cannot be translated into Turing machine operations, but all experience to date indicates that any process that we would agree to call a "computational method" can indeed be carried out on a Turing machine. This empirical observation has convinced most, if not all, that *every* computational method can be carried out by a Turing machine; this supposition is called the *Church–Turing thesis*.

10.3 Automata

The Turing machine was invented by Alan Mathison Turing (1912–1954) in the mid-1930s to address a problem in mathematical logic. Turing's seminal paper on the topic was published in 1937, a time when programmable electronic digital computers—computers as we now know them—did not exist. Nonetheless the idea of using machines to automate clerical processes was well developed.[2] Similarly, feedback mechanisms and automatic control using electro-mechanical systems had long been in use. By combining those simple elements that were known and used in the 1930s, Turing was able to design a machine that is believed capable of carrying out any step-by-step computational process.

The CPU. As our first step on the way to defining a Turing machine, we will look at its CPU or central processing unit. Any Turing machine works by processing the characters in some fixed, finite alphabet. The CPU of a Turing machine

- has finitely many *states* (for example, the states of an appliance might be on and off) and
- when an input character from the machine's alphabet is received, the machine undergoes a *transition* from its current state to another, or possibly the same state, as specified by a fixed set of *transition rules*.

Unless the contrary is stated, the CPU is assumed to be *deterministic*, meaning that, for every state of the machine, there is exactly one transition rule for every character of the alphabet. If the CPU is nondeterministic, then it might stop working because it receives a character for which there is no rule, or there might be several transition rules that apply and the machine could do more than one thing (and therefore cannot decide). In everyday life, such nondeterministic behavior is bad. A real machine exhibiting nondeterministic behavior may well be broken and ready for the scrap heap. Nonetheless, we will see later that, for theoretical purposes, nondeterminism can be useful and good.

Deterministic Finite Automata. If all you have is a CPU with no way to feed it some data, then nothing will happen. When we allow the CPU to read a string of finitely many characters from the alphabet, we get a *deterministic finite automaton*. In fact, you are already quite familiar with many deterministic finite automata. For example, many small appliances cycle through a number of settings using a pushbutton. In the simplest case,

---

[2]For instance, Herman Hollerith's machine to automate analysis of the census, a machine that used punch cards, was invented in 1890. Hollerith went on to found the company IBM.

10.3. Automata

the only settings are "ON" and "OFF." Even in this case, the appliance is an example of a deterministic finite automaton having two states, namely, on and off. The automaton's alphabet is the one character "push" read by the pushbutton, and the transition rules are the following:

- If the appliance is off, it will turn on when the button is pushed.
- If the appliance is on, it will turn off when the button is pushed.

The transition rules can be efficiently illustrated as in Fig. 10.2. To apply such a diagram, find the circle that corresponds to the current state of the automaton and follow the arrow labeled by the character read.

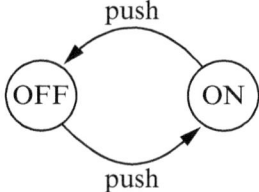

Figure 10.2 A simple appliance with a single pushbutton.

A more complicated appliance (such as a fan) may have additional settings; so, for example, the first push of the button switches the device on to "LO," the next push of the button switches the device to "HI," and a third push of the button switches the device back to "OFF." This appliance is an example of an automaton with three states, but still the one character alphabet. The transition rules are illustrated in Fig. 10.3.

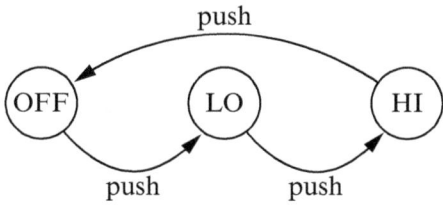

Figure 10.3 An appliance with two settings and a single pushbutton.

For an appliance with many settings, having only one button is not convenient. A small improvement can be made by adding a dedicated on/off button and another that chooses a speed. That setup is illustrated in Fig. 10.4.

There are two pushbuttons indicated by "B1" and "B2." Button B1 is the on/off switch. If the appliance is off and B1 is pushed, then the appliance comes on in its low setting, as is shown by the arrow starting at the circle labeled "OFF" and ending at the circle labeled "LO." The button B2 is used to change from low, to medium, to high, and back to low as indicated by the arrows labeled by B2. Another push on B1 turns the appliance off. This appliance is an example of an automaton with four states and a two-character alphabet.

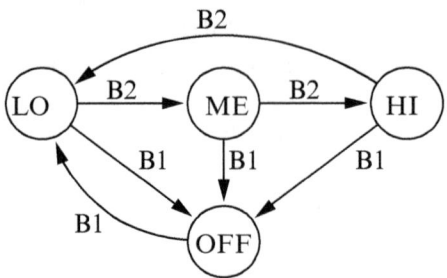

Figure 10.4 An appliance with two pushbuttons.

The appliance corresponding to Fig. 10.4 always comes on in its low setting. One might prefer that the appliance would come back on in the same setting as it was in when turned off. Such a setup is illustrated in Fig. 10.5. Again, there are two pushbuttons with button B1 being the on/off switch and button B2 selecting among low, medium, and high. What has changed is that we need six circles to keep track of what is happening, hence the automaton has six states. In the figure, F denotes "off," N denotes "on," L denotes "low," M denotes "medium," and H denotes "high."

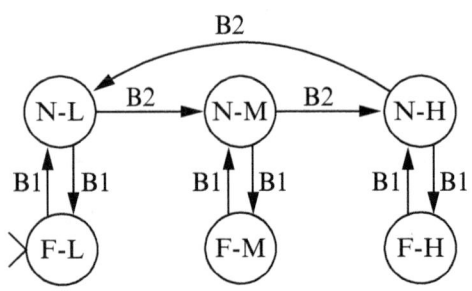

State Label	On or Off	Setting
F-L	off	low
F-M	off	medium
F-H	off	high
N-L	on	low
N-M	on	medium
N-H	on	high

Figure 10.5 Another appliance with two pushbuttons.

10.3. Automata

In the previous examples, one naturally assumes that the appliance is off when it is first plugged in, but in this example there is no "obvious" initial state. The initial state of an automaton is indicated by a ">" pointing at the appropriate circle, in this case, the F-L circle.

Formal Languages. The finite automata we have considered so far may make one doubt that a finite automaton could solve any interesting problem. In fact, finite automata can be used very naturally to decide whether or not particular words are in a language. We look at this application next.

A *formal language* is a set of finite strings of characters from some alphabet. The written English language is a certain set of strings of characters chosen from the Roman alphabet: "fish" is an element of the language and "qqxef" is not. We say that "fish" is *accepted* in English and "qqxef" is not accepted. As a living, natural language, English is constantly changing in unpredictable ways. A formal language is a *fixed* subset of the set of finite strings of characters in the given alphabet.

Some formal languages can be defined very succinctly. For example, suppose the alphabet has only the two characters a and b; then the language E consisting of all strings that end with the substring aa is easily defined. Recognizing whether a string is accepted in E is also easy; in fact, recognizing words in E is so easy that we can design a deterministic finite automaton to carry out the task.

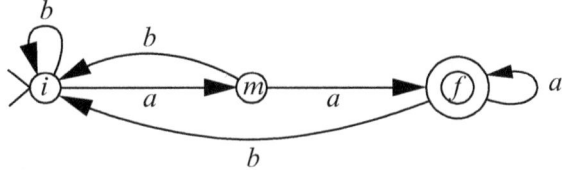

States: initial (i), middle (m), final (f)

Figure 10.6 An automaton that accepts strings ending in aa.

Figure 10.6 illustrates an automaton that reads strings in the alphabet $\{a, b\}$ and accepts a string if and only if it is a word in E. To understand the automaton represented by this figure, we first note that a string will be accepted if the automaton is in the double circled state when the last character in the string has been read and acted on. If, after the last character in the string has been read, the automaton is in a state indicated by a single circle, then the string is not accepted. As a matter of terminology, the double circled state is called a *final state*. It is permissible for there to be more than one final state, but we only need one for this example.

To see how the automaton in Fig. 10.6 functions, let us consider applying the automaton to the strings

$$abaa, \quad baaa, \quad aaba, \quad aaab.$$

We will keep track of things by writing ordered pairs consisting of the current unread part of the string as the first entry and a name for the state as the second entry. For this purpose, the states have been labeled i for "initial state," m for "middle state," and f for "final state." The symbol $\vdash$ indicates one step of the automaton, $\emptyset$ denotes the empty string, and each step uses up one character from the string.

$$(abaa, i) \;\vdash\; (baa, m) \;\vdash\; (aa, i) \;\vdash\; (a, m) \;\vdash\; (\emptyset, f)$$

$$(baaa, i) \;\vdash\; (aaa, i) \;\vdash\; (aa, m) \;\vdash\; (a, f) \;\vdash\; (\emptyset, f)$$

$$(aaba, i) \;\vdash\; (aba, m) \;\vdash\; (ba, f) \;\vdash\; (a, i) \;\vdash\; (\emptyset, m)$$

$$(aaab, i) \;\vdash\; (aab, m) \;\vdash\; (ab, f) \;\vdash\; (b, f) \;\vdash\; (\emptyset, i)$$

We see that the strings $abaa$ and $baaa$ are accepted as words in E and the strings $aaba$ and $aaab$ are not.

Nondeterministic Finite Automata. If a person were given a long string of as and bs and asked whether it is a word in E, he or she would *not* go through the same step-by-step process as the automaton in Fig. 10.6. Instead, a person would quickly skip to the end of the string and check whether it ends with aa. That more efficient approach is modeled by the *nondeterministic finite automaton* illustrated in Fig. 10.7. The nondeterminism in this automaton explicitly occurs in the initial state where there are *two* arrows labeled with the character a. Those two arrows mean that the automaton is allowed to go to either of two states when an a is read. Also, a nondeterministic automaton is allowed to have states that do not have transition rules for every character that might occur. In the diagram for a nondeterministic automaton, that sort of nondeterminism leads to states that do not have arrows for every character that might be read. For the automaton in Fig. 10.7, this is the case for the states m and f and the character b. A string is said to be *accepted* by a nondeterministic automaton if there exists *some* sequence of allowed steps that ends in the final state with all the characters in the string used up.

10.3. Automata

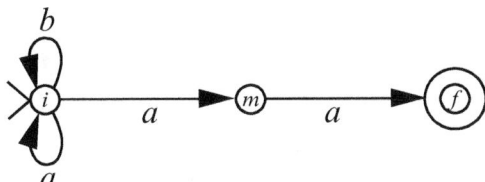

States: initial (i), middle (m), final (f)

Figure 10.7 A nondeterministic automaton that accepts strings ending in aa.

Let us apply the automaton in Fig. 10.7 to the string $abaa$ (which we already know is in E). When the first character a is read, there are two things the automaton can do: it can move to the state m or it can remain in the initial state. We write these two possibilities as follows:

$$(abaa, i) \vdash (baa, m) \tag{10.1}$$

$$(abaa, i) \vdash (baa, i) \tag{10.2}$$

If the automaton moves to the state m when the first a is read, then it will be unable to process the second character b, because there is no arrow labeled b leaving state m. We conclude that, if the transition (10.1) is followed, then the automaton will never get to the end of the input string. Thus (10.1) is useless for showing that $abaa \in E$.

On the other hand, the following sequence of allowed state transitions does end in the final state with all the characters in the string used up, thereby showing that $abaa$ is accepted by E:

$$(abaa, i) \vdash (baa, i) \vdash (aa, i) \vdash (a, m) \vdash (\emptyset, f)$$

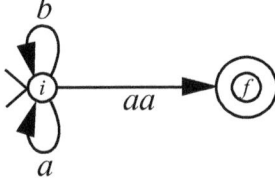

States: initial (i), final (f)

Figure 10.8 A nondeterministic automaton that accepts strings ending in aa.

Notice that the two arrows starting and ending at the initial state have the effect of allowing the automaton to skip over any number of initial characters in a string to get to the last two characters, which are the only two characters that really matter in deciding whether a string is in E. Also, notice that the middle state in this automaton is superfluous: we never want to go there unless there is a second a to allow the automaton to move on to the final state. A simplified version of the automaton is shown in Fig. 10.8, where the aa label tells us that the arrow may be followed when the substring aa is read.

A nondeterministic automaton is allowed to do everything a deterministic automaton is allowed to do, and more. For that reason, the family of nondeterministic automata, by definition, includes the deterministic automata, but not vice versa. Note that this terminology puts us in the situation where "nondeterministic" has a meaning different from "not deterministic."

Despite the fact that nondeterministic automata are allowed to do things that deterministic automata are not, the languages that can be defined by the two types of automata are the same. In fact, there is a theorem that tells us the following:

Theorem *If there is a nondeterministic finite automaton that accepts exactly the strings that are words in a language, then there exists a deterministic finite automaton that accepts exactly the words in the same language.*

The deterministic automaton that accepts the same words as a nondeterministic automaton will work with the same finite alphabet, but it will probably have more states. Also, the corresponding deterministic automaton is likely to be harder to understand from its diagram.

For example, the nondeterministic automaton in Fig. 10.9 accepts strings in the alphabet $\{a, b\}$ if and only if they begin with aa. We see from its diagram that it stops immediately if it does not see aa as the first two characters. If it does see aa first, it goes to the final state and it can then "throw away" any extra characters without leaving the final state. It seems more difficult to see that the deterministic automaton in Fig. 10.9 also accepts a string if and only if it begins with aa.

Limitations of Finite Automata. Finite automata provide a model for simple computers, but they are severely limited. If a deterministic finite automaton has only finitely many states and if it accepts a string with at least one more character than there are states in the automaton's CPU, then some substring in that accepted string must take the CPU on a round trip from a particular state back to itself. That substring can then be replaced with repeated copies of itself that make the CPU repeat its round trip over

10.3. Automata

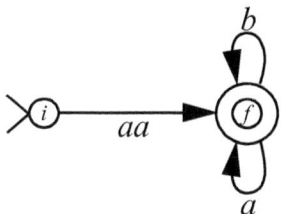

States: initial (i), final (f)
Nondeterministic

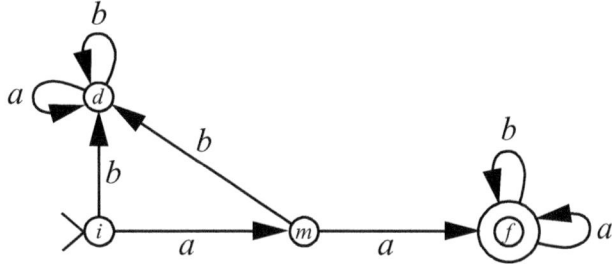

States: initial (i), dead-end (d), middle (m), final (f)
Deterministic

Figure 10.9 Two automata that accept strings beginning with aa.

and over again as many times as you want. But the resulting string will still be accepted because, once all the round trips among states are over, the ending of the string is unchanged and it takes the automaton to a final state. The argument in the preceding discussion is called a *pumping argument*. It proves the following pumping theorem:

Theorem *If a finite automaton accepts an infinite language (that is, a language containing infinitely many strings), then there are strings x, y, and z, where y is not the empty string, such that the language contains all strings of the form*

$$x \underbrace{y\,y \cdots y\,y}_{n \text{ copies}} z,$$

for every positive integer n. Here x and z each are allowed to be the empty string.

A consequence of this result is that, for the alphabet $\{a, b\}$, there is no finite automaton that accepts a string if and only if it consists of a number of a's followed by exactly the same number of b's, i.e., if the string is of the form

$$\underbrace{a\,a\cdots a\,a}_{n \text{ copies}} \underbrace{b\,b\cdots b\,b}_{n \text{ copies}} \tag{10.3}$$

To see this, ask yourself, "If a finite automaton accepted exactly strings of the form (10.3), then what string can play the role of y in the theorem?" It can't be a string of only a's, because an imbalanced number of a's and b's will be accepted. Likewise, y can't be a string of only b's for the same reason. Finally, if y contains both a's and b's, then repeating y two or more times will put a b before an a, and that is a pattern that is not supposed to be accepted.

10.4 Turing Machines

Finite automata have limited computing power because they have no memory. One consequence of this is the inability of any finite automaton to recognize strings of the form (10.3). Since Turing machines are intended to be, and are believed to be, capable of every computation that can be done on any computer, it will be essential to equip them with a memory.

Turing's original specifications called for the memory to be in the form of a doubly infinite tape divided into squares from which the machine could read and to which the machine could write. Here we will assume that our Turing machines have a tape that is infinitely long only to the right. This change in the design of the machines makes no fundamental difference to its computing power: anything that can be done with a Turing machine that has a doubly infinite tape can be reprogrammed to be done on a machine with a tape that is infinite in only one direction.

Basic Turing Machine Operations. To make the Turing machine's tape useful, we will need to supplement the basic state change function of a finite automaton with some tape handling functions. A Turing machine will have a *read/write head* that scans the contents of a square on the tape. Each square can be blank or it may contain a character from the machine's alphabet, but at any time all but finitely many squares must be blank. The alphabet must have at least two characters, and the blank is *not* allowed to be part of the

10.4. Turing Machines

alphabet. Depending on the state of the machine and the symbol currently being scanned by the read/write head, the machine

- transitions to a "new" state (which may be the current state),

- erases the symbol in the square, then leaves the square blank or writes a symbol in the square (which may be the same as the symbol scanned and erased), and

- moves the read/write head to the left or to the right or does not move the head.

If the read/write head moves past the left end of the tape, that constitutes an error condition and the machine will cease operation. If there is no action specified for the current state and scanned symbol, that also constitutes an error condition and the machine will cease operation. In both cases, we say the machine *hangs*.

While an automaton reads the next input character whenever there is one—automatically as it were—a Turing machine requires that the read/write head be directed to move. Since a Turing machine can write output to the tape, it is not necessary to use "final states" to give the results of the computation. Every Turing machine is required to have a *halted state* that signals the end of a computation. Halting is good, hanging is bad. Never halting and never hanging is also not good, because the computation never ends.

Computing with a Turing Machine. One way to use a Turing machine is to evaluate a function as we would do with a programmable calculator. We can do this by providing the values of the arguments of the function as strings on the tape at the start of the computation. We then let the Turing machine run until it halts, and read the value of the function as a string on the tape.

There are also at least two other ways we can put a Turing machine to work for us:

• We can use a Turing machine to solve decision problems. To do so, we add two special characters to the alphabet, one representing "Yes" and the other representing "No." The machine is given an input string and, when it reaches its halted state, we find the "Yes" character or the "No" character written on the tape. Thus the Turing machine makes a decision.

For instance, the Turing machine might be deciding whether the string represents a true statement in some axiomatic theory. Alternatively, the Turing machine might be deciding whether the string is a word in a particular formal language. If there exists a Turing machine that can decide whether any given string is a word in the language, then the language is said to be *Turing decidable*.

• Another way to use a Turing machine is to provide an input string and then see if the machine *ever* halts. If a formal language consists exactly of those strings for which the Turing machine halts, then the language is said to be *Turing acceptable*.

The perhaps surprising fact is that there are Turing acceptable languages that are not Turing decidable, but every Turing decidable language is Turing acceptable.

Working with Turing machines is easier if we impose precise requirements on the form of the input and output. For input consisting of a single string, we require the following:

(1) The machine is in its initial state.

(2) The input string contains no blanks.

(3) The input string is surrounded by a blank on each side.

(4) The input string is positioned as far to the left on the tape as possible.

(5) All squares to the right of the input string are blank.

(6) The read/write head is initially positioned so that it scans the blank adjacent to the rightmost character in the input string.

Output consisting of a single string is similarly restricted. We recognize that a string is the output resulting from a computation if the following hold:

(1) The machine is in its halted state.

(2) The tape contains one string of characters—the output string—and that string contains no blanks.

(3) The output string is surrounded by a blank on each side.

(4) The output string is positioned as far to the left on the tape as possible.

(5) All squares to the right of the output string are blank.

10.5. Examples of Turing Machines

(6) The read/write head is positioned so that it scans the blank adjacent to the rightmost character in the output string.

Notice that, except for the specific content of the string on the tape, the tape looks the same when the computation finishes as it did when the computation began. The difference is that the machine is in its initial state at the beginning of the computation and in its halted state at the end of the computation.

In the case of input consisting of multiple strings, we simply separate individual strings with blanks. For example, if the input is the three strings *abc*, *cba*, and *aa* then, at the start of the computation, the tape should be as follows, where we use # to indicate a blank square:

| # | a | b | c | # | c | b | a | # | a | a | # | # | ⋯ |

The same tape also could be the output from a computation.

Using the metaphor of the tape, we can imagine that any of the arithmetic operations that we learned in school could be performed by a Turing machine; the machine will keep all its work on one line, whereas we would work on multiple lines on a sheet of paper.

10.5 Examples of Turing Machines

Our understanding of Turing machines will be helped by looking at some specific examples.

The Incrementor. In this example, we design a Turing machine that takes as its input a binary number and returns as its output that number plus 1, again in binary representation. This Turing machine is called the *incrementor*. It works with the alphabet $\{0, 1\}$ consisting of the binary digits, or *bits*.

As an example of a binary number consider 1010. This binary number is converted to a decimal number by reading from right to left, so 1010 represents 0 ones, 1 two, 0 fours, and 1 eight for a total of ten. We will allow extra 0's on the left, so both 1010 and 0001010 will be legitimate binary representations of ten. (We will also allow the empty string as an alternative representation of zero.)

What the incrementor does is search the input string from right to left until a 0 is found. Once a 0 is found, it is changed to a 1, and then the read/write head moves back from left to right changing each 1 to a 0. For example, thirty-five has the binary representation 100011. Scanning from right to left, we see the first 0 is the third bit. Changing that 0 to a 1 and changing the first and second bits from 1s to 0s, we get 100100 which is the binary representation of 36.

If no 0 is found in searching from right to left, then the input string is of the form

$$\underbrace{11\cdots11}_{n \text{ copies}},\qquad(10.4)$$

where n may equal 0, so the output must be

$$1\underbrace{00\cdots00}_{n \text{ copies}}.$$

That output is produced by leaving the leftmost 1 unchanged, moving left to right changing each 1 to a 0, and finally changing the blank that marked the right end of the input to a 0. For example, thirty-one has the binary representation 11111, and when one is added to it, the result is thirty-two which has the binary representation 100000.

To carry out the process described above, the incrementor has the following eight states:

$$\text{Initial},\quad S_1,\quad S_2,\quad S_3,\quad S_4,\quad S_5,\quad S_6,\quad \text{Halted}.$$

The halted state requires no instructions. For the other states of the incrementor the instructions are as follows:

INITIAL STATE

read	write	move	go to
#	#	left	S_1

STATE S_1

read	write	move	go to	comment
#	#	right	S_2	input: ##... go to S_2 to write output
0	1	right	Halt	input: $\#d_1\ldots d_n 0\#\ldots$ output: $\#d_1\ldots d_n 1\#\ldots$
1	1	left	S_3	input: $\#d_1\ldots d_n 1\#\ldots$ go to S_3 to look to for a 0

STATE S_2

read	write	move	go to	comment
#	1	right	Halt	input: ##... output: #1#...

10.5. Examples of Turing Machines

STATE S_3

read	write	move	go to	comment
#	#	right	S_4	no 0 found: go to S_4 input: $\#1\ldots 1\#\ldots$
0	1	right	S_6	found a 0: go to S_6 input: $\#d_1\ldots d_n 011\ldots 1\#\ldots$ output: $\#d_1\ldots d_n 100\ldots 0\#\ldots$
1	1	left	S_3	continue looking for a 0

STATE S_4

read	write	move	go to	comment
1	1	right	S_5	keep the leading 1 on the left

STATE S_5

read	write	move	go to	comment
#	0	right	Halt	make the output one bit longer output: $\#100\ldots 0\#\ldots$
1	0	right	S_5	change all but leading 1 to 0's

STATE S_6

read	write	move	go to	comment
#	#	none	Halt	head has reached final position
1	0	right	S_6	moving head to its final position

If the input tape is properly formatted, only the blank (#) can be read in the initial state. If either of the other symbols is read while the machine is in the initial state, then the machine hangs.

Below we show the incrementor in action incrementing 7 to 8. Horizontal braces above and below a symbol indicate the symbol being scanned by the read/write head.

State	Tape contents
Initial		#	1	1	1	#	#	#	#	$\cdots$

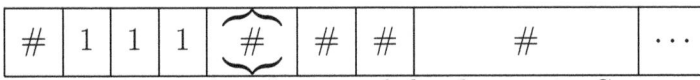
Instructions: write #, move left, change to S_1

Step 1

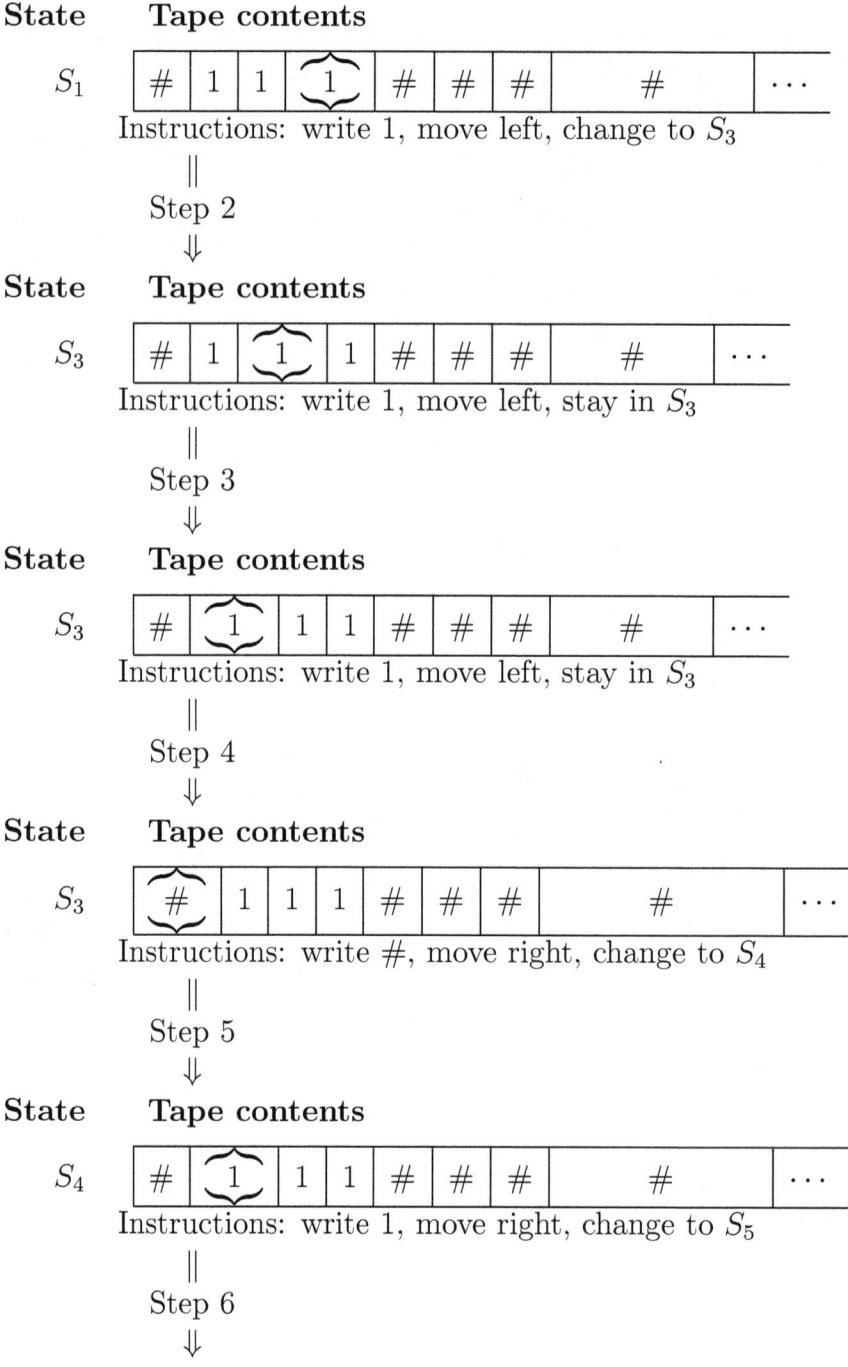

10.5. Examples of Turing Machines

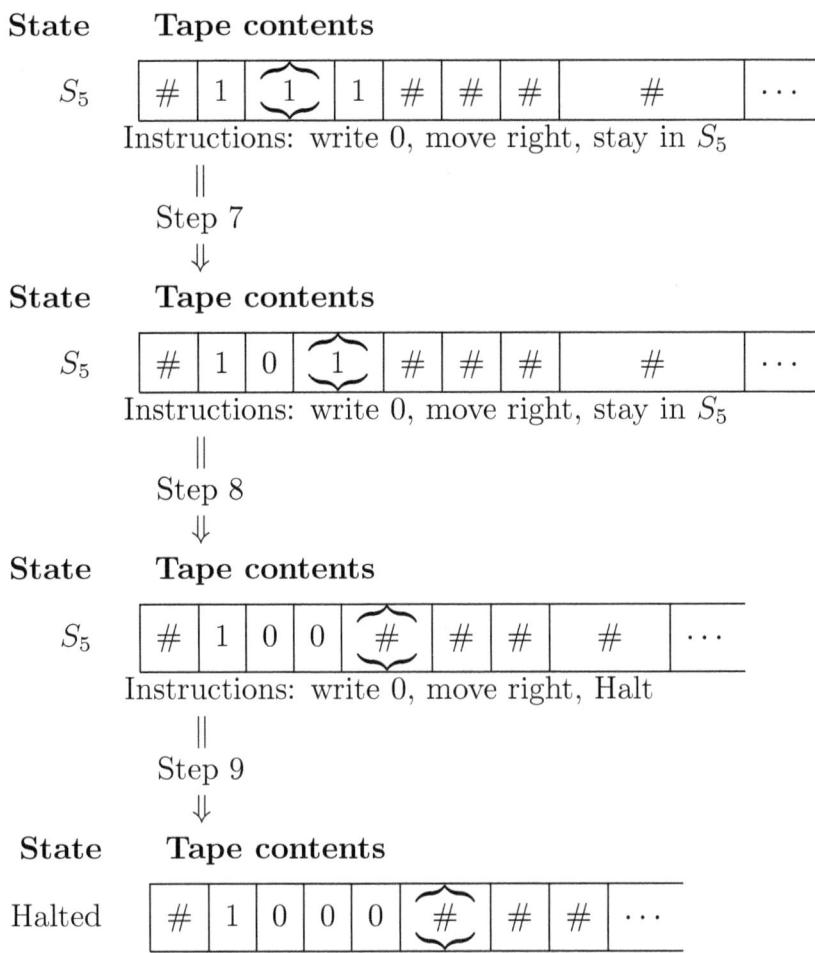

Notice that the incrementor takes nine steps to process the 3 bits of input. That those 3 bits happen to represent the number 7 does not change the fact that the input size is 3 symbols. Rating the size of the problem according to the number of characters required to provide the input is simply one of the rules of the game.

Because of the efficiency of binary notation, an input string of size n can represent a number of size $2^n - 1$. As a consequence, some computational tasks are more difficult than you might have thought. For instance, if the task is listing all nonnegative numbers less than a given number, you might think the time required is approximately the same magnitude as the input, but in fact it is exponential in the size of the input. Listing all nonnegative numbers less than the binary number 1,000,000,000 requires making a list with 512 entries, while the input required only 10 bits (the commas are for ease of reading, they do not count as input data). If we merely double the number of bits in the input to 20, then we face the task of listing all nonnegative numbers less than the binary number 10,000,000,000,000,000,000 and that requires making a list with with 524,288 entries.

Notice that, for an input string of n bits, it is the string in (10.4) that will require the most processing steps, namely, $3 + 2n$. When considering how many steps a Turing machine takes to process its input, the interesting question is, "How rapidly does the calculation grow as the length of the input string gets large?" The fact that it takes the incrementor three steps to process the empty string becomes unimportant when looking at large strings of bits.

The "big oh" notation of Landau provides a convenient way of expressing how rapidly the incrementor—or any Turing machine—processes long strings. The definition relevant to the behavior of functions when n approaches infinity is the following:

Definition Suppose that g is a real-valued function that is positive for large n. Then the real-valued function f is *big "O" of g* and we write

$$f(n) = \mathrm{O}(g(n))$$

in case there is a constant $0 < C < \infty$ such that

$$|f(n)| \le C\, g(n)$$

holds for all large enough n.

If the function g is given by a formula, for example, $g(n) = e^n$, then it is common practice to use the name of the function, g, interchangeably with the formula that defines it, e^n in this example, and to simply write $f(n) = \mathrm{O}(e^n)$.

Now let us consider the incrementor and the function $g(n) = n$. If we let $f(n)$ equal the maximum number of steps required by the incrementor to process an input string of n bits, then we know that $f(n) = 3 + 2n$. We see that $|f(n)| = 3 + 2n \le 3n = 3\,g(n)$ holds, whenever $n \ge 1$. Thus, we have $f(n) = \mathrm{O}(n)$. We conclude that the incrementor takes $\mathrm{O}(n)$ steps to process an input string with n bits.

The Decrementor. In this example, we design a Turing machine that takes as its input a binary number, i.e., a nonempty string of 0's and 1's,

10.5. Examples of Turing Machines

and returns as its output that number minus 1, unless the result would be negative, in which case it returns 0. Again we allow extra 0's on the left. We call this Turing machine the *decrementor*.

What the decrementor does is search the input string from right to left until a 1 is found. Once a 1 is found, it is changed to a 0, and then the machine moves back from left to right changing each 0 to a 1. If no 1 is found, then the input string represents zero and the tape is not to be changed.

To carry out the process described above, the decrementor has the following six states:

$$\text{Initial, } S_1, S_2, S_3, S_4, \text{ Halted.}$$

The halted state requires no instructions. For the other states of the decrementor the instructions are as follows:

INITIAL STATE

read	write	move	go to
#	#	left	S_1

STATE S_1

read	write	move	go to	comment
#	#	right	Halt	input: ##...
0	0	left	S_2	input: $\#d_1\ldots d_n0\#\ldots$ go to S_2 to look for a 1
1	0	right	Halt	input: $\#d_1\ldots d_n1\#\ldots$ output: $\#d_1\ldots d_n0\#\ldots$

STATE S_2

read	write	move	go to	comment
#	#	right	S_3	no 1 found: go to S_3 input: $\#0\ldots0\#\ldots$
0	0	left	S_2	continue looking for a 0
1	0	right	S_4	found a 1: go to S_4 input: $\#d_1\ldots d_n100\ldots0\#\ldots$ output: $\#d_1\ldots d_n011\ldots1\#\ldots$

STATE S_3

read	write	move	go to	comment
#	#	none	Halt	head has reached final position
0	0	right	S_3	moving head to its final position

STATE S_4

read	write	move	go to	comment
#	#	none	Halt	head has reached final position
0	1	right	S_4	moving head to its final position

As was true for the incrementor, if the input tape is properly formatted, only the blank (#) can be read in the initial state. If either of the other symbols is read while the machine is in the initial state, then the machine hangs.

Below we show the decrementor in action decrementing 4 to 3. The decrementor takes $O(n)$ steps to process an input string with n bits.

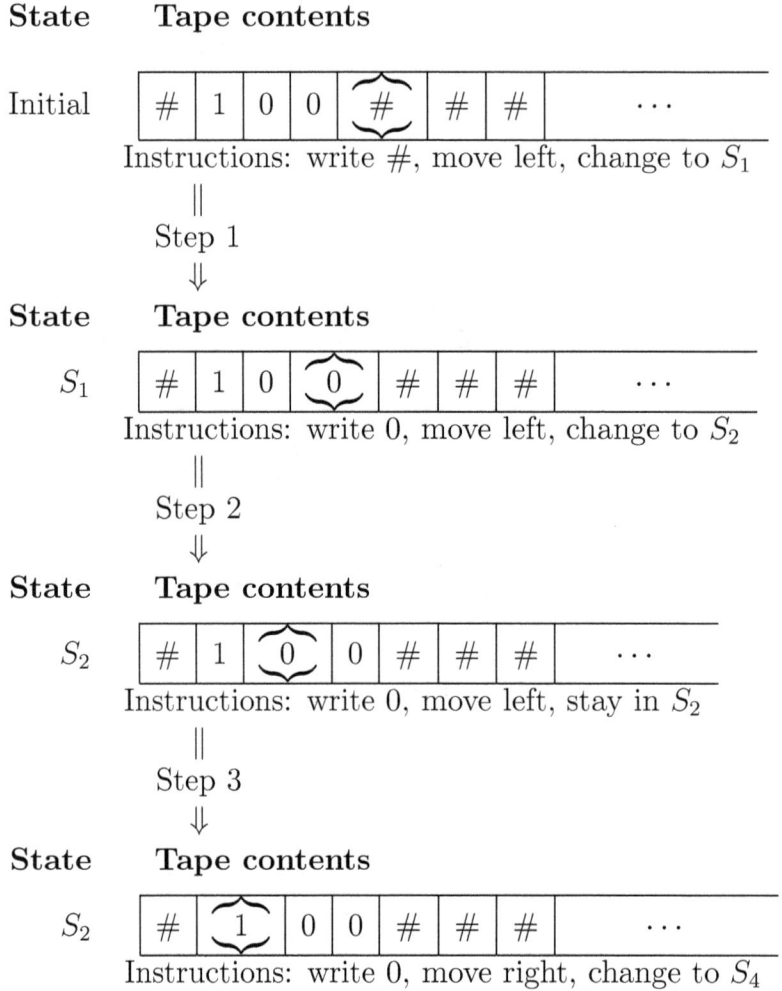

10.6. Bigger Calculations

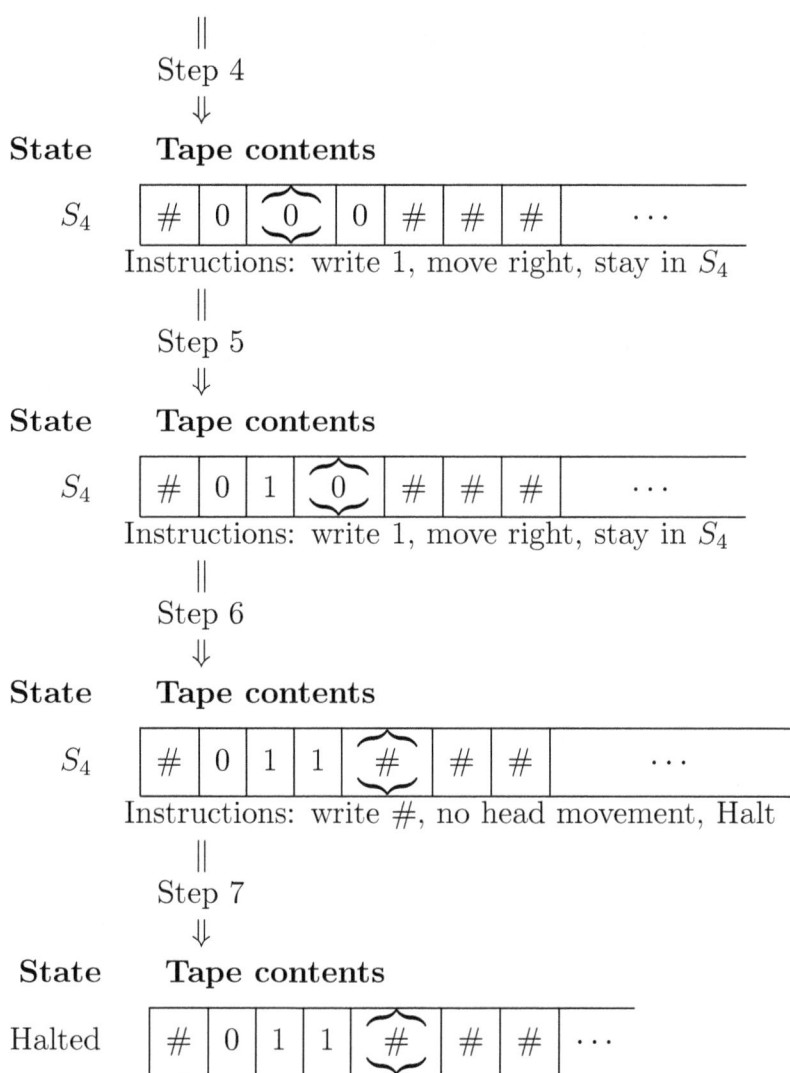

10.6 Bigger Calculations

The preceding section gave examples showing in complete detail the designs of the Turing machines we called the incrementor and the decrementor. The operations these machines perform are simple, but their designs are complicated. The prospect of designing a Turing machine to do a complicated computation is daunting.

Fortunately, we can chain together previously constructed Turing machines to make a bigger and better Turing machine. For example, if we want to add two binary numbers, we can put the two numbers on a tape and alternate decrementing and incrementing. When one number has been decremented to zero, the other number will have been incremented up to the sum. Some housecleaning will be needed to get the tape into proper form to be an output, but that is not conceptually difficult.

We see that we need to save all our Turing machine designs so we can combine them in the way we combine subroutines in computer programming. Also note that when a machine functioning as a subroutine is ready to pass control to another machine, a scanned symbol can be used to determine a branch in the process; thus different intermediate outputs can lead to different computations. The upshot is that any computation that can be programmed to run on an electronic digital computer can, in principle, be performed by an appropriate Turing machine. The Turing machine may be slow, but it will get the job done.

Computing the sum of two numbers by chaining together the incrementor and the decrementor will, as we said, get the job done, but it is a slow way to do addition. If both of the numbers being added are represented by n bits then, because a number represented in binary notation by n bits can be as large as $2^n - 1$, to do the addition, the incrementor and decrementor will each need to perform their operations as many as $2^n - 1$ times. Thus we might be led to think that the operation of addition of natural numbers requires exponential time.

In fact, addition is $O(n)$. One faster way to add two numbers is to use the binary version of the algorithm for addition (of decimal numbers) that we all learned in grade school. To obtain that efficiency, we would be need to design a new Turing machine instead of combining two that we already have. In general, two different Turing machines may perform the same calculation at vastly different speeds.

Finally, as an exercise to show definitively that Turing machines can do more than finite automata, you might try designing a Turing machine that works with the alphabet a, b, Y (representing "Yes"), and N (representing "No") and produces the output tape

$$\boxed{\#}\,\boxed{Y}\,\boxed{\#}\,\boxed{\#}\,\boxed{\#}\,\cdots$$

when the input is a string of the form

$$\underbrace{a\,a\cdots a\,a}_{n \text{ copies}}\underbrace{b\,b\cdots b\,b}_{n \text{ copies}} \tag{10.5}$$

10.7. Nondeterministic Turing Machines

and produces an output tape of the form

| # | N | # | # | # | ... |

when the input is a string of the form

$$\underbrace{a\,a\cdots a\,a}_{n \text{ copies}} \underbrace{b\,b\cdots b\,b}_{m\neq n \text{ copies}}$$

Recall that, earlier in this chapter when we discussed the limitations of finite automata, we showed that there is no finite automaton that accepts a string if and only if it is of the form (10.5).

10.7 Nondeterministic Turing Machines

The basic Turing machine can be modified in a number of ways. One simple change is to add a second tape. This seems like an obvious improvement, because we could provide input on one tape and read the output from the second tape. In fact, we might want to have additional tapes on which to write some of the intermediate steps in a calculation.

A Turing machine with additional tapes is better than a basic one-tape Turing machine, but the improvement is limited to speed. Any calculation that can be done with a multitape Turing machine can be done with a single tape Turing machine. While faster computation is always better, even the improvement in speed is not significant in the context of the P/NP problem, because any calculation that can be done in polynomial time on a multitape Turing machine can be done in polynomial time on a single tape Turing machine—though the degree of the polynomial may be larger, it is still a polynomial.

A modification in the design of a Turing machine that *might* make a fundamental improvement in the speed of calculation is to introduce nondeterminism. Indeed, the P/NP problem asks whether the introduction of nondeterminism into a Turing machine allows calculations to be done in polynomial time that otherwise could not.

Recall that a nondeterministic automaton is allowed to have several possible state changes that can be made when a particular input character is read. Similarly, a *nondeterministic Turing machine* is allowed to have more than one instruction that applies when a particular symbol is scanned by the read/write head while the machine is in a particular state. Nondeterminism in an actual computer would be bad; one especially unpleasant consequence of nondeterminism would be the possibility of different output from the same input. This consequence of nondeterminism is completely unacceptable in most contexts.

A nondeterministic Turing machine can be successfully used as a language acceptor. Recall that, with a language acceptor, you put an input string on the machine's tape, and if the machine halts, then the string is accepted in the language; but the specific contents of the tape when the machine halts are unimportant. The next example illustrates a nondeterministic Turing machine used as a language acceptor.

Example: A Nondeterministic Acceptor. Below we list the instructions for a nondeterministic Turing machine that halts when given an input tape containing a nonempty string that includes at least one a and at least one b. This nondeterministic Turing machine has the three states: initial, S_1, and halted. As always, the halted state requires no instructions. Notice that the state S_1 allows more than one transition as is indicated by the word "**OR**" in the instructions.

INITIAL STATE

read	write	move	new state
#	#	left	Initial
a	a	none	S_1
b	b	left	Initial

STATE S_1

read	write	move	new state
#	#	left **OR** right	S_1
a	a	left **OR** right	S_1
b	b	none	Halted

Described in words, the Turing machine finds an a if it can, and after an a has been found, it finds a b if it can. The only way to reach the halted state is to find an a and then find a b.

Next we show one sequence of states and tape configurations that demonstrate that this Turing machine would accept the string ab.

10.7. Nondeterministic Turing Machines

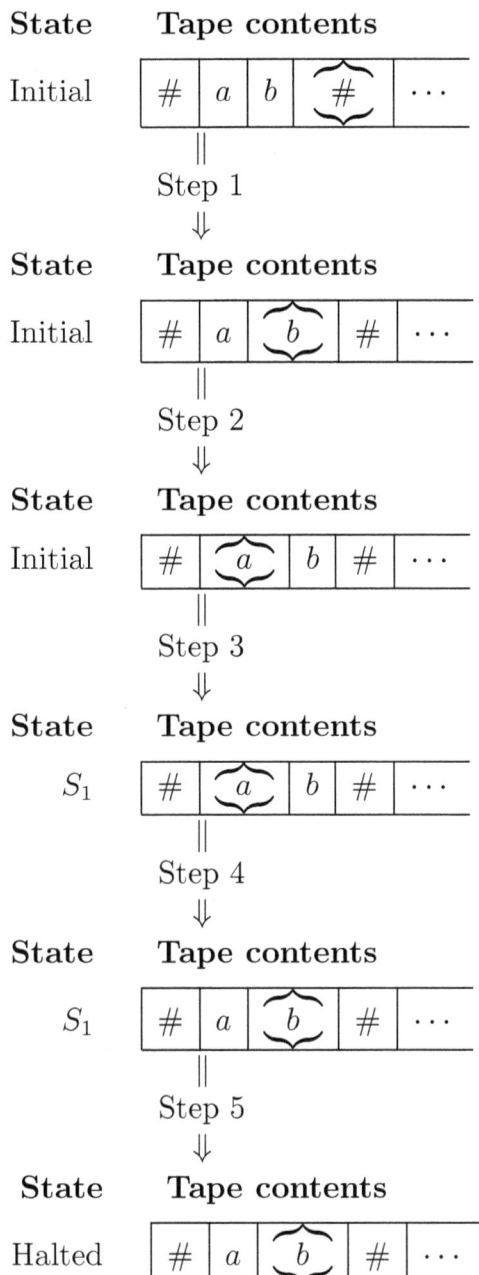

Notice that, when the Turing machine was in state S_1 and had read the symbol a, the read/write head moved to the right. Because the machine is nondeterministic, it could have "chosen" instead to move to the left to scan the leftmost blank square. So another way in which the Turing machine could accept the string *ab* is with the sequence of states and tape configurations shown below.

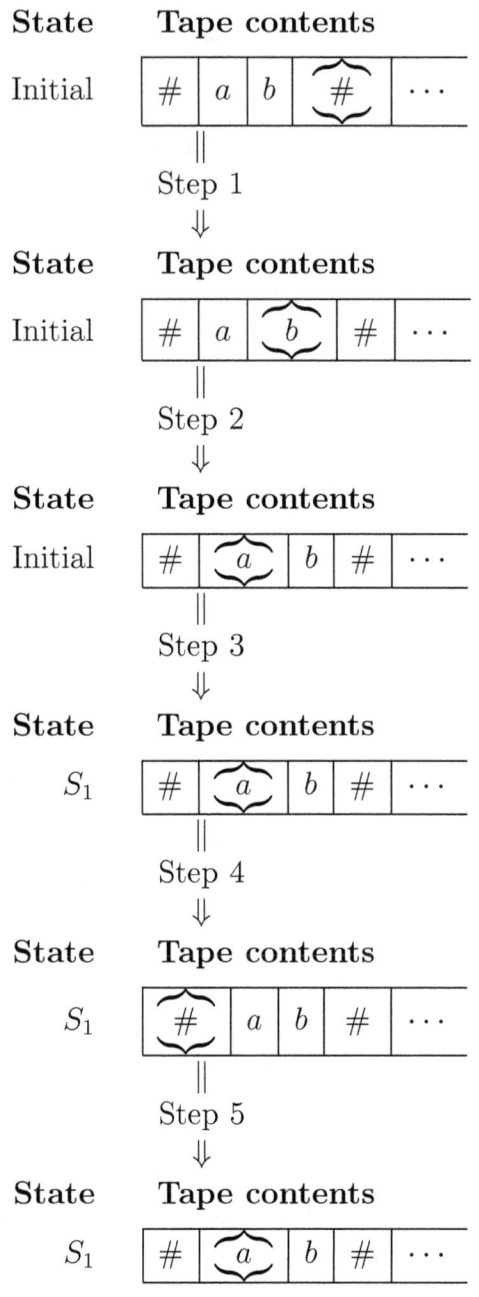

10.7. Nondeterministic Turing Machines

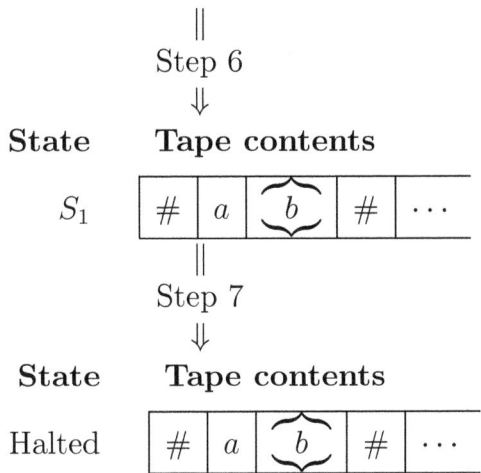

In fact, the machine could choose to have the read/write head bounce back and forth between reading the blank in the first square and the a in the second square. This bouncing back and forth could go on any number of times, so there are infinitely many ways that this nondeterministic Turing machine could deal with the string ab. Figure 10.10 shows more of the possibilities for the steps the Turing machine could take. An additional—foolish—possibility allowed by nondeterminism is for the machine to choose to run the read/write head past the left end of the tape and thus hang. As far as accepting the string, what matters is that there is a way for the machine to reach the halted state when given the input tape containing the string—the path to the halted state requiring the fewest steps is shown using thicker line segments.

The Value of Nondeterminism. While nondeterminism seems contrary to our notion of how computers work, nonetheless the nondeterministic language acceptor in the example does carry out an algorithmic process. One advantage of nondeterminism is that it can free us from describing some of the minute details about how the machine should behave.

It is also plausible that nondeterministic Turing machines have the potential to operate more quickly than do deterministic Turing machines. An analogy can be made with the problem of finding your misplaced reading glasses. One person searching each room of the house in turn until the glasses are found is the analog of the deterministic Turing machine. Several people searching different rooms for the glasses is the analog of the nondeterministic Turing machine. The several people correspond to the possible branches of the computation—like the branches in Fig. 10.10.

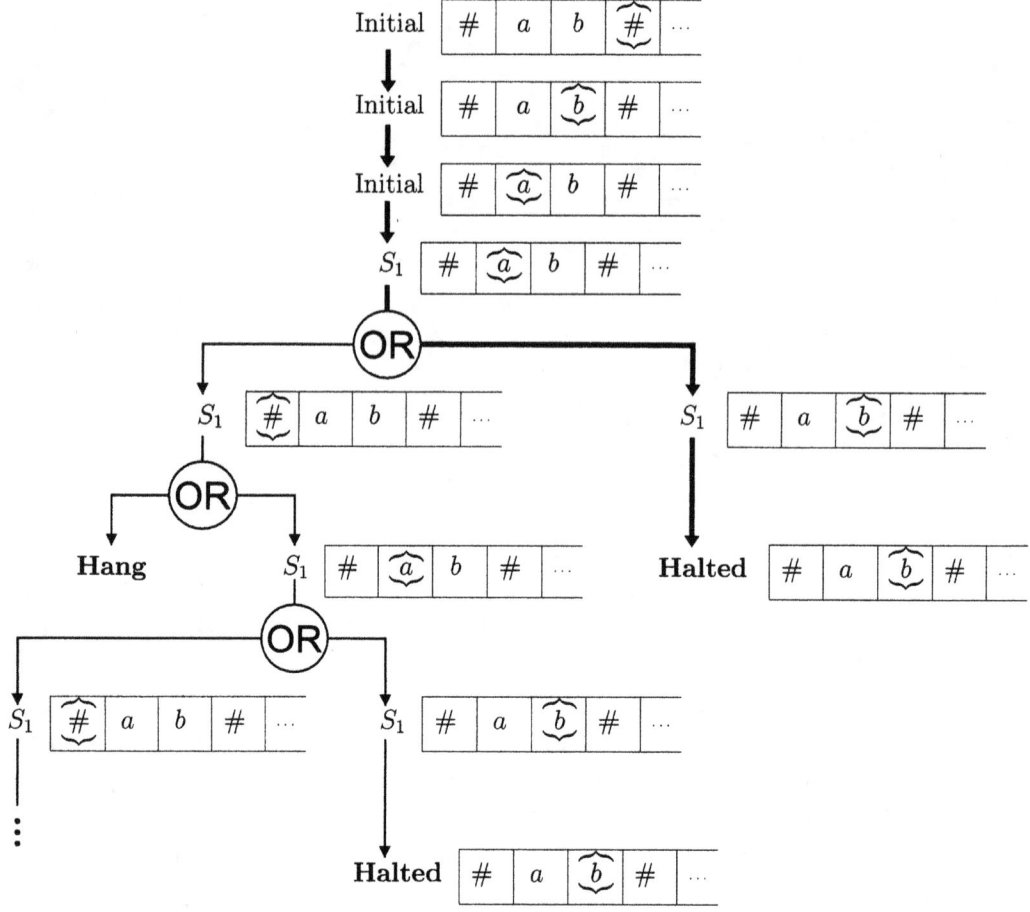

Figure 10.10 Nondeterministic processing of the string *ab*.

We might wonder whether nondeterminism allows any more powerful calculations than can be performed with a basic Turing machine serving as a language acceptor. In fact, we have the following theorem:

Theorem *If there is a nondeterministic Turing machine that accepts exactly the strings that are words in a language, then there exists a deterministic Turing machine that accepts exactly the words in the same language.*

The theorem tells us that, as was the case with finite automata, nondeterminism gives us convenience, but no additional power. The theorem does not tell us anything about speed.

The theorem is proved by a *dovetailing procedure*. In essence, dovetailing is what was done in constructing Fig. 10.10: each time there is a choice of actions, we construct a corresponding new computational path. More precisely, a deterministic Turing machine is designed to emulate the given nondeterministic machine. It is simplest to use a deterministic Turing machine with an unlimited number of tapes. Each time the nondeterministic Turing machine has a choice of instructions to follow, the deterministic Turing machine that emulates it opens a new computational path by making copies of the possible tape contents on currently blank tapes. Each step of the nondeterministic machine requires advancing the deterministic machine through one step of computation on every tape. If the nondeterministic machine halts after finitely many steps, then one of the deterministic machine's tapes will emulate that computation, and that will be the signal for the deterministic machine to halt. If the nondeterministic machine hangs, then the deterministic machine must emulate hanging by running a read/write head off the tape or by entering a state that allows no escape.

10.8 $\mathcal{P}$ and $\mathcal{NP}$

In this section we want to define and compare the notions of polynomial-time computation for deterministic and nondeterministic Turing machines. The notations $\mathcal{P}$ and $\mathcal{NP}$ indicate that we are concerned specifically with properties of formal languages. The appropriate definition for a deterministic Turing machine is as follows:

Definition A language in a given alphabet is said to be *polynomial-time decidable* or, more simply, in the class $\mathcal{P}$ if

• there is a Turing machine that can decide whether each string is in the language and

• there is a positive integer d and a constant $0 < C < \infty$ such that, for any string in the alphabet that is of length less than or equal to n, the Turing machine decides whether the string is in the language after no more than Cn^d steps, or in other words, for strings of length not exceeding n, the maximum of the number of steps required to process the string is $O(n^d)$.

Since a nondeterministic Turing machine is used as a language acceptor rather than as a decider, the definition of polynomial-time computation for a nondeterministic Turing machine is more subtle. We saw in Fig. 10.10 that there are many ways for our nondeterministic Turing machine to accept the string *ab*, but the fewest steps required is five, so we will say the machine accepts the string in five steps. The next definition generalizes that way of measuring the number of steps required to accept a string.

Definition If a language in a given alphabet is accepted by a nondeterministic Turing machine then, for each string in the language, the *number of steps* required for the nondeterministic Turing machine to accept the string is defined to be the number of steps in the shortest allowed computation that halts.

Using the preceding definition of the number of steps required to accept an individual string in the language, we can make the next definition.

Definition A language in a given alphabet is said to be *acceptable in nondeterministic polynomial-time* or, more simply, in the class $\mathcal{NP}$ if

• there is a nondeterministic Turing machine that accepts the strings in the language and

• there is a positive integer d and a constant $0 < C < \infty$ such that any string in the language that is of length less than or equal to n is accepted in no more than $C n^d$ steps; or in other words, for strings of length not exceeding n, the maximum of the minimum number of allowed steps required to accept the string is $\mathrm{O}(n^d)$.

If you have in hand a deterministic Turing machine that is a decider for a language, then you can convert it to an acceptor by having it hang or go into an inescapable state instead of halting with the output "No." Thus it is immediate that $\mathcal{P} \subset \mathcal{NP}$ holds. The following conjecture is at the heart of the P/NP Problem:

P/NP Conjecture. The class of languages $\mathcal{P}$ is a proper subset of the class of languages $\mathcal{NP}$. That is, there is at least one language that is acceptable in nondeterministic polynomial-time, but is not polynomial-time decidable.

Plausible as the conjecture may be, no one has been able to prove it nor anything close to it.

10.9 NP-Completeness

The P/NP Conjecture remains an open problem. To show that $\mathcal{P}$ is a proper subset of the class of languages $\mathcal{NP}$, one would need to prove the existence of a language in the class $\mathcal{NP}$ that can also be shown to not be in the class $\mathcal{P}$. Showing that something—in this case polynomial-time decidability—is impossible is typically difficult, but known proof techniques, such as diagonal arguments, could be tried.

To go the other way and refute the conjecture, one would need to show that *every* language in the class $\mathcal{NP}$ is in the class $\mathcal{P}$. Again, this seems a daunting prospect, because every language might need to be considered individually. The remarkable fact is that there are specific problems that are known to have the property of NP-*completeness*[3]; this means that if one of these NP-complete problems is shown to be solvable in polynomial time, then all the NP problems are solvable in polynomial time. Thus, to settle the conjecture, it would suffice to show one NP-complete problem is solvable in polynomial time.

The first example of an NP-complete problem was discovered by Stephen A. Cook (b. 1939) in 1971 (see [Coo 71]). Since that time, many problems have been shown to be NP-complete. Over 300 were already collected in Michael Garey and David Johnson's 1979 book *Computers and Intractability*.

Stephen A. Cook

Jiří Janíček
Stephen Cook

One would think academic life would be smooth sailing for a person like Dr. Cook who made a seminal contribution as he did with NP-completeness. Such is not the case. After getting his Ph.D. at Harvard in 1966, Cook joined the mathematics department of the University of California at Berkeley as an assistant professor, but instead of being fêted, he was denied tenure. He moved on to the University of Toronto.

---

[3]Here we use roman letters "P" and "NP" instead of the caligraphic letters "$\mathcal{P}$" and "$\mathcal{NP}$," because we have used the caligraphic letters for the restricted problem of language decidability.

A Look Back

Many people consider the **P/NP** to be the most difficult and important problem in the mathematical sciences. Field's medalist Stephen Smale, in the year 2000, wrote an article setting forth the important problems for mathematicians to think about in the twenty-first century. **P/NP** was at the top of his list.

The Clay Mathematics Institute has a list of seven important problems, for each of which it offers a $1 million bounty for the solution. Only one of them has been solved, and that is Grisha Perelman's solution of the Poincaré conjecture. But one of the problems on their list of 7 is **P/NP**.

This problem, as we have described and discussed in the present chapter, concerns the computational complexity of questions. It obviously relates in a rather direct way to our consideration of RSA encryption in Chap. 9. Now consider the factorization of a 150-digit integer n. Suppose that God knows that $n = p \cdot q$, where p and q are prime. It is straightforward to multiply p times q and verify that these are indeed the prime factors of n. So this problem is definitely **NP**. But, given existing techniques, it would take a great deal of computer time to begin with n and then to find p and q. So, at least as far as we can tell (but we are not sure), this problem is *not* **P**.

In a year 2002 poll of 100 top computer-science researchers, 61 believed that $\mathbf{P} \neq \mathbf{NP}$, 9 believed that $\mathbf{P} = \mathbf{NP}$, and 22 were undecided. Of these 22, 8 believed that the result may be independent of the currently accepted axioms in the subject. In the technical language of logic, this means that the question is undecidable: it can neither be proved nor disproved.

Cobham's thesis is that problems of type **P** are easy while problems not of type **P** are hard. While intuitively appealing, this view of the world is perhaps a bit naive. For a problem that can (at least in theory) be solved in polynomial time could in fact be governed by a polynomial that has huge coefficients, or that is of enormous degree. On the flip side, there can be problems or algorithms that are *not* of polynomial complexity from a theoretical point of view, but can still be tackled effectively and in good time in practice. An example of the latter is solving linear programming problems using the simplex method[4] (such problems are used to schedule airlines and to solve other common optimization problems). Theoretically, the simplex

---

[4] The linear programming problem is not solvable in polynomial time using the simplex method, but it is solvable in polynomial time by other methods, as was first shown by Leonid Khachiyan in 1979.

method is not polynomial time, but the specific problems that require more than polynomial time must be carefully contrived—it seems that all real world problems can be solved rapidly.

Another example of the latter phenomenon is the famous traveling salesman problem. Here a salesman is required to visit N different cities. Each road has a certain length and a certain cost associated to it. The problem is to find the most efficient route for the salesman to travel. This problem is known to be **NP**-complete. But there are many tricks (such as the well-known greedy algorithm) for obtaining "solutions" that are good enough for practical purposes.

A number of researchers have actually claimed solutions to the **P/NP** problem. Gerhard J. Woeginger has compiled a list of these. Among the noted attempts are that of Vinay Doelalikar. Although flaws have been found in his argument, the validity of his approach is still being debated. And he is endeavoring to formulate a more detailed and verifiable version of his proof.

The film *Travelling Salesman*, directed by Timothy Lanzone, is the story of four mathematicians hired by the U.S. government to solve the **P/NP** problem.

REFERENCES AND FURTHER READING

[**Coo 71**] Cook, S.A.: The complexity of theorem-proving procedures. Proceedings of the 3rd Annual ACM Symposium on Theory of Computing, pp. 151–158. Association for Computing Machinery, New York (1971)

[**Dan 57**] Dantzig, G.B.: On the significance of solving linear programming problems with some integer variables. Econometrica **28**, 30–44 (1957)

[**GJ 91**] Garey, M.R., Johnson, D.S.: Computers and Intractability: A Guide to the Theory of NP-Completeness. W. H. Freeman, San Francisco (1991)

[**Kar 72**] Karp, R.M.: Reducibility among combinatorial problems. In: Miller, R.E., Thatcher, J.W. (eds.) Complexity of Computer Computations, pp. 85–103. Plenum Press, New York (1972)

[**Kra 11**] Krantz, S.G.: The Proof Is in the Pudding. Birkhäuser, Boston (2011)

[**Law 76**] Lawler, E.L.: Combinatorial Optimization: Networks and Matroids. Holt, Rinehart, and Winston, New York (1976)

[**LP 98**] Lewis, H.E., Papadimitriou, C.H.: Elements of the Theory of Computation. Prentice-Hall, Upper Saddle River (1998)

[**NW 75**] Nijenhuis, A., Wilf, H.S.: Combinatorial Algorithms. Academic, New York (1975)

[**Tur 37**] Turing, A.M.: On computable numbers, with an application to the Entscheidungsproblem. Proceedings of the London Mathematical Society **42**, 230–265 (1936–37)

[**Wol 05**] Wolf, R.S.: A Tour Through Mathematical Logic. A Carus Monograph of the Mathematical Association of America. The Mathematical Association of America, Washington (2005)

Chapter 11
Primality Testing

11.1 Preliminary Concepts

A *prime number* is a whole number (an integer) with the property that its only divisors are 1 and itself. By custom we do not consider 1 to be a prime. Therefore the first several prime numbers are

$$2,\ 3,\ 5,\ 7,\ 11,\ 13,\ 17,\ 19,\ 23,\ 29,\ 31,\ 37,\ 41,\ 43,\ldots.$$

Prime numbers are of great interest because they are the building blocks of number theory. The *Fundamental Theorem of Arithmetic* tells us that every positive integer can be written in one and only one way as the product of primes. For example,

$$180 = 2^2 \cdot 3^2 \cdot 5$$

and

$$998{,}250 = 2 \cdot 3 \cdot 5^3 \cdot 11^3\,.$$

A number that is not prime is called *composite*. We have just seen explicitly that 180 and 998,250 are composite. A composite number is one that has nontrivial prime factors.

About 2300 years ago, Euclid proved the following result:

Theorem *There are infinitely many prime numbers.*

This theorem is remarkable for several reasons. One, the argument that he presented was one of the first "proofs by contradiction." Second, it was just about the first time that a result had been proved about the infinitude of a set. Third, the proof is simple and elegant and we can describe it in words in just a moment.

11.2 Euclid's Theorem

First, who was Euclid? Certainly one of the towering figures in the mathematics of the ancient world was Euclid of Alexandria (325 BCE–265 BCE). Although Euclid is not known as much (as were Archimedes and Pythagoras) for his original and profound insights, and although there are not many theorems named after Euclid, he has had an incisive effect on human thought. After all, Euclid wrote a treatise (consisting of thirteen Books)—now known as Euclid's *Elements*—which has been reproduced continuously for over 2000 years and has been through numerous editions. It is still studied in detail today, and continues to have a substantial influence over the way that we think about mathematics.

Not a great deal is known about Euclid's life, although it is fairly certain that he had a school in Alexandria. In fact "Euclid" was quite a common name in his day, and various accounts of Euclid, the mathematician's life, confuse him with other Euclids (one a prominent philosopher). One appreciation of Euclid comes from Proclus, one of the last of the ancient Greek philosophers:

> Not much younger than these [pupils of Plato] is Euclid, who put together the *Elements*, arranging in order many of Eudoxus's theorems, perfecting many of Theaetetus's, and also bringing to irrefutable demonstration the things which had been only loosely proved by his predecessors. This man lived in the time of the first Ptolemy; for Archimedes, who followed closely upon the first Ptolemy makes mention of Euclid, and further they say that Ptolemy once asked him if there were a shortened way to study geometry than the *Elements*, to which he replied that "there is no royal road to geometry." He is therefore younger than Plato's circle, but older than Eratosthenes and Archimedes; for these were contemporaries, as Eratosthenes somewhere says. In his aim he was a Platonist, being in sympathy with this philosophy, whence he made the end of the whole *Elements* the construction of the so-called Platonic figures.

As often happens with scientists and artists and scholars of immense accomplishment, there is disagreement, and some debate, over exactly who or what Euclid actually was. The three schools of thought are these:

11.2. Euclid's Theorem

- Euclid was an historical character—a single individual—who in fact wrote the *Elements* and the other scholarly works that are commonly attributed to him.

- Euclid was the leader of a team of mathematicians working in Alexandria. They all contributed to the creation of the complete works that we now attribute to Euclid. They even continued to write and disseminate books under Euclid's name after his death.

- Euclid was not an historical character at all. In fact "Euclid" was a *nom de plume*—an allonym if you will—adopted by a group of mathematicians working in Alexandria. They took their inspiration from Euclid of Megara (who *was* in fact an historical figure), a prominent philosopher who lived about 100 years before Euclid the mathematician is thought to have lived.

Most scholars today subscribe to the first theory—that Euclid was certainly a unique person who created the *Elements*. But we acknowledge that there is evidence for the other two scenarios. Certainly Euclid had a vigorous school of mathematics in Alexandria, and there is little doubt that his students participated in his projects.

It is thought that Euclid must have studied in Plato's Academy in Athens, for it is unlikely that there would have been another place where he could have learned the geometry of Eudoxus and Theaetetus on which the *Elements* are based.

Another famous story and quotation about Euclid is this. A certain pupil of Euclid, at his school in Alexandria, came to Euclid after learning just the first proposition in the geometry of the *Elements*. He wanted to know what he would gain by putting in all this study, doing all the necessary work, and learning the theorems of geometry. At this, Euclid called over his slave and said, "Give him threepence since he must needs make gain by what he learns."

What is important about Euclid's *Elements* is the paradigm it provides for the way that mathematics should be studied and recorded. He begins with several definitions of terminology and ideas for geometry, and then he records five important postulates (or axioms) of geometry. From this preliminary material he then derived theorems. This is still the way that mathematics is practiced today—with definitions and axioms and theorems. It is a tried and true method that makes mathematics travel well, and guarantees that our methodology and the truth of our results will stand the test of time.

Now let us return to Euclid's theorem about the infinitude of primes. Suppose, to the contrary, that there are only finitely many primes. Call them $p_1, p_2, \ldots, p_k$. That is alleged to be an exhaustive list of all the prime numbers. *There are no others.* Now consider the number

$$\mathcal{P} = p_1 \cdot p_2 \cdot p_3 \cdot \cdots \cdot p_k + 1.$$

This new number $\mathcal{P}$ cannot be composite. For, if it were composite, then it would be divisible by some prime. But when we divide $\mathcal{P}$ by p_1 we get the remainder 1. When we divide $\mathcal{P}$ by p_2 we get the remainder 1. And so forth up through p_k. In fact $\mathcal{P}$ is not divisible by any prime. And we have tried all the primes—there is just a finite list. It follows that $\mathcal{P}$ itself must be a prime. But that shows that our original exhaustive list of primes was incomplete. This is a contradiction. There are infinitely many primes.

11.3 The Sieve of Eratosthenes

Prime numbers are the nuts and bolts of number theory. Number theory is arguably the pinnacle of abstract mathematical reasoning. Understanding the properties of the integers is a dazzling and intoxicating line of thought, and one that has attracted the greatest minds in mathematics—Carl Friedrich Gauss, Niels Henrik Abel, Johann Peter Gustav Lejeune Dirichlet, Evariste Galois, David Hilbert, and many others.

The first thing you might want to have when embarking on a study of prime numbers is a good list of examples of prime numbers. For instance, the *CRC Standard Math Tables* (a reference widely used during the second half of the twentieth century) included a table listing all the primes in the range from 2 up to 100,000. The table occupied 8 pages. Such a table of primes can be made more easily and rapidly than you might imagine.

The *sieve of Eratosthenes* is an algorithm that rapidly finds all the prime numbers in the range from 2 up to a given number. To illustrate, suppose we want to find all the primes in the range from 2 up 105. First, we write down all the numbers in that range in some convenient fashion, such as in Table 11.1.

The idea behind the sieve is that we will eliminate all composite numbers from the table. To make the process systematic, we eliminate composite numbers in order according to each composite number's smallest prime factor. Among composite numbers with the same smallest prime factor, we will

11.3. The Sieve of Eratosthenes

	2	3	4	5	6	7	8	9	10	11	12	13	14	15
16	17	18	19	20	21	22	23	24	25	26	27	28	29	30
31	32	33	34	35	36	37	38	39	40	41	42	43	44	45
46	47	48	49	50	51	52	53	54	55	56	57	58	59	60
61	62	63	64	65	66	67	68	69	70	71	72	73	74	75
76	77	78	79	80	81	82	83	84	85	86	87	88	89	90
91	92	93	94	95	96	97	98	99	100	101	102	103	104	105

Table 11.1 The numbers from 2 up to 105.

	2	3	~~4~~	5	~~6~~	7	~~8~~	9	~~10~~	11	~~12~~	13	~~14~~	15
~~16~~	17	~~18~~	19	~~20~~	21	~~22~~	23	~~24~~	25	~~26~~	27	~~28~~	29	~~30~~
31	~~32~~	33	~~34~~	35	~~36~~	37	~~38~~	39	~~40~~	41	~~42~~	43	~~44~~	45
~~46~~	47	~~48~~	49	~~50~~	51	~~52~~	53	~~54~~	55	~~56~~	57	~~58~~	59	~~60~~
61	~~62~~	63	~~64~~	65	~~66~~	67	~~68~~	69	~~70~~	71	~~72~~	73	~~74~~	75
~~76~~	77	~~78~~	79	~~80~~	81	~~82~~	83	~~84~~	85	~~86~~	87	~~88~~	89	~~90~~
91	~~92~~	93	~~94~~	95	~~96~~	97	~~98~~	99	~~100~~	101	~~102~~	103	~~104~~	105

Table 11.2 Multiples of 2, but not 2 itself, have been crossed out.

eliminate them in the usual order from smallest to largest. Conveniently, it will happen that as we proceed the primes will be revealed in increasing order.

We now apply this procedure to the numbers in Table 11.1. The number 2 is the smallest prime number. So we eliminate all the composite numbers that have 2 as a prime factor. This is easily done by crossing out all the multiples of 2, but not 2 itself, that appear in Table 11.1. The result is shown in Table 11.2. In Table 11.2, we have also emphasized "2" to help keep track of the fact that we have eliminated any composite number in the table that has 2 as its smallest prime factor.

At each stage, the next prime will be the smallest number that (a) has not been crossed out and (b) is also larger than all the primes located so far. In Table 11.2, we thus identify 3 as the next prime after 2. (Of course we knew all along that 3 is prime, but we are establishing the pattern to be followed in this process.) We then eliminate all the composite numbers that have 3 as a prime factor. This is easily done by crossing out all the multiples of 3, but not 3 itself, that appear in Table 11.2. One can do this by going through the table in groups of 3 numbers—crossed out or not. That's why we cross out numbers rather than erase them. The result is shown in Table 11.3.

2	3	~~4~~	5	~~6~~	7	~~8~~	~~9~~	~~10~~	11	~~12~~	13	~~14~~	~~15~~
~~16~~	17	~~18~~	19	~~20~~	~~21~~	~~22~~	23	~~24~~	25	~~26~~	~~27~~	~~28~~	29	~~30~~
31	~~32~~	~~33~~	~~34~~	35	~~36~~	37	~~38~~	~~39~~	~~40~~	41	~~42~~	43	~~44~~	~~45~~
~~46~~	47	~~48~~	49	~~50~~	~~51~~	~~52~~	53	~~54~~	55	~~56~~	~~57~~	~~58~~	59	~~60~~
61	~~62~~	~~63~~	~~64~~	65	~~66~~	67	~~68~~	~~69~~	~~70~~	71	~~72~~	73	~~74~~	~~75~~
~~76~~	77	~~78~~	79	~~80~~	~~81~~	~~82~~	83	~~84~~	85	~~86~~	~~87~~	~~88~~	89	~~90~~
91	~~92~~	~~93~~	~~94~~	95	~~96~~	97	~~98~~	~~99~~	~~100~~	101	~~102~~	103	~~104~~	~~105~~

Table 11.3 Multiples of 2 and multiples of 3 have been crossed out.

The smallest number, after 3, that has not been crossed out in Table 11.3 is 5. Thus 5 is the next prime after 3. We then cross out all the multiples of 5, but not 5 itself, that appear in Table 11.3. The result is shown in Table 11.4.

2	3	~~4~~	5	~~6~~	7	~~8~~	~~9~~	~~10~~	11	~~12~~	13	~~14~~	~~15~~
~~16~~	17	~~18~~	19	~~20~~	~~21~~	~~22~~	23	~~24~~	~~25~~	~~26~~	~~27~~	~~28~~	29	~~30~~
31	~~32~~	~~33~~	~~34~~	~~35~~	~~36~~	37	~~38~~	~~39~~	~~40~~	41	~~42~~	43	~~44~~	~~45~~
~~46~~	47	~~48~~	49	~~50~~	~~51~~	~~52~~	53	~~54~~	~~55~~	~~56~~	~~57~~	~~58~~	59	~~60~~
61	~~62~~	~~63~~	~~64~~	~~65~~	~~66~~	67	~~68~~	~~69~~	~~70~~	71	~~72~~	73	~~74~~	~~75~~
~~76~~	77	~~78~~	79	~~80~~	~~81~~	~~82~~	83	~~84~~	~~85~~	~~86~~	~~87~~	~~88~~	89	~~90~~
91	~~92~~	~~93~~	~~94~~	~~95~~	~~96~~	97	~~98~~	~~99~~	~~100~~	101	~~102~~	103	~~104~~	~~105~~

Table 11.4 Multiples of 2, 3, or 5 have been crossed out.

The smallest number, after 5, that has not been crossed out in Table 11.4 is 7. Thus 7 is the next prime after 5. We then cross out all the multiples of 7, but not 7 itself, that appear in Table 11.4. The result is shown in Table 11.5.

The smallest number, after 7, that has not been crossed out in Table 11.5 is 11. Thus 11 is the next prime after 7. We would cross out all the multiples of 11, but not 11 itself, that appear in Table 11.5, but there are none. That is because any composite number that has 11 as its smallest prime factor must be at least $11 \cdot 11 = 121$, and thus is out of our range of numbers. In fact, no further crossing out of numbers can happen in our table, and all the

	2	3	4̶	5	6̶	7	8̶	9̶	1̶0̶	11	1̶2̶	13	1̶4̶	1̶5̶
1̶6̶	17	1̶8̶	19	2̶0̶	2̶1̶	2̶2̶	23	2̶4̶	2̶5̶	2̶6̶	2̶7̶	2̶8̶	29	3̶0̶
31	3̶2̶	3̶3̶	3̶4̶	3̶5̶	3̶6̶	37	3̶8̶	3̶9̶	4̶0̶	41	4̶2̶	43	4̶4̶	4̶5̶
4̶6̶	47	4̶8̶	4̶9̶	5̶0̶	5̶1̶	5̶2̶	53	5̶4̶	5̶5̶	5̶6̶	5̶7̶	5̶8̶	59	6̶0̶
61	6̶2̶	6̶3̶	6̶4̶	6̶5̶	6̶6̶	67	6̶8̶	6̶9̶	7̶0̶	71	7̶2̶	73	7̶4̶	7̶5̶
7̶6̶	7̶7̶	7̶8̶	79	8̶0̶	8̶1̶	8̶2̶	83	8̶4̶	8̶5̶	8̶6̶	8̶7̶	8̶8̶	89	9̶0̶
9̶1̶	9̶2̶	9̶3̶	9̶4̶	9̶5̶	9̶6̶	97	9̶8̶	9̶9̶	1̶0̶0̶	101	1̶0̶2̶	103	1̶0̶4̶	1̶0̶5̶

Table 11.5 Multiples of 2, 3, 5, or 7 have been crossed out.

numbers in Table 11.5 that have not been crossed out are prime. Thus the process has ended with the prime numbers

$$2 \quad 3 \quad 5 \quad 7 \quad 11 \quad 13 \quad 17 \quad 19 \quad 23 \quad 29 \quad 31 \quad 37 \quad 41 \quad 43$$
$$47 \quad 53 \quad 59 \quad 61 \quad 67 \quad 71 \quad 73 \quad 79 \quad 83 \quad 89 \quad 97 \quad 101 \quad 103$$

sifted out from among all the numbers from 2 up to 105.

11.4 Recognition of Composite Numbers

In today's world, an essential use of prime numbers is in cryptography (see our chapter on RSA encryption). The efficacy of some of the most important cryptographic techniques hinges on the fact that it is calculationally infeasible to factor very large integers into their prime components. To factor a 150-digit integer—even on a very fast digital computer—can take years. Thus it is a matter of great interest to be able to look at an integer and to have an efficient means of determining whether it is prime. If the integer is called n, then we want to be able to do so in a number of steps that is a relatively small number (as a function of n). [In practice, as we shall see below, we do not deal with n but rather with the number of digits in n—which is $\log_{10} n$. Thus the "size" of the number 10,000 is 5.]

This is a problem worth pondering. The most naive algorithm for determining whether a given n is prime is to divide n by $2, 3, 4, \ldots, n-1$. If n is divisible by any of these numbers, then it is composite; otherwise it is prime. This extremely naive algorithm takes $n - 2$ steps—each step requiring a long division.

But we can improve the algorithm almost instantly if we apply just a little analytical reasoning. Notice that $n = \sqrt{n} \cdot \sqrt{n}$. Of course $\sqrt{n}$ may not be an integer, but that observation is not salient to the point that we now wish to make. We see by this formula that n can be written as the product of two factors that are equally balanced: they are of the same size. It follows that, if we write $n = k \cdot \ell$ for *any* two factors k and ℓ, then one of these will be $\leq \sqrt{n}$ and the other will be $\geq \sqrt{n}$. As a result, a composite number n will always have one factor of size not exceeding $\sqrt{n}$. We conclude then that, in order to test n for compositeness, we need only divide by $2, 3, 4, \ldots, \sqrt{n}$.[1] Thus the problem can be analyzed in at most $\sqrt{n} - 1$ steps. This is an order-of-magnitude improvement over the estimate in the last paragraph.

But we can do even better. In fact, when we are dividing n by the numbers $2, 3, 4, \ldots, \sqrt{n}$, we can skip all the even numbers. Since if n is divisible by an even number then it is divisible by 2. So we can just test $2, 3, 5, 7, 9, \ldots, \sqrt{n}$. That already cuts the number of calculations down by about half.

Another useful observation is this. Any integer n can be written in the form $6k + j$ for some quotient k and some remainder $j = -1, 0, 1, 2, 3, 4$ (this is just the Euclidean algorithm: divide n by 6 and obtain the quotient k and remainder j). Note that $6k + 3$ and $6k$ are divisible by 3 while $6k + 2$ and $6k + 4$ are divisible by 2. Thus any prime must have the form $6k \pm 1$. Very interesting. So when we want to test the primality of n, we can first try dividing by 2 or 3, and then just test by dividing by the numbers $6k \pm 1$ which are less than n. This is three times as fast as the methods described above.

Using more sophisticated methods we can cut down significantly on the number of steps needed to test for primality. As noted, the only numbers between 2 and $\sqrt{n}$ that we need test as divisors for n are the prime numbers in the list. For n is composite if and only if it is divisible by a prime. So if we have a very large list of primes covering the range from 2 up to $\sqrt{n}$, then we can test whether n is prime. Let's see how well this will work when n has 150 digits.

The celebrated Prime Number Theorem of Hadamard (1865–1963) and de la Vallée Poussin (1866–1962) tells us that, if we let $\pi(x)$ denote the number of primes less than or equal to a number x, then

---

[1]More precisely, we need to divide by $2, 3, 4, \ldots, \lfloor \sqrt{n} \rfloor$, where $\lfloor x \rfloor$ denotes the *floor function* (also called the *greatest integer function*) equal to the largest integer less than or equal to x. We prefer to keep the notation cleaner by suppressing the floor function.

$$\pi(x) \sim \frac{x}{\ln x},$$

where $\ln x$ is the natural logarithm of x and the symbol $\sim$ means the relative error (i.e., the difference of the quantities divided by x) goes to 0 as x goes to ∞. For example, the prime number theorem gives us the estimate $434{,}294{,}482$ for $\pi(10^{10})$, whereas the exact value is $455{,}052{,}511$. For very rough calculations, it may be adequate to approximate $\pi(x)$ by x divided by the number of digits in x. This simpler estimate applied to $\pi(10^{10})$ yields 10^9.

Returning to primality testing for a number n with 150 digits, we note that $n \approx 10^{150}$. Consequently, $\sqrt{n} \approx 10^{75}$, and the prime number theorem tells us that there are about 5.8×10^{72} prime numbers in the range from 2 up to $\sqrt{n}$. That is a very large list indeed. In fact, there are more than 10^{15} entries on the list for every proton and neutron in the sun.[2]

The last discussion illustrates a very important point about methods of testing for primality. We want a method that is not only theoretically fast, but that can be programmed into a computer for quick execution. It is not clear that the last indicated method will be so efficacious, because how are we to generate and store the needed list of primes?

11.5 Speed of Algorithms

An algorithm for solving a particular type of problem is considered to be fast based on how many steps are required for larger and larger instances of the problem. An algorithm for testing whether n is prime is fast when it does well for large values of n. One might reasonably think that for this purpose the size of n *is* n. But that is *not* the way it is done. In studying the speed of algorithms for primality testing, the measure of the size of n is the *number of digits* in n. Of course that number of digits is $\log_{10} n$—if we use base 10 notation. So ideally we would like to have an algorithm that can say whether n is prime after a computation requiring a number of steps bounded by some fixed power of $\log_{10} n$. Of course most computers use base 2 notation, so we might want to replace $\log_{10} n$ by $\log_2 n$. In fact choosing a specific base for the logarithm is silly, because logarithms using two distinct bases differ merely by a numerical factor. Thus, for simplicity, we shall adopt the custom of just writing $\log n$—omitting the base of the logarithm function.

---

[2]The mass of the sun is about 1.99×10^{30} kilograms. The mass of a proton or neutron is about 1.67×10^{-27} kilograms. Thus there are about 1.2×10^{57} protons and neutrons in the sun.

Since the speed of an algorithm for testing the primality of n is determined by how the number of steps required grows as a function of $\log n$, we see that the methods of the last section based on *trial division* require exponential time. Only a decade ago it was believed by many experts that primality testing inherently required exponential time. Thus it came as quite a surprise when, in the year 2002, a trio of Indian computer scientists led by Manindra Agrawal found a way to determine whether a positive integer is prime or composite using a method that is computer-implementable and that takes a number of steps bounded by a fixed power of $\log n$. Moreover, the techniques that this team used to establish their result are quite elementary—in fact they date back 370 years to Pierre de Fermat. Let us say a few words about that esteemed figure in the history of number theory.

11.6 Pierre de Fermat

unknown artist

Pierre de Fermat

Pierre de Fermat(1601–1665) was one of the most remarkable mathematicians who ever lived. He spent his entire adult life as a magistrate or judge in the city of Toulouse, France. His career was marked by prudence, honesty, and scrupulous fairness. He led a quiet and productive life. His special passion was for mathematics. Fermat was perhaps the most talented amateur mathematician in history.

Fermat is remembered today by a large statue that is in the basement of the Hôtel de Ville (i.e., city hall) in Toulouse. The statue depicts Fermat, dressed in formal attire, and seated. There is a sign, etched in stone and part of the statue, that says, "Pierre de Fermat, the father of differential calculus." Seated in Fermat's lap is a scantily clad muse showing her ample appreciation for Fermat's powers.

Pierre Fermat had a brother and two sisters and was almost certainly brought up in the town of his birth (Beaumont-de-Lomagne). Although there is little evidence concerning his school education it must have been at the local Franciscan monastery.

He attended the University of Toulouse before moving to Bordeaux in the second half of the 1620s. In Bordeaux he began his first serious mathematical research and in 1629 he gave a copy of his restoration of Apollonius's *Plane*

11.6. Pierre de Fermat

loci to one of the mathematicians there. Certainly in Bordeaux he was in contact with Beaugrand and during this time he produced important work on maxima and minima which he gave to Étienne d'Espagnet who clearly shared mathematical interests with Fermat.

From Bordeaux, Fermat went to Orléans where he studied law at the University. He received a degree in civil law and he purchased the offices of councillor at the parliament in Toulouse. So by 1631 Fermat was a lawyer and government official in Toulouse and, because of the rank he then held, he became entitled to change his name from Pierre Fermat to Pierre de Fermat.

For the remainder of his life Fermat lived in Toulouse but, as well as working there, he also worked in his home town of Beaumont-de-Lomagne and a nearby town of Castres. The plague struck the region in the early 1650s, meaning that many of the older men died. Fermat himself was struck down by the plague and in 1653 his death was wrongly reported, then corrected:

> I informed you earlier of the death of Fermat. He is alive, and we no longer fear for his health, even though we had counted him among the dead a short time ago.

The period from 1643 to 1654 was one when Fermat was out of touch with his scientific colleagues in Paris. There are a number of reasons for this development. First, pressure of work kept him from devoting so much time to mathematics. Secondly, the Fronde, a civil war in France, took place and from 1648 Toulouse was greatly affected. Finally, there was the plague of 1651 which must have had great consequences both on life in Toulouse and of course its near fatal consequences for Fermat himself. However it was during this time that Fermat worked on the theory of numbers.

Fermat is best remembered for his work in number theory, in particular for Fermat's Last Theorem. This theorem states that the equation

$$x^n + y^n = z^n$$

has no nonzero integer solutions x, y and z when the integer exponent $n > 2$. Fermat wrote, in the margin of Bachet's translation of Diophantus's *Arithmetica*,

> I have discovered a truly remarkable proof which this margin is too small to contain.

These marginal notes only became known after Fermat's death, when his son Samuel published (in 1670) an edition of Bachet's translation of Diophantus's

Arithmetica that included his father's notes. See our chapter on Fermat's Last Theorem for more about this result which challenged mathematicians for 320 years.

11.7 Fermat's Little Theorem

We briefly recall some of our earlier discussion of modular arithmetic from the chapter on RSA encryption.

When we write $k \bmod n$ we mean simply the remainder when k is divided by n. When we write $k \equiv \ell \bmod n$, we mean $k \bmod n = \ell \bmod n$ or, what is the same, $k - \ell$ is evenly divisible by n.

It is an important, but easily checked, fact that modular arithmetic respects sums and products. That is, if

$$a \equiv c \bmod n \quad \text{and} \quad b \equiv d \bmod n,$$

then

$$a + b \equiv c + d \bmod n,$$

and likewise if

$$a \equiv c \bmod n \quad \text{and} \quad b \equiv d \bmod n,$$

then

$$a \cdot b \equiv c \cdot d \bmod n.$$

The most basic enunciation of Fermat's little theorem is below. The result is a special case of Euler's theorem which we discussed in the chapter on RSA encryption.

Theorem *If p is prime, then for any integer $a = 1, 2, 3 \ldots, p - 1$,*

$$a^{p-1} \equiv 1 \bmod p$$

or, what is the same, $a^{p-1} - 1$ is divisible by p.

11.7. Fermat's Little Theorem

Since $1^{p-1} = 1$ holds even when p is not prime, the nontrivial cases are when $a = 2, 3, \ldots, p - 1$. But now look at the following facts about various powers of 2:

$$\begin{aligned}
\mathbf{3:} & \quad \mathbf{2^2} = \mathbf{4} \equiv \mathbf{1 \bmod 3} \\
4: & \quad 2^3 = 8 \equiv 0 \bmod 4 \\
\mathbf{5:} & \quad \mathbf{2^4} = \mathbf{16} \equiv \mathbf{1 \bmod 5} \\
6: & \quad 2^5 = 32 \equiv 2 \bmod 6 \\
\mathbf{7:} & \quad \mathbf{2^6} = \mathbf{64} \equiv \mathbf{1 \bmod 7} \\
8: & \quad 2^7 = 128 \equiv 0 \bmod 8 \\
9: & \quad 2^8 = 256 \equiv 4 \bmod 9 \\
10: & \quad 2^9 = 512 \equiv 2 \bmod 10 \\
\mathbf{11:} & \quad \mathbf{2^{10}} = \mathbf{1024} \equiv \mathbf{1 \bmod 11} \\
12: & \quad 2^{11} = 2048 \equiv 8 \bmod 12 \\
\mathbf{13:} & \quad \mathbf{2^{12}} = \mathbf{4096} \equiv \mathbf{1 \bmod 13} \\
14: & \quad 2^{13} = 8192 \equiv 2 \bmod 14 \\
15: & \quad 2^{14} = 16{,}384 \equiv 4 \bmod 15 \\
16: & \quad 2^{15} = 32{,}768 \equiv 0 \bmod 16 \\
\mathbf{17:} & \quad \mathbf{2^{16}} = \mathbf{65{,}536} \equiv \mathbf{1 \bmod 17} \\
18: & \quad 2^{17} = 131{,}072 \equiv 14 \bmod 18 \\
\mathbf{19:} & \quad \mathbf{2^{18}} = \mathbf{262{,}144} \equiv \mathbf{1 \bmod 19} \\
20: & \quad 2^{19} = 524{,}288 \equiv 8 \bmod 20
\end{aligned}$$

We notice that, in this last list, $2^{n-1} \equiv 1 \bmod n$ holds precisely when n is prime. This should at least make us wonder if perhaps we could test for primality by merely checking whether $2^{n-1} \equiv 1 \bmod n$. Alas, things are not that simple. It turns out that $2^{340} \equiv 1 \bmod 341$, but $341 = 11 \times 31$ is composite. Since 341 acted like a prime number when $a = 2$, we say that 341 is a *pseudoprime to the base* 2.

Nonetheless Fermat's little theorem is a powerful tool. For a suspected prime n, we say n *passes the Fermat test* with base a if $a^{n-1} \equiv 1 \bmod n$; otherwise we say n *fails the Fermat test*. While the pseudoprime 341 passes the Fermat test with the base 2, it is quickly unmasked as a composite if we test it with the next base of 3. In fact, $3^{340} \equiv 56 \bmod 341$ holds.

Unfortunately, a number n can be a pseudoprime to every base $a = 2, 3, \ldots, n - 1$ and still be composite; there are infinitely many of these miscreants. Such a number is called a *Carmichael number*. An example of a Carmichael number is $561 = 3 \times 11 \times 17$. Because of the existence of Carmichael numbers, all we can say with certainty is that *if n fails the Fermat test, then n is* **not** *prime*.

11.8 The Strong Pseudoprimality Test and Probabilistic Primality Testing

The RSA method for encryption has been in wide use for at least two decades. The implementation of the RSA method *requires* the availability of large prime numbers, but we have also noted that it was long believed that primality testing is an insurmountably difficult problem. These two facts would appear to be in conflict. Fortunately there are probabilistic methods for primality testing—due to Miller and Rabin and to Solovay and Strassen. These probabilistic methods only tell us that the number n being tested is composite with a certain probability (3/4 for Miller–Rabin and 1/2 for Solovay–Strassen). Repetition of the technique increases the probability that the method is correct. Any degree of certainty (as close to 1 as we please) can be achieved with a sufficient number of repetitions. We will describe one of these methods next.

The *strong pseudoprimality test* is a procedure that can be applied to a number n that you suspect might be prime. It will have one of two outcomes:

Fail: n is definitely not prime.

Pass: The probability that n is prime is at least 3/4, but we still don't know for certain that n is prime.

The strong pseudoprimality test is based on Fermat's little theorem and the following observation:

If p is prime, then $a^2 \equiv 1 \bmod p$ holds only when $a \equiv \pm 1 \bmod p$. $(\star)$

This last fact is seen to be true by factoring the difference of two squares $a^2 - 1$ as follows:
$$a^2 - 1 = (a-1)(a+1).$$
Because p is prime, if it divides a product, then it must divide one of the factors. So either p divides $a - 1$ which tells us that $a \equiv 1 \bmod p$ or p divides $a + 1$ which tells us that $a \equiv -1 \bmod p$. □

11.8. Probabilistic Primality Testing

On the other hand, there are many instances in which n is composite and $a^2 \equiv 1 \bmod n$, while a is congruent to neither $+1$ nor -1 modulo n. For instance, $5^2 \equiv 1 \bmod 12$.

The Strong Pseudoprimality Test

Let n be a number that you suspect may be prime. We already know that 2 is the only even prime, so n may be assumed to be odd. If you have a table of primes available, then we may also assume that n is larger than any of the primes in your table.

(1) Do trial divisions by all the primes in your table to make sure none of them divides n. If one of the primes in your table does divide n, then n is obviously composite and n has **failed**. (Doing a fixed number of trial divisions is harmless, so we would be remiss if we did not make this easy check for compositeness.)

(2) Make a random choice of base a with $2 \leq a \leq n-1$.

(3) Compute the greatest common divisor of a and n (that is, the greatest number that divides both a and n). If the greatest common divisor is greater than 1, then n is not prime. The test is over and n has **failed**. (The greatest common divisor of a and n can be rapidly computed, so again we would be remiss if we did not make this easy check for compositeness.)

(4) Since n is odd, $n-1$ is even and thus $n-1$ contains some highest power of 2 as a factor. That is, we can write

$$n - 1 = 2^k m, \text{ where } m \text{ is odd.}$$

Now we proceed to the heart of the test (also illustrated in Fig. 11.1):

(5) If $a^m \equiv \pm 1 \bmod n$, then we know that $a^{n-1} \equiv 1 \bmod n$ holds as it must when n is prime. So the test is over and n has **passed**. Otherwise set $A \equiv a^m \bmod n$ and go to the next step.

(6) Square A modulo n, and keep squaring it modulo n as many as $k-1$ times. Check the following conditions after each squaring:

(a) If you get neither -1 nor $+1$, then square the number again.

(b) If you get a -1, then the next squaring will give $+1$, so n will pass the Fermat test and the fact $(\star)$ above will not have been violated. The test is over and n has **passed**.

(c) If you get a $+1$, then n has passed the Fermat test, but that is not good enough. You got to this point by squaring a number that was neither congruent to $+1$ nor to -1 modulo n. Thus fact $(\star)$ above has been violated. The test is over and n has **failed**.

(7) After you have done the $k-1$ squarings called for in step **(6)** without getting a -1 or a $+1$, the test is over and n has **failed**. The reason n has failed is that you have computed $a^{2^{k-1}m}$ and found that it is not congruent to -1 modulo n. Squaring once more will give you $a^{2^k m} = a^{n-1}$, and it will be impossible to satisfy the Fermat test without violating the fact $(\star)$ above.

Example If we apply the strong pseudoprimality test to the Carmichael number 561—without doing the trial divisions that would quickly reveal it to be composite—we see that

$$560 = 2^4 \cdot 35$$

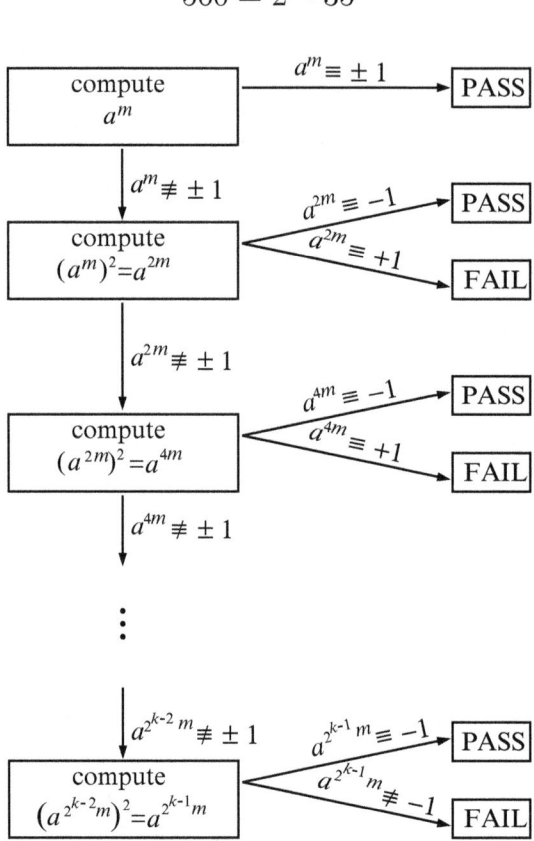

Figure 11.1 The main algorithm in the strong pseudoprimality test.

11.9. The AKS Primality Test

and we compute

$$2^{35} \equiv 263 \mod 561$$
$$2^{70} \equiv 166 \mod 561$$
$$2^{140} \equiv 67 \mod 561$$
$$2^{280} \equiv 1 \mod 561$$

In the process of squaring, we obtain $+1$ before -1, so 561 fails the test. □

The strong pseudoprimality test was first considered by Gary Miller in the context of a deterministic test for primality, but he needed to make an additional assumption of the validity of the extended Riemann hypothesis—which remains an unproven conjecture.[3] The probabilistic test was justified in 1980 when Michael Rabin proved the following theorem:

Theorem *If $n > 4$ is composite, then n will pass the strong pseudoprimality test for no more than $1/4$ of the choices of a, which means it will fail for at least $3/4$ of the choices of a.*

Based on Rabin's theorem, if a number n passes the strong pseudoprimality test for five random choices of the base a, then we should feel 99.9 % certain that n is prime (just because 4 to the 5th power is about 1,000). Of course, n either is prime or it is not, and we still don't know which is true. Nonetheless, n is informally called an *industrial grade prime*, and we would use it in practical applications as if it were prime—until something convinces us otherwise.

11.9 The AKS Primality Test

Manindra Agrawal and his students Neeraj Kayal and Nitin Saxena—who were then candidates for bachelor's degrees at the Indian Institute of Technology at Kanpur—based their work on an interesting generalization of Fermat's result to polynomials. This is an important tool in mathematics: to take a

---

[3]In fact the Riemann hypothesis, first formulated by B. Riemann in 1859, is probably the most important unsolved problem in pure (abstract) mathematics. In its simplest form, the Riemann hypothesis (affectionately known to experts as RH) asserts that the prime numbers are randomly distributed. The more technical forms of the hypothesis are very concrete and technical estimates about the distribution of primes.

known idea and spruce it up for a new application. Their idea was to apply the following result (the proof of which is at the level of an exercise for students in an advanced undergraduate course in number theory or algebra):

If n and a are relatively prime (i.e., their greatest common divisor is 1), then n is prime if and only if the congruence of polynomials

$$(x-a)^n \equiv (x^n - a) \bmod n \tag{11.1}$$

holds (meaning all their coefficients are congruent).

As examples of the above result the reader may verify the following:

$$
\begin{aligned}
n=3,\ a=2: &\quad (x-2)^3 \equiv x^3 - 2 &&\bmod 3 \\
n=4,\ a=3: &\quad (x-3)^4 \equiv x^4 + 2x^2 + 2x + 1 &&\bmod 4 \\
n=5,\ a=2: &\quad (x-2)^5 \equiv x^5 - 2 &&\bmod 5 \\
n=5,\ a=3: &\quad (x-3)^5 \equiv x^5 - 3 &&\bmod 5 \\
n=5,\ a=4: &\quad (x-4)^5 \equiv x^5 - 4 &&\bmod 5 \\
n=6,\ a=5: &\quad (x-5)^6 \equiv x^6 + 3x^4 + 2x^3 + 3x^2 + 1 &&\bmod 6
\end{aligned}
$$

Notice that, for a prime n, Fermat's little theorem is the special case that arises by substituting 0 for x in (11.1). The crucial difference between this last result and Fermat's little theorem is that, while Fermat's little theorem only goes in one direction, the above result goes both ways, making it a deterministic test for primality. Unfortunately it is a slow test because of the large computation time required to calculate the polynomial $(x-a)^n$.

Two more ideas were needed to turn this into a polynomial time test. One was to divide both sides of (11.1) by $x^r - 1$, where r is a specially chosen prime, and look at congruence of the remainders (which are polynomials of degree smaller than r) after that division. The second idea was to vary a.

What has now become known as the AKS (acronym for the names of the authors) technique actually consists of a body of algorithms. In their original paper, Agrawal, Kayal, and Saxena showed that their method can test the primality of a positive integer n in $(\log n)^{13}$ steps. Not long afterward in July 2005, Carl Pomerance and Henrik Lenstra [LP] improved the estimate to $(\log n)^6$. Research and refinements continue to appear.

It should be stressed that the AKS algorithm is revolutionary in that it tells us whether a positive integer n is prime or composite in polynomial time. In the case that the number is composite, it does *not* tell us how to factor the number.

While AKS is revolutionary, any algorithm that tells us how to factor a large integer (say 150 digits) would be moderately catastrophic. Many modern cryptographic methods that are currently in use—both in the financial sector and in the defense world—are based on the apparent fact that very large numbers *cannot* be factored in a reasonable amount of time. The film *Sneakers* starring Robert Redford is predicated on the idea that a scientist has produced a machine that *can* factor a large integer in polynomial time. The drama in that film only begins to suggest (somewhat humorously) all the consequences of such a discovery.

It is still not known with certainly that the problem of factoring a large integer is of exponential complexity. But we *think* it is. Our current, cutting-edge ideas about cryptoanalysis are all based on that supposition. But current research is vigorous and new results appear every month. We do not know what possibly shocking developments are around the corner.

A Look Back

The lore of the prime numbers is broad and deep. The primes are the building blocks of the integers. All questions about the structure of the integers, about number-theoretic relations, about congruences and equivalences, boil down to questions about the primes.

Historically, the first great result about the primes is due to Euclid. He showed that there are infinitely many primes. The proof is fairly simple, but the information is of the utmost importance. Moreover, the proof is by contradiction—and this is one of the first instances of that technique.

Another great result in the theory of primes is the Prime Number Theorem. No less an eminence than Carl Friedrich Gauss (1777–1855) formulated and studied the question. He wanted to know the density of the prime numbers in the full set of integers. When he was young, Gauss spent many hours staring at tables of primes and formulating his conjecture. He ended up enunciating the Prime Number Theorem, but was never able to prove it.

Bernhard Riemann had a great interest in the Prime Number Theorem. His celebrated 1859 paper on the distribution of primes, the one that introduced the Riemann zeta function and led to the Riemann hypothesis, was primarily devoted to developing tools for proving the Prime Number Theorem. Unfortunately Riemann did not live long enough to achieve that goal.

However, in 1896, Charles de la Vallée Poussin (1866–1962) and Jacques Hadamard (1865–1963) simultaneously and independently proved the Prime Number Theorem. The statement is that if, $\pi(N)$ is the number of primes less than or equal to N, then

$$\lim_{N \to \infty} \frac{\pi(N)}{N/\log N} = 1 \, .$$

This is a dramatic and precise result, and all the more fascinating because the proof is not number-theoretic. Both mathematicians had to bring in techniques from complex analysis to complete their arguments.

In fact there is an interesting history involved in this problem. In the early 1950s, Atle Selberg at the Institute for Advanced Study in Princeton announced that he had an "elementary" proof of the Prime Number Theorem. This meant that his proof did not use any complex analysis; it only used basic ideas from number theory. It did *not* mean that the proof was easy or accessible. Only that it did not use advanced techniques from other parts of mathematics. The only glitch in Selberg's program was that there was a particular lemma that he was sure was true, but he couldn't prove it. Paul Erdős was in residence at the Institute at the time, and he was fascinated by this development. He went home and thought hard about Selberg's putative lemma. Two days later he had a proof.

Well, Erdős was one of history's great collaborators. He wrote papers with everyone. And naturally he assumed that now he and Selberg would write a wonderful paper about the elementary proof of the Prime Number Theorem. But Selberg never had any collaborators in his whole career, and he did not plan to take one on now. He instead went home and found his own proof of the critical lemma. In this fashion he was able to rationalize rejecting Erdős's contribution and to write up the result on his own.

The end result is that there are now two papers on the elementary proof of the Prime Number Theorem—one by Erdős and one by Selberg. Both men are given considerable credit for the result. The vituperation and hard feelings connected with this misunderstanding lived on for many years, and much has been said about the matter. A colleague of mine once went to Erdős and asked him, "Paul, you have always said that mathematics is for sharing. You cast your bread upon the waters and you reap what comes forth. Why are you engaged in this dreadful controversy with Selberg?" Erdős's reply was, "Ah, yes. But this is the Prime Number Theorem."

REFERENCES AND FURTHER READING

[**AKS 04**] Agrawal, M., Kayal, N., Saxena, N.: PRIMES is in P. Annals of Mathematics **160**, 781–793 (2004)

[**Bor 03**] Bornemann, F.: PRIMES is in P: a breakthrough for everyman. Notices of the American Mathematical Society, vol. 50, pp. 545–552. (2003)

[**Car 12**] Carmichael, R.D.: On composite numbers p which satisfy the Fermat congruence $a^{p-1} \equiv 1 \mod p$. American Mathematical Monthly **19**, 22–27 (1912)

[**LP**] Lenstra, Jr. H.W., Pomerance, C.: Primality testing with Gaussian periods. www.math.dartmouth.edu/~carlp/PDF/complexity12.pdf. Accessed July (2005)

[**Mil 76**] Miller, G.L.: Riemann's hypothesis and a test for primality. Journal of Computer and System Sciences **13**, 300–317 (1976)

[**Rab 80**] Rabin, M.O.: Probabilistic algorithm for testing primality. Journal of Number Theory **12**, 128–138 (1980)

[**SS 77**] Solovay, R.M., Strassen, V.: A fast Monte-Carlo test for primality. SIAM Journal on Computing **6**, 84–85 (1977)

Chapter 12
The Foundations of Mathematics

12.1 The Evolution of the Concept of Proof

Mathematics has a long and distinguished history. Babylonian tablets as old as 1800 BCE show clear evidence of sophisticated mathematical thinking. More elementary mathematics for keeping track of commerce and land area dates back even further.

In the early days, mathematics was largely an intuitive endeavor. The practitioners combined measurements, intuition, and deliberation to come up with solutions to problems. Logic was used only intermittently, and the concepts of "theorem" and "proof" were unknown.

The earliest mathematics was phenomenological. If one could draw a plausible picture, or give a compelling description, then that was all the justification that was needed for a mathematical "fact." Sometimes one argued by analogy. Or by invoking the gods. The notion that mathematical statements could be *proved* was not yet an idea that had been developed. There was no standard for the concept of proof. The logical structure, the "rules of the game," had not yet been created.

Thus we are led to ask: What is a proof? Heuristically, a proof is a rhetorical device for convincing someone else that a mathematical statement is true or valid. And how might one do this? A moment's thought suggests that a natural way to prove that something new (call it **B**) is true is to relate it to something old (call it **A**) that has already been accepted as true. Thus

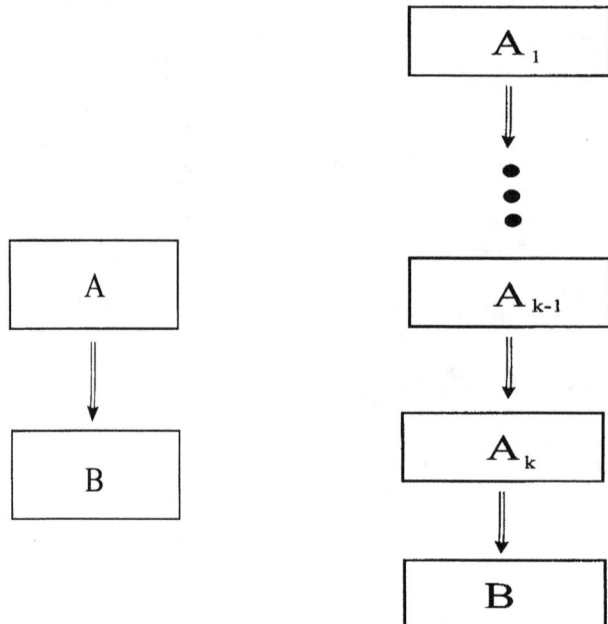

Figure 12.1 Logical derivation.

arises the concept of *deriving* a new result from an old result. See Fig. 12.1. The next question then is, "How was the old result verified?" Applying this regimen repeatedly, we find ourselves considering a chain of reasoning as in Fig. 12.1. But then one cannot help but ask: "Where does the chain begin?" This is a fundamental issue.

It will not do to say that the chain has no beginning: it extends infinitely far back into the fogs of time. Because if that were the case it would undercut our thinking of what a proof should be. We are endeavoring to justify new mathematical facts in terms of old mathematical facts. But if the reasoning regresses infinitely far back into the past, then we cannot in fact ever grasp a basis or initial justification for our reasoning. As we shall see below, the answer to these questions is that the mathematician puts into place certain definitions and axioms before beginning to explore the firmament, to determine what is true, and then to prove it. Considerable discussion will be required to put this paradigm into context.

As a result of these questions, ancient mathematicians had to think hard about the nature of mathematical proof. Thales (640 BCE–546 BCE), Eudoxus (408 BCE–355 BCE), and Theaetetus of Athens (417 BCE–369

12.1. The Evolution of the Concept of Proof

BCE) actually formulated theorems. Thales definitely proved some theorems in geometry (and these were later put into a broader context by Euclid—see also Chaps. 6 and 11). A theorem is the mathematician's formal enunciation of a fact or truth. Eudoxus is to be admired for the rigor of his work (his ideas about comparing incommensurable quantities—comparable to our modern method of cross-multiplication—are to be particularly noted), but he generally did not prove theorems. His work had a distinctly practical bent, and he was notably fond of calculations. Hippocrates of Chios (470 BCE–410 BCE) proved a theorem about ratios of areas of circles that he put to especially good use in calculating the areas of lunes (moon-shaped crescents).

It was Euclid of Alexandria who first formalized the way that we now think about mathematics. Euclid had definitions and axioms and then theorems—in that order. There is no denying the assertion that Euclid set the paradigm by which we have been practicing mathematics for 2300 years. This was mathematics done right. Now, following Euclid, in order to address the issue of the infinitely regressing chain of reasoning, we begin our studies by putting into place a set of *Definitions* and a set of *Axioms*.

What is a definition? A definition explains the meaning of a piece of terminology. There are logical problems with even this simple idea, for consider the first definition that we are going to formulate. Suppose that we wish to define a *rectangle*. This will be the first piece of terminology in our mathematical system. What words can we use to define it? Suppose that we define rectangle in terms of points and lines and planes and right angles. That begs the questions: What is a point? What is a line? What is a plane? How do we define "angle"? What is a right angle?

Thus we see that our *first* definition(s) must be formulated in terms of commonly accepted words that require no further explanation. Aristotle (384 BCE–322 BCE) insisted that a definition must describe the concept being defined in terms of other concepts already known. This is often quite difficult. As an example, Euclid defined a *point* to be that which has no part. Thus he is using words *outside of mathematics*, that are a commonly accepted part of everyday argot, to explain the precise mathematical notion of "point."[1] Once "point" is defined, then one can use that term in later definitions—for example, to define "line." And one will also use everyday language that

---

[1] It is quite common, among those who study the foundations of mathematics, to refer to terms that are defined in nonmathematical language—that is, which cannot be defined in terms of other mathematical terms—as *undefined terms*. The concept of "point" is an undefined term.

does not require further explication. That is how we build up our system of definitions.

The definitions give us then a language for doing mathematics. We formulate our results, or *theorems*, by using the words that have been established in the definitions. But wait, we are not yet ready for theorems. Because we have to lay logical cornerstones upon which our reasoning can develop. That is the purpose of axioms.

What is an axiom? An axiom[2] (or postulate[3]) is a mathematical statement of fact, formulated using the terminology that has been defined in the definitions, that is taken to be self-evident. An axiom embodies a crisp, clean mathematical assertion. One does not *prove* an axiom. One takes the axiom to be given, and to be so obvious and plausible that no proof is required. Following an ancient precept called Occam's Razor, we seek to have as few axioms as possible.

Generally speaking, in any subject area of mathematics, one begins with a brief list of definitions and a brief list of axioms. Once these are in place, and are accepted and understood, then one can begin proving theorems.[4] And what is a proof? A proof is a rhetorical device for convincing another mathematician that a given statement (the theorem) is true. Thus a proof can take many different forms. The most traditional form of mathematical proof is that it is a tightly knit sequence of statements linked together by strict rules of logic. Today, a proof could (and often does) take the traditional form that goes back 2300 years to the time of Euclid. But it could also consist of a computer calculation. Or it could consist of constructing a physical model. Or it could consist of a computer *simulation* or *model*. Or it could consist of a computer algebra computation using `Mathematica` or `Maple` or `MatLab`. It could also consist of an agglomeration of these various techniques.

12.2 Kurt Gödel and the Birth of Uncertainty

Mathematics today follows the model set by Euclid and his school. In formal, recorded mathematics, we state theorems and we prove them. This paradigm gives mathematics a permanence and a portability that is unknown in other disciplines. Mathematicians proudly archive their work, because they know

---

[2]The word "axiom" derives from the Greek *axios*, meaning "something worthy."

[3]The word "postulate" derives from a medieval Latin word *postulatus* meaning "to nominate" or "to demand."

[4]The word "theorem" derives from the Greek *theōrein*, meaning "to look at."

that future generations will use it confidently, knowing that the mathematics of yesterday is just as valid as any new mathematics being developed at the moment. Similar statements cannot be made for computer science or medicine. In the late nineteenth century it was taken as a given that any sensible mathematical statement can be proved or disproved. This was the intrinsic nature of the mathematical program. David Hilbert (1862–1943), arguably the spokesman for the field for over 50 years, formulated a program for all of mathematics to be worked out rigorously and axiomatically. There was to be no room for doubt or equivocation.

It came as something of a shock when, in 1932, the Austrian mathematician Kurt Gödel proved that Hilbert's program was impossible. More precisely, Gödel showed that, in any logical system sophisticated enough to include arithmetic, there will be sensible statements that can neither be proved nor disproved. Nobody had ever imagined such a situation, much less demonstrated its reality. Gödel's ideas shook the world. Today he is considered to have been one of the greatest logicians in history.

Gödel went on to develop the theory of recursive functions, and to study the structure and shape of the real number system (in the form of the continuum hypothesis). We shall describe his contributions below. First we begin with a consideration of Gödel the man.

12.3 Origins of Kurt Gödel

Rudolf Gödel was from Vienna. He managed a textile firm. Marianne Gödel was from the Rhineland, also from a textile family. The Gödel's had two sons: Rudolf and Kurt. Kurt was the younger, born in 1906. He had a happy childhood, and was very close to his mother. He suffered rheumatic fever at the age of six, but his recovery was direct and he resumed normal childhood activities thereafter. On the other hand, this serious bout with illness at an early age seems to have been the seed of Gödel's adult hypochondria. He soon began reading medical books to learn about the causes of rheumatic fever. Gödel attended school in Brno (see Fig. 12.2), in what is now the Czech Republic, completing his studies in 1923. German speakers, as Gödel was, call that city Brünn. Gödel excelled in both mathematics and Latin. He mastered much of university mathematics while still at the Gymnasium.

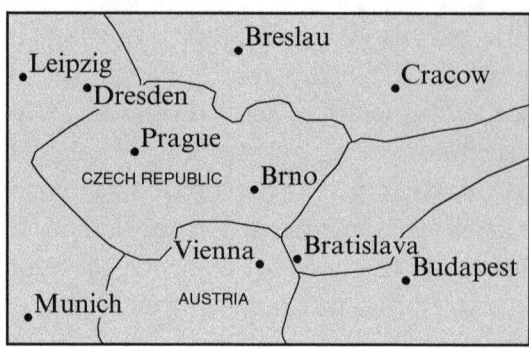

Figure 12.2 The location of Brno, known as Brünn in German.

At the University of Vienna Gödel studied mathematics and theoretical physics. He was particularly taken by the lectures of Furtwängler, in part because the man was a brilliant scholar and in part because he was paralyzed from the waist down. Furtwängler's lectures were delivered verbally by the famous professor, but written on the board by an assistant. Since Gödel was intensely conscious of his own health, he took particular fascination with a professor having a disability. During this time Gödel was exposed to Bertrand Russell's book on mathematical philosophy, and was quite taken with the subject matter. This set the course for his future studies. He determined to master logic.

Gödel wrote his doctoral dissertation in 1929 and became a faculty member at the University of Vienna in 1930. Gödel published his celebrated incompleteness theorem in 1931. It showed that the program initiated by Hilbert, and also by Russell and Whitehead's *Principia Mathematica* [WR 62], was doomed to failure. It was a seminal result.

Hitler came to power in 1933. Gödel had no particular interest in politics, so was not immediately affected by these developments. He went to Princeton in 1934 to give some lectures on his big results. Stephen Kleene took notes, and these were subsequently published (Gödel himself published little during his lifetime). Unfortunately, on his return to Europe, Gödel suffered a nervous breakdown. This was to be the first of many bouts of mental illness that plagued Gödel for his entire adult life. Gödel recovered in a sanitorium. His mathematics continued to progress. He proved important independence results for the axioms of set theory. Unfortunately, in 1936, Gödel's mentor Schlick was murdered by the Nazis; Gödel suffered another mental breakdown.

12.3. Origins of Kurt Gödel

Gödel always liked older women. In 1927 he met, in a nightclub, a dancer named Adele Porkert who was 6 years his senior. They married in 1938. He again visited Princeton in 1938–1939. In March 1938, Austria had become part of Germany, but Gödel was not much interested and carried on his life much as normal.

But events began to trouble Gödel, and he was forced to consider his life in painful detail. Kurt Gödel's subsequent move from Vienna to the United States was abrupt but necessary for his survival. In March of 1939 the Nazis abolished university lectureships and created a new position called *Dozent neuer Ordnung* (Lecturer of the New Order). There were no such positions for Jews. Kurt Gödel was in fact *not* Jewish, but he was commonly thought to be a Jew, and he suffered for it. Thus Gödel was out of a job, and he also received a letter ordering him to report to the German army for a physical examination. This was a man who had recently spent quite a lot of time in sanatoria for overwork and depression, and now on top of losing his job, he faced the prospect of becoming a grunt in the army.

When Gödel next got a bill from his cleaning lady, he found that it listed the total due (which was *Reichsmark 6.80*) and then, underneath, a neatly typed *Heil Hitler!*. For Gödel this was the last straw. He decided that he and his new wife had to move to Princeton (where he had an offer). They decided that traveling by boat across the Atlantic was too risky, so they took the long way around—across Russia, the Pacific, and the continent of North America. It can truly be said that, for Gödel, the Institute for Advanced Study, in Princeton, New Jersey, was (in founder Abraham Flexner's words) "a paradise, a haven where scholars and scientists could regard the world and its phenomena as their laboratory, without being carried off into the maelstrom of the immediate."

Kurt Gödel was an eccentric and unworldly man who frequently needed to be protected from the vicissitudes of life by his friends. After Gödel had lived in the United States for many years, he was persuaded to become an American citizen. He therefore began studying for the citizenship exam.

Unfortunately, as soon as Gödel began reading the U.S. Constitution, he discovered troubling logical loopholes. This insight cast him into deep distress. John von Neumann (1903–1957)—Gödel's colleague at the Institute for Advanced Study—was finally called in to convince Gödel that, if you looked at things the right way, then there would be no logical inconsistency.

Gödel became a U.S. citizen in 1948. Albert Einstein (1879–1955) and the economist Oskar Morgenstern (1902–1976) were the ones who chaperoned Gödel to the hearing for his citizenship application. The judge was overwhelmed by this opportunity to talk to Einstein, and they conversed at length about events in Nazi Germany. Finally, as an afterthought, the judge turned to Gödel and said, "But of course from your reading of the Constitution you now know that nothing like that could happen here." "As a matter of fact, ..." Gödel began—but then, under the table, Morgenstern kicked Gödel. So Gödel got his citizenship after all. (By some accounts, Morgenstern had to do a heck of a lot more than just kick Gödel. Fortunately, he was equal to the task.)

Einstein and Gödel were great friends but they could hardly have been more different. The former was famous for dressing in rumpled clothing that was just one step above the ragpile; the latter typically sported a white linen suit and matching fedora. Einstein was a gregarious character who loved to have fun and tell jokes. Gödel was solemn, solitary, pessimistic, and ultimately paranoid. Einstein enjoyed heavy German cooking, and he freely indulged his appetite for wienerschnitzel and sachertorte. Gödel subsisted on a strange combination of butter, baby food, and laxatives. Gödel kept a careful diary of his food intake and daily health status; he used a strange and unusual form of German shorthand that is now extinct.

Einstein enjoyed the company of other people. He was a lively presence at parties. Gödel believed in ghosts; he had a morbid dread of being poisoned by refrigerator gases; he refused to go out when certain distinguished mathematicians were in town, apparently because he feared that they would try to kill him. Gödel said, "Every chaos is a wrong appearance."

But colleague Freeman Dyson observed that "Gödel was ... the only one of our colleagues who walked and talked on equal terms with Einstein." Gödel was undaunted by Einstein's reputation and was not afraid to challenge his ideas. Einstein found this refreshing. Of course they conversed in German, and both found that to be cathartic.

Both men were generally perceived to be on a plane higher than the rest of us. Gödel biographer Rebecca Goldstein relates that "I once found the philosopher Richard Rorty standing in a bit of a daze in Davidson's food market. He told me in hushed tones that he'd just seen Gödel in the frozen food aisle." As a result of their rarification, both men were lonely. United by a shared sense of intellectual isolation, they found comfort in each other's company. One member of the Institute community noted that the two men did not seem to want to speak to anyone else; they only wanted to converse with each other. Gödel was born in 1906, the year after the publication of

Einstein's four great papers that shook the world of physics. So Einstein was old enough to be Gödel's father. But they were close personal friends.

Kurt Gödel's favorite movie was *Snow White and the Seven Dwarfs*. Gödel said, "Only fables present the world as it should be and as if it had meaning." He never convinced his friend Albert Einstein to see the movie. Likewise, Einstein never convinced Gödel to sample Beethoven or Mozart. It is not known which was Gödel's favorite dwarf.

In his later years, Gödel became a recluse. He rarely left his house, and he stopped eating because he feared that people were trying to poison him.

A slight person and very fastidious, Gödel was generally worried about his health and did not travel or lecture widely in later years. He had no doctoral students, but through correspondence and personal contact with the constant succession of visitors to Princeton, many people benefited from his extremely quick and incisive mind. Friend to Einstein, von Neumann and Morgenstern, he particularly enjoyed philosophical discussion. He died in 1978 from malnutrition and debilitation.

12.4 Elements of Formal Logic

We begin here with an introduction to the elements of formal logic. This will lead to an appreciation of Gödel's contribution to the subject.

Strictly speaking, our approach to logic is "intuitive" or "naive." Whereas, in ordinary conversation, these emotion-charged words may be used to downgrade the value of that which is being described, our use of these words here is more technical. What is meant is that we shall prescribe in this chapter certain rules of logic that are to be followed in the rest of the discussion. They will be presented to you in such a way that their validity should be intuitively appealing and self-evident. We cannot *prove* these rules. The rules of logic are the point where our learning begins. A more advanced course in logic will explore other logical methods. The ones that we present here are universally accepted in mathematics and science.

We shall begin with sentential logic and elementary connectives. This material is called the *propositional calculus* (to distinguish it from the predicate calculus, which will be treated later). In other words, we shall be discussing *propositions*—which are built up from atomic statements and connectives. The elementary connectives include "and," "or," "not," "if–then," and "if and only if." Each of these will have a precise meaning and will have exact relationships with the other connectives. Later we shall discuss the completeness of this system of elementary sentential logic, although we shall not present the *proof* of completeness (see [Sto 61, p. 147 ff.] for a discussion of the work of Frege, Whitehead and Russell, Bernays, and Gödel in this regard).

An *atomic statement*, also called *elementary statement*, is a sentence with a subject and a verb (and sometimes an object) but no connectives ("and," "or," "not," "if–then," "if-and-only-if"). For example,

John is good.

Mary has bread.

Ethel reads books.

are all atomic statements. We build up sentences, or propositions, from atomic statements using connectives.

Later we shall consider the quantifiers "for all" and "there exists" and their relationships with the connectives from the last paragraph. The quantifiers will give rise to the so-called *predicate calculus*. Connectives and quantifiers will prove to be the building blocks of all future statements in this discussion, indeed in all of mathematics.

In everyday conversation, people sometimes argue about whether a statement is true or not. In mathematics there is nothing to argue about. In practice a sensible statement in mathematics is either true or false, and there is no room for opinion about this attribute. How do we determine which statements are true and which are false?

The modern methodology in mathematics works as follows.

- We *define* certain terms.

- We *assume* that these terms have certain properties or truth attributes (these assumptions are called axioms).

We specify certain rules of logic.

Every sensible statement in our logical system will have a truth value, that is, it will either be true or false, but not both. The interesting question is: How do we show, or verify, that a given statement is true?

Any statement that can be derived from the axioms, using the rules of logic, is understood to be true. In practice, this is our method for verifying

12.4. Elements of Formal Logic

that a statement is true. On the other hand, a statement is false if it is inconsistent with the axioms and the rules of logic. That is, a statement is false if the assumption that it is true leads to a contradiction. Alternatively, a statement **P** is false if the negation of **P** can be established or proved.

Two questions naturally arise:

(1) Could our system ever lead us to conclude that a statement is both true and false?

(2) Is it the case that every true statement can be derived from the axioms?

It is essential that the answer to the first question be "No." A system of logic that allows one to prove that a statement is both true and false is said to be *inconsistent*, and an inconsistent system is useless. A system that allows one to prove only true statements is said to be *consistent*.

It is desirable that the answer to the second question be "Yes." A system of logic that allows one to prove every true statement is said to be *complete*. If there are true statements that cannot be proved, then the system is said to be *incomplete*. One can tolerate incompleteness, but the goal is a consistent and complete logical system. Gödel showed us that our mathematics, the mathematics that we practice every day, is incomplete.

The point of view being described here is special to mathematics. While it is indeed true that mathematics is used to model the world around us—in physics, engineering, and in other sciences—the subject of mathematics itself is a man-made system. Its internal coherence is guaranteed by the axiomatic method that we have just described.

It is reasonable to ask whether mathematical truth is a construct of the human mind or an immutable part of nature. For instance, is the assertion that "the area of a circle is π times the radius squared" actually a fact of nature just like Newton's inverse square law of gravitation? Our point of view is that mathematical truth is relative. The formula for the area of a circle is a logical consequence of the axioms of mathematics, nothing more. The fact that the formula seems to describe what is going on in nature is convenient, and is part of what makes mathematics useful. But that aspect is something over which we as mathematicians have no control. Our concern is with the internal coherence of our logical system.

It can be asserted that a "proof" (a concept to be discussed later) is a psychological device for convincing the reader that an assertion is true. However our view in this section is more rigid: a proof is a sequence of applications of the rules of logic to derive the assertion from the axioms. There is no room for opinion here. The axioms are plain. The rules are rigid. A proof is like a sequence of moves in a game of chess. If the rules are followed, then the proof is correct. Otherwise not.

12.4.1 Definitions of Connectives

Let **A** and **B** be atomic statements such as "Chelsea is smart" or "The earth is flat." The statement

$$\mathbf{A} \text{ and } \mathbf{B}$$

means that both **A** is true *and* **B** is true. For instance,

$$\textbf{Arvid is old and Arvid is fat.}$$

means both that Arvid is old *and* Arvid is fat. If we meet Arvid and he turns out to be young and fat, then the statement is false. If he is old and thin then the statement is false. Finally, if Arvid is *both* young and thin then the statement is false. The statement is *true* precisely when both properties—oldness and fatness—hold. We may summarize these assertions with a *truth table*. We let

$$\mathbf{A} = \text{Arvid is old.}$$

and

$$\mathbf{B} = \text{Arvid is fat.}$$

The expression

$$\mathbf{A} \wedge \mathbf{B}$$

will denote the phrase "**A** and **B**." We call this statement the *conjunction* of **A** and **B**. The letters "T" and "F" denote "True" and "False" respectively. Then we have

A	B	A ∧ B
T	T	T
T	F	F
F	T	F
F	F	F

This table is to be read line-by-line. For example, the first line tells us that if **A** is true and **B** is true, then the conjunction $\mathbf{A} \wedge \mathbf{B}$ is true. Notice that

12.4. Elements of Formal Logic

we have listed all possible truth values of **A** and **B** and the corresponding values of the *conjunction* **A** ∧ **B**.

In a restaurant the menu often contains phrases such as

soup or salad

This means that we may select soup *or* select salad, but we may not select both. This use of "or" is called the *exclusive* "or"; it is not the meaning of "or" that we use in mathematics and logic. In mathematics we instead say that "**A or B**" is true provided that **A** is true or **B** is true or *both* are true. This is the *inclusive* "or." If we let **A** ∨ **B** denote "**A or B**" then the truth table is

A	B	A ∨ B
T	T	T
T	F	T
F	T	T
F	F	F

We call the statement **A** ∨ **B** the *disjunction* of **A** and **B**.

We see from the truth table that the only way that "**A or B**" can be false is if *both* **A** is false and **B** is false. For instance, the statement

Hillary is beautiful or Hillary is poor.

means that Hillary is either beautiful or poor or both. In particular, she will not be both ugly and rich. Another way of saying this is that if she is ugly she will compensate by being poor; if she is rich she will compensate by being beautiful. *But she could be both beautiful and poor.*

EXAMPLE 12.4.1 The statement

$$x > 2 \text{ and } x < 5$$

is true for the number $x = 3$ because this value of x is both greater than 2 *and* less than 5. It is false for $x = 6$ because this x value is greater than 2 but not less than 5. It is false for $x = 1$ because this x is less than 5 but not greater than 2.

EXAMPLE 12.4.2 The statement

$$x \text{ is odd and } x \text{ is a perfect cube}$$

is true for $x = 27$ because both assertions hold. It is false for $x = 7$ because this x, while odd, is not a cube. It is false for $x = 8$ because this x, while a cube, is not odd. It is false for $x = 10$ because this x is neither odd nor is it a cube.

EXAMPLE 12.4.3 The statement

$$x < 3 \text{ or } x > 6$$

is true for $x = 2$ since this x is < 3 (even though it is not > 6). It holds (that is, it is true) for $x = 9$ because this x is > 6 (even though it is not < 3). The statement fails (that is, it is false) for $x = 4$ since this x is neither < 3 nor > 6.

EXAMPLE 12.4.4 The statement

$$x > 1 \text{ or } x < 4$$

is true for every real x.

EXAMPLE 12.4.5 The statement $(\mathbf{A} \vee \mathbf{B}) \wedge \mathbf{B}$ has the following truth table:

A	B	A ∨ B	(A ∨ B) ∧ B
T	T	T	T
T	F	T	F
F	T	T	T
F	F	F	F

Notice in Example 12.4.5 that the statement $(\mathbf{A} \vee \mathbf{B}) \wedge \mathbf{B}$ has the same truth values as the simpler statement $\mathbf{B}$. In what follows we shall call such pairs of statements (having the same truth values) *logically equivalent*.

The words "and" and "or" are called *connectives*: their role in sentential logic is to enable us to build up (or to connect together) pairs of statements. The idea is to use very simple statements, like "**Jennifer is swift**" as building blocks; then we compose more complex statements from these building blocks by using connectives.

12.4. Elements of Formal Logic

In the next paragraphs we will become acquainted with the other two basic connectives "not" and "if–then."

The statement "not **A**," written $\sim \mathbf{A}$, is true whenever **A** is false. We also call "not **A**" the *negation* of **A**. For example, the statement

Charles is not happily married.

is true provided the statement "Charles is happily married" is false. The truth table for $\sim \mathbf{A}$ is as follows:

A	$\sim \mathbf{A}$
T	F
F	T

Greater understanding is obtained by combining connectives:

EXAMPLE 12.4.6 Here is the truth table for $\sim (\mathbf{A} \wedge \mathbf{B})$:

A	**B**	$\mathbf{A} \wedge \mathbf{B}$	$\sim (\mathbf{A} \wedge \mathbf{B})$
T	T	T	F
T	F	F	T
F	T	F	T
F	F	F	T

EXAMPLE 12.4.7 Now we look at the truth table for $(\sim \mathbf{A}) \vee (\sim \mathbf{B})$:

A	**B**	$\sim \mathbf{A}$	$\sim \mathbf{B}$	$(\sim \mathbf{A}) \vee (\sim \mathbf{B})$
T	T	F	F	F
T	F	F	T	T
F	T	T	F	T
F	F	T	T	T

Notice that the statements $\sim (\mathbf{A} \wedge \mathbf{B})$ and $(\sim \mathbf{A}) \vee (\sim \mathbf{B})$ have the *same truth values*. As previously noted, such pairs of statements are called *logically equivalent*.

The logical equivalence of $\sim (\mathbf{A} \wedge \mathbf{B})$ with $(\sim \mathbf{A}) \vee (\sim \mathbf{B})$ makes good intuitive sense: the statement $\mathbf{A} \wedge \mathbf{B}$ fails precisely when either **A** is false *or* **B** is false. Since in mathematics we cannot rely on our intuition to establish facts, it is important to have the truth table technique for establishing logical equivalence.

One of the main reasons that we use the *inclusive* definition of "or" rather than the exclusive one is so that the connectives "and" and "or" have the nice relationship just discussed. Similarly, it is also the case that $\sim (\mathbf{A} \vee \mathbf{B})$ and $(\sim \mathbf{A}) \wedge (\sim \mathbf{B})$ are logically equivalent. The two facts

$$\sim (\mathbf{A} \wedge \mathbf{B}) \text{ is logically equivalent to } (\sim \mathbf{A}) \vee (\sim \mathbf{B})$$

and

$$\sim (\mathbf{A} \vee \mathbf{B}) \text{ is logically equivalent to } (\sim \mathbf{A}) \wedge (\sim \mathbf{B})$$

are referred to as *de Morgan's laws*.

A statement of the form "If **A** then **B**" asserts that, whenever **A** is true, then **B** is also true. This assertion (or "promise") is tested when **A** is true, because it is then claimed that something else (namely **B**) is true as well. *However*, when **A** is false then the statement "If **A** then **B**" *claims nothing*. Using the symbols $\mathbf{A} \Rightarrow \mathbf{B}$ to denote "If **A** then **B**," we obtain the following truth table:

A	B	$\mathbf{A} \Rightarrow \mathbf{B}$
T	T	T
T	F	F
F	T	T
F	F	T

The connective $\Rightarrow$ is called the *material conditional* and the sentence $\mathbf{A} \Rightarrow \mathbf{B}$ is called a *conditional statement*.

Notice that we use here an important principle of Aristotelian logic: every sensible statement is either true or false. There is no "in between" status. When **A** is false we can hardly assert that $\mathbf{A} \Rightarrow \mathbf{B}$ is false. For $\mathbf{A} \Rightarrow \mathbf{B}$ asserts that "whenever **A** is true then **B** is true," and **A** is not true!

Put in other words, when **A** is false then the statement $\mathbf{A} \Rightarrow \mathbf{B}$ is not tested. It therefore cannot be false. So it must be true.

EXAMPLE 12.4.8 The statement "If $2 = 4$, then Calvin Coolidge was our greatest president." is true. This is the case no matter what you think of Calvin Coolidge.

The statement "If fish have hair, then chickens have lips." is true.

The statement "If $9 > 5$, then dogs don't fly." is true.

The statement "If $2+2=4$, then Newt Gingrich is ten feet tall." is false.

12.4. Elements of Formal Logic

Notice that the "if" part of the sentence and the "then" part of the sentence need not be related in any intuitive sense. The truth or falsity of an "if–then" statement is simply a fact about the logical values of its hypothesis and of its conclusion.

EXAMPLE 12.4.9 The "if–then" statement $\mathbf{A} \Rightarrow \mathbf{B}$ is logically equivalent with $(\sim \mathbf{A}) \vee \mathbf{B}$. For the truth table for the latter is

A	B	$\sim$ A	$(\sim \mathbf{A}) \vee \mathbf{B}$
T	T	F	T
T	F	F	F
F	T	T	T
F	F	T	T

which is the same as the truth table for $\mathbf{A} \Rightarrow \mathbf{B}$.

You should think for a bit to see that $(\sim \mathbf{A}) \vee \mathbf{B}$ *says the same thing* as $\mathbf{A} \Rightarrow \mathbf{B}$. To wit, assume that the statement $(\sim \mathbf{A}) \vee \mathbf{B}$ is true. Now suppose that $\mathbf{A}$ is true. Then, according to the disjunction, $\mathbf{B}$ must be true. But that says that $\mathbf{A} \Rightarrow \mathbf{B}$. For the converse, assume that $\mathbf{A} \Rightarrow \mathbf{B}$ is true. This means that if $\mathbf{A}$ holds, then $\mathbf{B}$ must follow. But that just says $(\sim \mathbf{A}) \vee \mathbf{B}$. So the two statements are equivalent, i.e., they say the same thing.

Once you believe that assertion, then the truth table for $(\sim \mathbf{A}) \vee \mathbf{B}$ gives us another way to understand the truth table for $\mathbf{A} \Rightarrow \mathbf{B}$.

There are in fact infinitely many pairs of logically equivalent statements. But just a few of these equivalences are really important in practice—most others are built up from these few basic ones.

EXAMPLE 12.4.10 The statement

$$\text{If } x \text{ is negative, then } -5 \cdot x \text{ is positive.}$$

is true. Because if $x < 0$, then $-5 \cdot x$ is indeed > 0; while if $x \geq 0$, then the statement is unchallenged.

EXAMPLE 12.4.11 The statement

$$\text{If } \{x > 0 \text{ and } x^2 < 0\}, \text{ then } x \geq 10.$$

is true since the hypothesis "$x > 0$ and $x^2 < 0$" is never true.

EXAMPLE 12.4.12 The statement

$$\text{If } x > 0, \text{ then } \{x^2 < 0 \text{ or } 2x < 0\}.$$

is false since the conclusion "$x^2 < 0$ *or* $2x < 0$" is false whenever the hypothesis $x > 0$ is true.

EXAMPLE 12.4.13 Let us construct a truth table for the statement

$$(\mathbf{A} \vee (\sim \mathbf{B})) \Rightarrow ((\sim \mathbf{A}) \wedge \mathbf{B})$$

A	B	$\sim$ A	$\sim$ B	$(\mathbf{A} \vee (\sim \mathbf{B}))$	$((\sim \mathbf{A}) \wedge \mathbf{B})$
T	T	F	F	T	F
T	F	F	T	T	F
F	T	T	F	F	T
F	F	T	T	T	F

$(\mathbf{A} \vee (\sim \mathbf{B})) \Rightarrow ((\sim \mathbf{A}) \wedge \mathbf{B})$
F
F
T
F

Notice that the statement $(\mathbf{A} \vee (\sim \mathbf{B})) \Rightarrow ((\sim \mathbf{A}) \wedge \mathbf{B})$ has the same truth table as $\sim (\mathbf{B} \Rightarrow \mathbf{A})$. Can you comment on the logical equivalence of these two statements?

12.4.2 Logical Syllogisms

Perhaps the most commonly used logical syllogism is the following. Suppose that we know the truth of **A** and of $\mathbf{A} \Rightarrow \mathbf{B}$. We wish to conclude **B**. Examine the truth table for $\mathbf{A} \Rightarrow \mathbf{B}$. The only line in which both **A** is true and $\mathbf{A} \Rightarrow \mathbf{B}$ is true is the line in which **B** is true. That justifies our reasoning. In logic texts, the syllogism we are discussing is known as *modus ponendo ponens* (often shortened to *modus ponens*).

The statement

A if and only if B

12.4. Elements of Formal Logic

is a brief way of saying

If A then B *and* **If B then A**

We abbreviate **A if and only if B** as **A ⇔ B** or as **A iff B**. Here is a truth table for **A ⇔ B**.

A	B	A ⇒ B	B ⇒ A	A ⇔ B
T	T	T	T	T
T	F	F	T	F
F	T	T	F	F
F	F	T	T	T

The connective ⇔ is called the *material biconditional* and the sentence **A ⇔ B** is called a *biconditional statement*.

Notice that we can say that **A ⇔ B** is true only when both **A ⇒ B** and **B ⇒ A** are true. An examination of the truth table reveals that **A ⇔ B** is true precisely when **A** and **B** are either both true or both false. Thus **A ⇔ B** means precisely that **A** and **B** are logically equivalent. One is true when and *only when* the other is true.

EXAMPLE 12.4.14 The statement

$$x > 0 \Leftrightarrow 2x > 0$$

is true. Because if $x > 0$, then $2x > 0$; and if $2x > 0$, then $x > 0$.

EXAMPLE 12.4.15 The statement

$$x > 0 \Leftrightarrow x^2 > 0$$

is false. Because $x > 0 \Rightarrow x^2 > 0$ is certainly true, while $x^2 > 0 \Rightarrow x > 0$ is false ($(-3)^2 > 0$ but $-3 \not> 0$).

EXAMPLE 12.4.16 The statement

$$\{\sim (A \vee B)\} \Leftrightarrow \{(\sim A) \wedge (\sim B)\} \qquad (*)$$

is true because the truth table for $\sim(A \vee B)$ and that for $(\sim A) \wedge (\sim B)$ are the same. Thus they are logically equivalent: one statement is true precisely when the other is. Another way to see the truth of $(*)$ is to examine the truth table:

A	B	$\sim(A \vee B)$	$(\sim A) \wedge (\sim B)$	$\sim(A \vee B) \Leftrightarrow \{(\sim A) \wedge (\sim B)\}$
T	T	F	F	T
T	F	F	F	T
F	T	F	F	T
F	F	T	T	T

Some classical treatments augment the concept of *modus ponens* with the idea of *modus tollendo tollens* (more often shortened to *modus tollens*). It is in fact logically equivalent to *modus ponens*. *Modus tollens* says

If $\sim B$ and $A \Rightarrow B$, then $\sim A$.

Modus tollens actualizes the fact that $\sim B \Rightarrow \sim A$ is logically equivalent to $A \Rightarrow B$. The first of these implications is of course the *contrapositive* of the second. You can check the logical equivalence of an implication with its contrapositive by using a truth table.

12.4.3 Quantifiers

The mathematical statements that we will encounter in practice will use the *connectives* "and," "or," "not," "if–then," and "iff." They will also use *quantifiers*. The two basic quantifiers are "for all," the *universal quantifier*, and "there exists," the *existential quantifier*.

EXAMPLE 12.4.17 Consider the statement

All automobiles have wheels.

This statement makes an assertion about *all* automobiles. It is true, because every automobile does have wheels.

Compare this statement with the next one:

There exists a woman who is blonde.

This statement is of a different nature. It does not claim that all women have blonde hair—merely that there exists *at least one* woman who does. Since that is true, the statement is true.

EXAMPLE 12.4.18 Consider the statement

All positive real numbers are integers.

12.4. Elements of Formal Logic

This sentence asserts that something is true for all positive real numbers.[5] It is indeed true for *some* positive numbers, such as 1 and 2 and 193. However it is false for at least one positive number (such as $1/10$ or π), so the entire statement is false.

Here is a more extreme example:

The square of any real number is positive.

This assertion is *almost* true—the only exception is the real number 0: $0^2 = 0$ is not positive. But it only takes one exception to falsify a "for all" statement. So the assertion is false. This example illustrates the principle that the negation of a "for all" statement is a "there exists" statement.

EXAMPLE 12.4.19 Look at the statement

There exists a real number which is greater than 5.

In fact there are lots of numbers which are greater than 5; some examples are $7, 42, 2\pi$, and $97/3$. Other numbers, such as 1, 2, and $\pi/6$, are not greater than 5. Since there is *at least one* number satisfying the assertion, the assertion is true.

EXAMPLE 12.4.20 Consider the statement

There is a man who is at least ten feet tall.

This statement is false. To *verify* that it is false, we must demonstrate that *there does not exist a man who is at least ten feet tall.* In other words, we must show that all men are shorter than ten feet.

The negation of a "there exists" statement is a "for all" statement.

A somewhat different example is the sentence

There exists a real number which satisfies the equation
$$x^3 - 2x^2 + 3x - 6 = 0.$$

There is in fact only one real number that satisfies the equation, and that is $x = 2$. Yet this information is sufficient to show that the statement is true.

---

[5]Here an integer is a whole number such as -3 or 2 or 9. A real number is any number with a decimal expansion. This latter includes both rational and irrational numbers.

We often use the symbol $\forall$ to denote "for all" and the symbol $\exists$ to denote "there exists." The assertion

$$\forall x, \ x + 1 < x$$

claims that for every x, the number $x + 1$ is less than x. If we take our universe to be the standard real number system, then this statement is false. The assertion

$$\exists x, \ x^2 = x$$

claims that there is a number whose square equals itself. If we take our universe to be the real numbers, then the assertion is satisfied by $x = 0$ and by $x = 1$. Therefore the assertion is true.

In all the examples of quantifiers that we have discussed thus far, we were careful to specify our *universe*. That is, "There is a woman such that ..." or "All positive real numbers are ..." or "All automobiles have ...". The quantified statement makes no sense unless we specify the universe of objects from which we are making our specification. In the discussion that follows, we will always interpret quantified statements in terms of a universe. Sometimes the universe will be explicitly specified, while other times it will be understood from the context.

Quite often we will encounter $\forall$ and $\exists$ used together. The following examples are typical:

EXAMPLE 12.4.21 The statement

$$\forall x \ \exists y, \ y > x$$

claims that for any number x, there is a number y which is greater than it. In the realm of the real numbers this is true. In fact $y = x + 1$ will always do the trick.

The statement

$$\exists x \ \forall y, \ y > x$$

has quite a different meaning from the first one. It claims that there is an x which is less than *every* y. This is absurd. For instance, x is *not* less than $y = x - 1$.

EXAMPLE 12.4.22 The statement

$$\forall x \ \forall y, \ x^2 + y^2 \geq 0$$

12.4. Elements of Formal Logic

is true in the realm of the real numbers: it claims that the sum of two squares is always greater than or equal to zero. This statement happens to be *false* in the realm of the complex numbers. This example underscores the fact that when we interpret a logical statement, it will always be important to understand the context, or universe, in which we are working.

The statement
$$\exists x \, \exists y, \, x + 2y = 7$$
is true in the realm of the real numbers: it claims that there exist x and y such that $x + 2y = 7$. Certainly the numbers $x = 3, y = 2$ will do the job (although there are many other choices that work as well).

We conclude by noting that $\forall$ and $\exists$ are closely related. The statements
$$\forall x, \, \boldsymbol{B}(\boldsymbol{x}) \qquad \text{and} \qquad \sim \exists x, \, \sim \boldsymbol{B}(\boldsymbol{x})$$
are logically equivalent. The first asserts that the statement $B(x)$ is true for all values of x. The second asserts that there exists no value of x for which $B(x)$ fails, which is the same thing.

Likewise, the statements
$$\exists x, \, \boldsymbol{B}(\boldsymbol{x}) \qquad \text{and} \qquad \sim \forall x, \, \sim \boldsymbol{B}(\boldsymbol{x})$$
are logically equivalent. The first asserts that there is some x for which $B(x)$ is true. The second claims that it is not the case that $B(x)$ fails for every x, which is the same thing. The books [Hal 62] and [GH 98] explore the algebraic structures inspired by these quantifiers.

A "for all" statement is something like the conjunction of a very large number of simpler statements. For example, the statement

For every nonzero integer n, $n^2 > 0$

is actually an efficient way of saying that $1^2 > 0$ and $(-1)^2 > 0$ and $2^2 > 0$, etc. It is not feasible to apply truth tables to "for all" statements, and we usually do not do so.

A "there exists" statement is something like the disjunction of a very large number of statements (the word "disjunction" in the present context means an "or" statement). For example, the statement

There exists an integer n such that $P(n) = 2n^2 - 5n + 2 = 0$

is actually an efficient way of saying that "$P(1) = 0$ or $P(-1) = 0$ or $P(2) = 0$, etc.". It is not feasible to apply truth tables to "there exist" statements, and we usually do not do so.

It is common to say that *first-order logic* consists of the connectives $\wedge$, $\vee$, $\sim$, $\Rightarrow$, $\Leftrightarrow$, the equality symbol $=$, and the quantifiers $\forall$ and $\exists$, together with an infinite set of variables $x, y, z, \ldots, x', y', z', \ldots$ and, finally, parentheses $(\,,\,)$ to keep things readable. The word "first" here is used to distinguish the discussion from second-order and higher-order logics. In first-order logic the quantifiers $\forall$ and $\exists$ always range over elements of the domain M of discourse. Second-order logic, by contrast, allows us to quantify over subsets of M and functions F mapping $M \times M$ into M. Third-order logic treats sets of function and more abstract constructs. The distinction among these different orders is often moot.

12.5 Truth and Provability

An elementary statement like

$$\mathbf{A} = \text{``George is tall."}$$

has a truth value assigned to it. It is either true or false. From the point of view of mathematics, there is nothing to prove about this statement. Likewise for the statement

$$\mathbf{B} = \text{``Barbara is wise."}$$

On the other hand, the statement

$$\mathbf{A} \vee \mathbf{B}$$

is subject to mathematical analysis. Namely, it is true if at least one of $\mathbf{A}$ or $\mathbf{B}$ is true. Otherwise it is false.

Any statement which is true regardless of the truth value of its individual components is called a *tautology*. An example of a tautology is

$$\mathbf{B} \Rightarrow (\mathbf{A} \vee \sim \mathbf{A})$$

This statement is true all the time—regardless of the truth values of $\mathbf{A}$ and $\mathbf{B}$. Set up a truth table to satisfy yourself that this is the case.

12.5. Truth and Provability

Another example of a tautology is

$$(A \Rightarrow B) \Leftrightarrow (\sim A \vee B)$$

Again, you may verify that this is a tautology by setting up a truth table.

So we have two ways to think about whether a certain statement is valid all the time: **(i)** to substitute in all possible truth values, and **(ii)** to prove the statement from elementary principles. We have seen two examples of **(i)**. Now let us think about method **(ii)**.

In order to provide an example of a provable statement, we must isolate in advance the syllogisms we assume to be true, and the rules of logic that are allowed. In a formal treatment of logic, such as [Sto 61], we would begin on page 1 of the book with these syllogisms and rules of logic and then proceed rigidly, step-by-step. At each stage we would have to check which rule or syllogism is being applied. The present treatment is *not* a formal treatment of logic. It is in fact a more intuitive approach. For the remainder of the section, however, we lapse into the formal mode so that we may learn more carefully to distinguish truth from provability.

It is natural to ask whether every tautology is provable (that every provable statement is a tautology is an elementary corollary of our logical structure, or see [Sto 61, p. 152]). That this is so is Frege's theorem. We shall revisit Frege in Sect. 12.6. This statement is summarized by saying that *elementary sentential logic is complete*.

In fact Gödel proved in 1930 that *the first-order predicate calculus is consistent and complete*. The first-order predicate calculus is essentially the logic we have described in this chapter: it includes elementary connectives, the quantifiers $\forall$ and $\exists$, and statements P with one or more (but finitely many) variables $x_1, \ldots, x_k$. Thus, according to Gödel, any provable statement in this logic is true—consistency—and, more profoundly, any true statement is provable—completeness. Gödel went on to construct a model for any consistent system of axioms.

Gödel's more spectacular contribution to modern thought is that in any logical system that is complex enough to contain arithmetic, there are sensible statements that cannot be proved either true or false. For example, Peano's arithmetic (the standard arithmetic of the positive whole numbers) contains statements that cannot be proved either true or false. A rigorous discussion of this celebrated "incompleteness theorem" is beyond the scope of this book. Suffice it to say that Gödel's proof consists of making an (infinite) list of all provable statements, enumerating them with a system of "Gödel numbers," and then constructing a new statement that differs from each of these. Since the constructed statement could not be on the list, it also cannot be provable. For further discussion of Gödel's ideas, see [Dav 64], [NN 58], [Smu 92b].

Theoretical computer scientists have shown considerable interest in the incompleteness theorem. For a computer language—even an expert system—can be thought of as a logical theory. Gödel's theorem says, in effect, that given any sufficiently complex language there will be statements that can be formulated in the language that cannot be established through a sequence of logical steps from first principles. For more on this matter, see [Kar 76], [Sho 67], [Sto 61].

12.6 Cracks in the Edifice

Even as early as 1906 there were indications that mathematics and formal reasoning were perhaps not as rock solid as Hilbert and others had hoped. In 1902, Gottlob Frege (1848–1925) was enjoying the fact that the second volume of his definitive work *The Basic Laws of Arithmetic* [Fre 64b] was at the printer when he received a polite and modest letter from Bertrand Russell offering the following paradox.[6] We begin by noting that in mathematics, a *set* is a collection of objects. We may discuss the set of all positive rational numbers, or the set of solutions of the equations $x^2 + x + 1 = 0$ or the set of algebraic equations of degree 2. Now we have

---

[6]This paradox was quite a shock to Frege. After considerable correspondence with Russell, he modified one of his axioms and added an Appendix explaining how the modification addresses Russell's concerns. Unfortunately this modification nullified several of the results in Frege's already-published Volume 1. Frege's second volume *did* ultimately appear (see [Fre 64b]). Frege was somewhat disheartened by his experience, and his research productivity definitely went into a decline. His planned third volume never appeared.

Lesniewski proved, after Frege's death, that the resulting axiom system that appears in print is inconsistent. Frege is nonetheless remembered as one of the most important figures in the foundations of mathematics. He was one of the first to formalize the rules by which mathematicians operate, and in that sense he was a true pioneer. Many scholars hold that his earlier work, *Begriffsschrift und andere Aufsätze* [Fre 64a], is the most important single work ever written in logic. It lays the very foundations of modern logic. A more recent 1995 paper of G. Boolos [Boo 95] makes considerable strides in rescuing much of Frege's original program that is represented in the two-volume work [Fre 64b].

12.6. Cracks in the Edifice

> Let S be the collection of all sets that are not elements of themselves. Can S be an element of S?

Why is this a paradox?[7]

Here is the problem. If $S \in S$ (the symbol "$\in$" means "is an element of") then, by the way that we defined S, S is *not* an element of S. And if S is *not* an element of S, then, by the way that we defined S, it follows that S *is* an element of S. Thus we have a contradiction no matter what.

Of course we are invoking Archimedes's law of the excluded middle. It *must* be the case that either $S \in S$ or $S \notin S$ (the symbol "$\notin$" means "is not an element of"), but in fact either situation leads to a contradiction. And that is Russell's Paradox. Frege had to rethink his book, and make notable revisions, in order to address the issues raised by Russell's paradox.[8]

Now that we have had about a century to think about Russell's paradox, we realize that what it teaches us is that we cannot allow sets that are *too large*. The set S described in Russell's paradox is unallowably large. In a rigorous development of set theory, there are very specific rules for which sets we are allowed to consider and which not. In particular, modern set theory does not allow us to consider a set that is an element of itself. We cannot indulge in the details here.

It turns out that Russell's paradox is only the tip of the iceberg. Nobody anticipated what Kurt Gödel (1906–1978) would teach us 30 years later. In informal language, what Gödel showed us is that—in any sufficiently complex logical system (i.e., at least as complex as arithmetic)—there will be a

---

[7]The thoughtful reader may well wonder whether it is actually possible for a set to be an element of itself. This sounds like a form of mental contortionism that is implausible at best. But consider the set S described by

> The collection of all sets that can be described in fewer than fifty words.

Notice that S is a set, and S can be described in fewer than fifty words. Aha! So S is certainly an element of itself.

[8]A popular version of Russell's Paradox goes like this. A barber in a certain town agrees to shave every man in the town who does not shave himself. He will not touch any man who ever deigns to shave himself. Who shaves the barber? If the barber shaves himself, then he himself is someone whom the barber has agreed not to shave. If instead the barber does not shave himself, then he himself is someone whom the barber must shave. A contradiction either way.

true statement that we cannot prove within that system.[9] This is Gödel's incompleteness theorem. It came as an unanticipated bombshell, and has completely altered the way that we think of our subject. It should be stressed however that the statement that Gödel found is not *completely unprovable*. If one transcends the specified logical system, and works instead in a larger and more powerful system, then one *can* create a proof. Further discussion occurs later in the chapter.

Just as quantum mechanics taught us that nature is not completely deterministic—we cannot know everything about a given physical system, even if we have a complete list of all the initial conditions—so it is the case that Gödel has taught us that there will always be statements in mathematics that are "undecidable."

It is safe to say that Gödel's ideas shook the foundations of mathematics. They have profound implications for the logical basis for our subject, and for what we can expect from it. There are also serious consequences for theoretical computer science—just because the computer scientist wants to know where any given programming language (which is certainly a logical system) will lead and what it can produce.

12.7 The Gödel Incompleteness Theorem

Gödel's incompleteness theorem has been hailed as a milestone in human thought. Gödel himself has been characterized as the greatest logician since Aristotle. His seminal paper appeared in 1931, and it was actually several years before the ideas were fully understood and accepted. The great mathematician John von Neumann (1903–1957) played a decisive role in promoting the significance of Gödel's work. As we shall see below, the reasoning behind Gödel's theorem is rather simple—as mathematical theorems go. What was revolutionary in the work is that Gödel was the first person to consider, and to find ways to mathematically manipulate, an entire formal language at once. He did not simply write sentences and connect them with chains of logic. He in fact considered the collection of *all* sentences, and performed mathematical operations on that collection. The results were original, profound, and far-reaching.

---

[9]Gödel even went so far as to show that the consistency of arithmetic itself is not provable. Certainly arithmetic is the most fundamental and widely accepted part of basic mathematics. The notion that we can never be certain of what we are doing is distinctly unsettling.

12.7. The Gödel Incompleteness Theorem

A statement of Gödel's incompleteness theorem is this:

Theorem *Let T be any logical system that is sufficiently sophisticated that it contains the whole numbers and the operations of arithmetic. Then there is a statement P in the system T that is true but which cannot be proved.*

This statement bears some thought.

Our standard means for confirming that a statement is true is by supplying a proof. A proof is a sequence of logical steps, beginning with the axioms and definitions of the logical system at hand, which leads to the desired conclusion. It was part of David Hilbert's program, formulated at the dawn of the twentieth century, that all statements in mathematics should be provable from a fixed, finite set of axioms. In the technical language of formal logic, Hilbert believed that mathematics was *complete*. Here completeness means simply that every true statement can be proved. The important book *Principia Mathematica* by Whitehead and Russell (see [WR 62]) was meant to be a first (albeit lengthy and complex) step in carrying out Hilbert's program. Gödel's theorem shows that the Hilbert program is doomed to failure. Not only is mathematics incomplete, but essentially *any* logical system is incomplete.

It has been argued that Gödel's theorem has even more far-reaching consequences. One train of thought—see [Pen 89]—is that a computer can never be as smart as a human being because a computer is limited by its programming language, which is essentially a fixed set of axioms and definitions (in other words, a logical system). So, according to Gödel, there are limitations to what the computer can do. There will be true statements that it cannot verify. A human being, on the other hand, can make logical leaps and discover unexpected truths.

The verification of Gödel's theorem is an elucidating exercise in formal reasoning. Let the symbol G denote the sentence

Our logical system T will never prove that this sentence is true.[10]

So G says that G is not provable. We cannot help but note how Gödel's idea here is clearly inspired by Russell's Paradox!! Let us consider whether the statement G is provable.

---

[10] It is in the rigorous logical formulation of this sentence that the ideas of arithmetic come in—see [Coh 66].

If the sentence G is provable, then it is certainly true. But the sentence says that G is not provable. That is a contradiction. We conclude therefore that G must not be provable.

Is the sentence G true? If it is false, then it is not the case that G is not provable. So in fact G is provable. Therefore it must be true. But that contradicts the assumption that G is false. We conclude that G must be true.

In summary, the sentence G is a true statement that cannot be proved in our logical system. For a more complete and formal treatment of Gödel's incompleteness theorem, see [Coh 66].

A Look Back

Interest in the foundations of mathematics goes back to the ancient Greeks. Very early mathematics contained no enunciations of theorems and no proofs. It was mostly calculation and measurement—in many cases inspired by questions of land management. The Babylonian tablet Plimpton 322 has long been considered an important source of information about ancient mathematics. It dates to 1800 BCE, so precedes the ancient Greeks. Some say that it contains the first proof, but this is not the case. It contains sequences of numbers that look like Pythagorean triples (i.e., triples of numbers like 3, 4, 5 or 5, 12, 13 that satisfy the Pythagorean identity $x^2 + y^2 = z^2$), suggesting that the Babylonians were at least aware of the Pythagorean theorem. But no proofs.

It is arguable that the very first proof was due to Thales around 600 BCE. This really changed the nature of mathematics forever. Aristotle discusses different methods of proof in his *Physics*. But the truly remarkable and revolutionary creation of ancient times was Euclid's *Elements*. Although remembered primarily for its treatment of geometry, this 13-volume work recorded most of the known mathematics of the time—including number theory and 3-dimensional analysis. But the main feature of Euclid's work was the *form* in which it recorded mathematics. Namely, it enunciated definitions and axioms, and then it derived theorems from those axioms. This is the way that we still do mathematics today. It is what makes mathematics reliable and robust, and is the reason that it travels so well.

Not much happened with mathematics during the dark ages. This was partly because those 1000 years were a time of ignorance and religious superstition. But it was also because people were wedded to the use of Roman numerals, and it is just about impossible to do any intelligent mathematics using Roman numerals.

Fortunately, the Renaissance saw a great blossoming of mathematics. Fermat, Descartes, Newton, Leibniz, and many others contributed profoundly to our understanding of planar geometry and calculus. But these new, somewhat modern, mathematicians paid scant attention to logic and the theory of rigor. They were too busy discovering new mathematical truths.

In fact mathematics was being done in different countries with rather different styles, and there was scant communication among different groups of mathematicians in far-flung places. Terms were defined differently in different places, and the standard of what constitutes a proof varied considerably. In the late nineteenth century some of the mathematical leaders concluded that mathematics needed a dose of rigor and structure. Some rules needed to be laid down. Some paradigms needed to be set.

In Germany, David Hilbert set the standard. He wanted all of mathematics to be derived rigorously from a compact and clearly enunciated set of axioms. Later in France, the secret society that operated under the name of Nicolas Bourbaki wrote a set of textbooks designed to set all the basic areas of mathematics in stone for future reference.

Unfortunately Gödel's incompleteness theorem, discussed in the present chapter, knocked this program for a rigorous foundation of mathematics on its head. We are still recovering from the impact of Gödel's ideas.

REFERENCES AND FURTHER READING

[**Boo 95**]   Boolos, G.: Frege's theorem and the Peano postulates. Bulletin of Symbolic Logic **1**, 317–326 (1995)

[**Coh 66**]   Cohen, P.J.: Set Theory and the Continuum Hypothesis. W. A. Benjamin, Inc., New York (1966)

[**Dav 64**]   Davis, A.S.: Gödel's Theorem. University of Oklahoma. preprints (1964)

[**Fre 64a**]   Frege, G.: Begriffsschrift und Andere Aufsätze. G. Olms, Hildesheim (1964)

[Fre 64b]   Frege, G.: Grundgesetze der Arithmetik (two volumes in one). G. Olms, Hildesheim (1964)

[GH 98]   Givant, S., Halmos, P.R.: Logic as Algebra. Mathematical Association of America, Washington, D.C. (1998)

[Gol 05]   Goldstein, R.: Incompleteness: the proof and paradox of Kurt Gödel. W. W. Norton & Company, New York (2005)

[Hal 62]   Halmos, P.R.: Algebraic Logic. Chelsea, New York (1962)

[Kar 76]   Karp, R.M.: The probabilistic analysis of some combinatorial search problems. In: Algorithms and Complexity. Proceedings Symposium Carnegie-Mellon University, Pittsburgh, pp. 1–19. Academic, New York (1976)

[NN 58]   Nagel, E., Newman, J.R.: Gödel's Proof. New York University Press, New York (1958)

[Pen 89]   Penrose, R.: The Emperor's New Mind. Oxford University Press, Oxford (1989)

[Sho 67]   Shoenfield, J.: Mathematical Logic. Addison-Wesley, Reading (1967)

[Smu 87]   Smullyan, R.: Forever Undecided: A Puzzle Guide to Gödel. Alfred Knopf, New York (1987)

[Smu 92b]   Smullyan, R.: Gödel's Incompleteness Theorems. Oxford University Press, New York (1992)

[Smu 92a]   Smullyan, R.: The Lady or the Tiger? and Other Logic Puzzles. Times Books, New York (1992)

[Sto 61]   Stoll, R.R.: Sets, Logic, and Axiomatic Theories. W.H. Freeman and Company, San Francisco (1961)

[WR 62]   Whitehead, A.N., Russell, B.: Principia Mathematica. Cambridge University Press, Cambridge (1962)

Chapter 13
Fermat's Last Theorem

13.1 Introduction

Sometime during 1637 Pierre de Fermat (1601–1665) wrote in the margin of a book the assertion that he had found a truly marvelous proof of the following result:

> **Fermat's Last Theorem.** *It is impossible to find nonzero integers a, b, and c that satisfy the equation*
> $$a^n + b^n = c^n,$$
> *when $n \geq 3$ is also an integer.*

Unfortunately, the book margin was too small for Fermat to write his proof there, nor did he record it anywhere else. In consequence, nobody knows for sure what Fermat had in mind at that moment in 1637 when he penned his marginal note.

For the impossibility of $a^4 + b^4 = c^4$, we do know Fermat's proof, because that one he did write down. As far as Fermat's claim of having a proof that works for all $n \geq 3$, there is no hard evidence, but the consensus seems to be that, in addition to the case $n = 4$, Fermat probably also had a valid proof for the case $n = 3$, but that he mistakenly thought his proofs for $n = 3$ and $n = 4$ generalized to all larger n.

It was not until three hundred fifty eight years later that a proof of Fermat's last theorem was established. That proof embodied the work of many mathematicians, but the capstone was the pair of papers "Modular elliptic curves and Fermat's last theorem" and "Ring-theoretic properties of certain Hecke algebras." The first paper was written by Andrew John Wiles (b. 1953) and the second paper was coauthored by Wiles and Richard Lawrence Taylor (b. 1962). Both papers appeared in the *Annals of Mathematics* in 1995,[1] after thorough review by other experts.

Diophantus of Alexandria

Diophantus of Alexandria was one of the great figures of Greek mathematics, but we know almost nothing about his life. A credible guess is that he was born around 200 BCE. It is thought that he lived to age 84, but that figure is also a guess. This latter guess is based on the assumption that the following puzzle problem reflects the actual facts about Diophantus's life:

> ...his boyhood lasted 1/6th of his life; he married after 1/7th more; his beard grew after 1/12th more, and his son was born 5 years later; the son lived to half his father's age, and the father died 4 years after the son.

The solution is that Diophantus married at the age of 26 and had a son who died at the age of 42, four years before Diophantus himself died aged 84.

Despite our limited knowledge of Diophantus's personal history, we do know about his work: At least six of the 13 books of his major work *Arithmetica* have survived. One notable edition of *Arithmetica* was prepared by Claude Gaspar Bachet de Méziriac (1581–1638). This edition was bilingual; it included a Latin translation along with the original Greek. The jurist and amateur mathematician Pierre de Fermat owned and studied a copy of Bachet's edition of *Arithmetica*.

Arithmetica is a collection of 130 problems. Problem 8 of book II asks how to split a square number into two squares. Figure 13.1a represents the square number 25 using a 5 by 5 square. In Fig. 13.1b, the 25 small squares are split into two colors. The group of 9 small blue squares obviously represents the square number 9, and, in Fig. 13.1c, the small yellow squares are rearranged into a 4 by 4 square. Thus the square number 25 has been split into the square numbers 9 and 16.

It was in his copy of Diophantus's *Arithmetica*, near Problem II.8—the square splitting problem—that Fermat wrote his marginal note claiming that he had found the truly marvelous proof of the result now known as Fermat's last theorem.

---

[1] This is the most prestigious journal in mathematics.

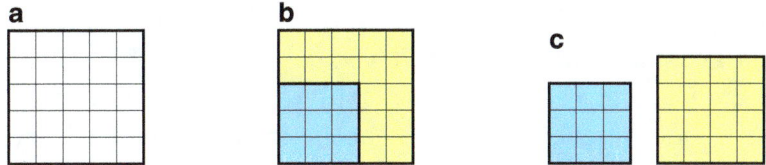

Figure 13.1 Splitting the square number 25 into two squares.

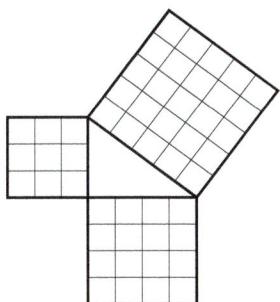

Figure 13.2 The Pythagorean triple $(3, 4, 5)$.

13.2 Splitting Square Numbers

Any triple (a, b, c) of positive integers that satisfy $a^2 + b^2 = c^2$ represents a splitting of a square number into two other square numbers. Such a triple is called a *Pythagorean triple*, because the Pythagorean theorem tells us that a and b are the side lengths of a right triangle whose hypotenuse has length c. For the Pythagorean triple $(3, 4, 5)$ this right triangle is illustrated in Fig. 13.2.

The reader may note that $(5, 12, 13)$ is a Pythagorean triple because $5^2 + 12^2 = 13^2$ and $(28, 45, 53)$ is a Pythagorean triple because $28^2 + 45^2 = 53^2$. There are infinitely many Pythagorean triples.

Neither Fig. 13.1 nor Fig. 13.2 gives us a method for creating Pythagorean triples, but there is a geometric construction that will do so.[2] Let the square number b^2, greater than 1, be given. If possible, write b^2 as the product of two unequal numbers u and v such that both are even or both are odd, that is,

$$b^2 = u \cdot v \text{ with } u < v \text{ and } v - u \text{ divisible by } 2.$$

---

[2]Euclid gave this construction in Book X, Prop. 28, Lemma 1 of the *Elements*.

For example, we can write $4^2 = 16$ as $16 = 2 \cdot 8$, so $b = 4$, $u = 2$, and $v = 8$. In Fig. 13.3a, we illustrated the rectangle with width $v = 8$ and height $u = 2$. Then in Fig. 13.3b, we split that rectangle into three pieces: a square with side length $u = 2$ on the far right and two rectangles with width $(v - u)/2 = 3$. Next in Fig. 13.3c, we formed a square with side length $(v - u)/2 = 3$ that sits above the blue rectangle. Finally in Fig. 13.3d, we illustrated the fact that the magenta rectangle is the perfect shape to go above the yellow square with side length $u = 2$ and thus complete a square with side length $(v + u)/2 = 5$.

The construction illustrated in Fig. 13.3 works whenever we write a square number b^2 in the form $b^2 = u \cdot v$ with $u < v$ and with $v - u$ divisible by 2. The result is the construction of two numbers a and c so that (a, b, c) is a Pythagorean triple. In algebraic notation, we have $a = (v - u)/2$ and $c = (v + u)/2$. In fact, the construction in Fig. 13.3 works in reverse, so every Pythagorean triple can be found using this construction due to Euclid.

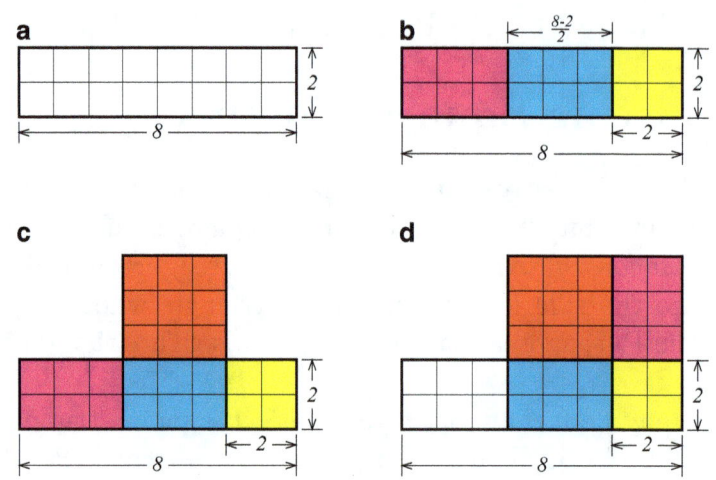

Figure 13.3 Splitting the square number 25 into two squares.

13.3 Pythagorean Triples: Another Construction

If we look at the first few Pythagorean triples generated by Euclid's construction that was described in the last section and illustrated in Fig. 13.3, we obtain the following list:

13.3. Pythagorean Triples: Another Construction

- $(3, 4, 5)$,
- $(5, 12, 13)$,
- $(6, 8, 10)$,
- $(7, 24, 25)$,
- $(8, 15, 17)$,
- $(9, 12, 15)$,
- $(9, 40, 41)$,
- $(10, 24, 26)$.

Not all these triples are equally interesting. In fact, because $(6, 8, 9)$ and $(9, 12, 15)$ are multiples of $(3, 4, 5)$ and because $(10, 24, 26)$ is a multiple of $(5, 12, 13)$, it makes sense to group the triples in the following way:

- $(3, 4, 5)$, $(6, 8, 10)$, $(9, 12, 15)$,
- $(5, 12, 13)$, $(10, 24, 26)$,
- $(7, 24, 25)$,
- $(8, 15, 17)$,
- $(9, 40, 41)$,

In each row, the leftmost Pythagorean triple (a, b, c) is made up of three numbers a, b, and c that have no common factor (other than 1). The triples to the right in a row are simply multiples of the leftmost triple.

Pythagorean triples in which the three numbers have no common factor are called *primitive Pythagorean triples*. One way to avoid common factors and find only primitive Pythagorean triples is to divide both sides of the equation $a^2 + b^2 = c^2$ by c^2, write x for a/c, y for b/c, and then express the rational numbers x and y in the natural way as fractions without common factors in the numerator and denominator. Thus the primitive Pythagorean triples correspond to the rational points on the unit circle (i.e., the circle with radius one and center the origin)

$$x^2 + y^2 = 1, \qquad (13.1)$$

where by a *rational point* we mean a point in the (x, y)-plane for which both the x-coordinate and the y-coordinate are rational numbers.

It may seem at first that finding the rational solutions of (13.1) will be extremely difficult, but that is not the case. There are two key facts that make finding the rational solutions of (13.1) easy:

(1) We already know a rational solution to (13.1). Indeed, we know many, but we only need one.

(2) If a quadratic equation $Ax^2 + Bx + C = 0$ with rational coefficients has one rational solution, then it has a second rational solution. (Those two rational solutions might be equal.)

The first fact is evident from Euclid's construction, and additionally, the points $(\pm 1, 0)$, $(0, \pm 1)$ satisfy (13.1). The second fact is true for the following reason: If r is one rational solution of the equation $Ax^2 + Bx + C = 0$, then $x - r$ divides the polynomial $Ax^2 + Bx + C$ with remainder 0. Moreover, the quotient polynomial can be found by an algorithm (namely, synthetic division) that involves only addition, subtraction, multiplication, and division of the coefficients. In fact, if we write the quotient in the form $cx + d$ (the letter c used here does *not* refer to its earlier use as the third number in a Pythagorean triple), then we have $c = A$, $d = B + Ar$, and the second solution is the rational number $-d/c = -r - B/A$.

Now let us see how the two facts above allow us to find all the rational points on the unit circle. We start from the existence of the rational point $(-1, 0)$ on the unit circle:

• On the one hand, if we are given a rational point (x, y), other than $(-1, 0)$, on the unit circle, then the line through $(-1, 0)$ and (x, y) will have the rational slope $m = (y + 1)/x$. So every rational point on the unit circle, other than $(-1, 0)$, gives us a rational slope.

• On the other hand, let a rational slope m be given. Consider the line with slope m that passes through the point $(-1, 0)$. That line consists of the points (x, y) for which x and y satisfy the equation

$$y = m(x + 1). \qquad (13.2)$$

The intersection of the line given by (13.2) with the unit circle $x^2 + y^2 = 1$ can be found by eliminating y (i.e., replace y in the equation $x^2 + y^2 = 1$ by what (13.2) says it is equal to). We obtain the equation

$$x^2 + m^2(x + 1)^2 = 1. \qquad (13.3)$$

Because the line and the unit circle intersect at $(-1, 0)$, the Eq. (13.3) has the rational solution $x = -1$. Since $x = -1$ is one rational solution of (13.3) and since (13.3) has rational coefficients, there is a second rational solution of (13.3). That second rational solution of (13.3) corresponds to a second rational point of intersection of the line $y = m(x + 1)$ and the unit circle $x^2 + y^2 = 1$ (because the Eq. (13.3) has distinct solutions, that second point is distinct from $(-1, 0)$). So each rational slope gives us a rational point, other than $(-1, 0)$, on the unit circle.

The two procedures just described are inverse to each other, so we have established a correspondence between the rational numbers (the slopes of the lines) and the rational points, other than $(-1, 0)$, on the unit circle.

Figure 13.4 illustrates the preceding situation with the line through $(-1, 0)$ having slope $3/7$. The second point of intersection of the line and the unit circle is $(20/29, 21/29)$. The rational point $(20/29, 21/29)$ on the unit circle corresponds to the primitive Pythagorean triple $(20, 21, 29)$.

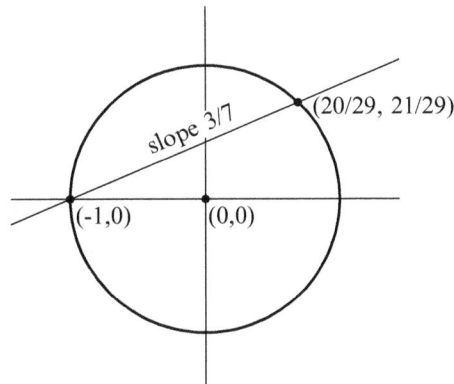

Figure 13.4 Constructing rational points on the unit circle.

In this section we have seen how the theory of a particular algebraic curve (a curve defined by a polynomial equation), namely the unit circle, could be used to find the integers that satisfy the equation $a^2 + b^2 = c^2$. We will return to this theme later.

13.4 Kummer's Criterion

Fermat's last theorem asserts the impossibility of solving $a^n + b^n = c^n$ in integers with $abc \neq 0$, where $n \geq 3$ is also an integer. Even in the case $n = 3$, showing this impossibility is nontrivial. Geometrically, finding positive integers a, b, and c so that $a^3 + b^3 = c^3$ holds would represent a splitting of a cube of side length c into cubes with side lengths a and b. One would presume that mathematicians of Euclid's time must have considered this problem, but the earliest known mention of this problem is of an attempted, but defective, proof of impossibility attributed to the tenth century Persian astronomer Abu Mahmud Hamid ibn al-Khidr Al-Khujandi. Fermat may well have known a proof for the case $n = 3$, but the first documented correct proof is due to Euler.

In the case $n = 4$, Fermat had a valid proof and we know what his proof was. Based on the fact that Fermat proved the result for $n = 4$, we can see that to prove Fermat's last theorem it suffices to consider

$$a^n + b^n = c^n \text{ when } n \text{ is an odd prime.}$$

For example, the case

$$a^{15} + b^{15} = c^{15}$$

can be rewritten as

$$(a^5)^3 + (b^5)^3 = (c^5)^3,$$

so reduces to the case of exponent 3.

The early results on Fermat's last theorem proceeded incrementally until the work of Ernst Eduard Kummer (1810–1893) appeared in the mid-1800s. Kummer's method involves deep work in algebra beyond the scope of this book, but one result we can describe here tells us the following:

> **Kummer's Criterion.** *If p is an odd prime and if p does not divide the numerator in any of the Bernoulli numbers[3] B_2, $B_4, \ldots, B_{p-3}$, then Fermat's last theorem holds for p.*

The Bernoulli numbers can be generated recursively,[4] so Kummer's criterion can be checked for a given odd prime. The first twelve nonzero Bernoulli number numerators are

$$\begin{aligned}
N_0 &= 1, & N_1 &= -1 & N_2 &= 1 \\
N_4 &= -1, & N_6 &= 1 & N_8 &= -1 \\
N_{10} &= 5, & N_{12} &= -691, & N_{14} &= 7, \\
N_{16} &= -3617, & N_{18} &= 43867, & N_{20} &= -174,611.
\end{aligned}$$

---

[3] The Bernoulli numbers, named after Jakob Bernoulli (1654–1705), are special rational numbers that have an important role in number theory.

[4] The Bernoulli numbers satisfy $B_0 = 1$ and $\sum_{j=0}^{k} \binom{k+1}{j} B_j = 0$.

13.5. Fields

Since 5, 7, 691, 3617, and 43867 are prime and since 23 does not divide 174,611, Kummer's criterion tells us that Fermat's last theorem is true for

$$p = 5,\ 7,\ 11,\ 13,\ 17,\ 19,\ 23\,.$$

Based on Kummer's criterion and other more refined criteria, by the year 1992 the smallest p for which Fermat's last theorem could fail was pushed beyond 4,000,000.

Each more refined extension of Kummer's criterion pushed the frontier of Fermat's last theorem higher, but no end to the process could be foreseen. To take care of all the odd primes at once required a different approach. The argument that finally proved Fermat's last theorem was based on understanding the deep connection between two apparently unrelated mathematical objects: elliptic curves and modular forms. Before we can discuss either of these structures, we need to introduce the notion of a field.

13.5 Fields

In elementary school you put in a lot of effort in learning how to add, subtract, multiply, and divide. Along the way you probably learned quite a few tricks that were given important sounding names such as the *commutative law*, the *associative law*, and the *distributive law*. A *field* is any mathematical structure in which you can add, subtract, multiply, and divide and in which the commutative, associative, and distributive laws hold for those operations.

The fundamental example of a field is the set of rational numbers, denoted $\mathbb{Q}$, consisting of all the fractions m/n, where m is any integer—positive, negative, or zero—and n is a positive integer. It was the arithmetic of $\mathbb{Q}$ that you mastered when you learned to add, subtract, multiply, and divide with fractions. As a visual aid, we can plot the rational numbers on a "number line" by picking points to represent 0 and 1 and then positioning the rest of the rational numbers proportionally as illustrated in Fig. 13.5.

After the rational numbers have been plotted on the number line, there will be many, many points on the line that don't correspond to rational numbers: for instance, whenever the positive integer n is not a perfect square, then $\sqrt{n}$ is not a rational number. We know that $\sqrt{n}$ can be constructed as a length using constructions in plane geometry, so it should be a legitimate number. By letting every point on the line represent a number, we can fill in the missing numbers, including $\sqrt{n}$ and many many more. The resulting

Figure 13.5 The number line.

larger collection of numbers is the set of real numbers, which is denoted by $\mathbb{R}$. Since every irrational number ("irrational" means "not rational") can be approximated as closely as we wish by rational numbers, we can use those approximations to extend the arithmetic of the rational numbers to the real numbers $\mathbb{R}$. The real numbers with the arithmetic so defined also form a field.

13.5.1 Complex Numbers

Having identified the points in a line with the elements of the field of real numbers, we might ask if we can similarly identify the points in a plane with the elements of a field. Indeed that can be done and the result is the field of complex numbers, denoted $\mathbb{C}$. To construct the complex numbers, fix a plane and a line in the plane. The points on the chosen line will represent the real numbers within the complex numbers. Each point in the plane represents a complex number z. The distance from the point to 0 is called the *modulus of z* and it is written $|z|$. The angle that the ray from 0 through z makes with the positive real axis is called the *argument of z* and it is written $\arg z$. It is important that the argument be an oriented angle: The argument is positive when measured counterclockwise from the positive real axis and it is negative if measured in the clockwise direction.

To add z_1 and z_2 in the complex plane, add them as vectors, that is, form the parallelogram whose sides are the segments from 0 to z_1 and from 0 to z_2, then the diagonal of the parallelogram is the segment from 0 to $z_1 + z_2$ (Fig. 13.6).

To multiply z_1 and z_2 in the complex plane, construct $z_1 z_2$ so that $|z_1 z_2| = |z_1| |z_2|$ and $\arg(z_1 z_2) = (\arg z_1 + \arg z_2)$ (Fig. 13.7). With the operations so defined the complex numbers form a field.

There is a unique complex number with modulus 1 and argument $\pi/2$; we denote that particular number by i. Since $i \cdot i$ has modulus $1 \cdot 1 = 1$ and $\arg(i \cdot i) = \pi$, we see that $i \cdot i = -1$ (Fig. 13.8).

Every complex number can be written in the form $a + ib$ and the complex number $a + ib$ corresponds to the point in the plane with horizontal coordinate a and vertical coordinate b. This identification is known as the

13.5. Fields

Argand diagram.[5] The line determined by 0 and i is called the *imaginary axis*.

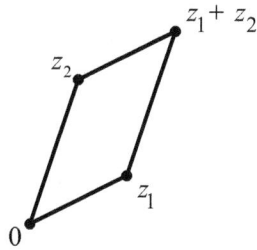

Figure 13.6 Addition of complex numbers.

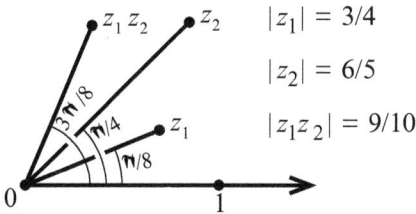

Figure 13.7 Multiplication of complex numbers.

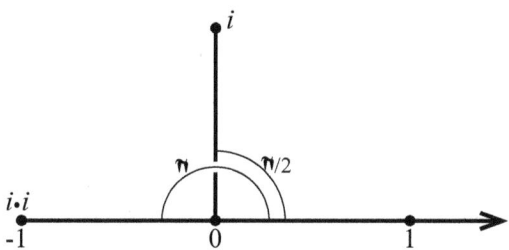

Figure 13.8 The square root of minus one.

It is natural to wonder if the geometric construction of larger and larger fields can be continued beyond the plane. While a well-behaved invertible operation of addition can always be defined using the parallelogram construction, defining a well-behaved (i.e., commutative, associative, and distributive) multiplication, with inverses, is not possible in higher dimensions. In four

---
[5] Jean-Robert Argand (1768–1822).

dimensions, a noncommutative multiplication, with inverses, can be defined and the resulting structure is called the *quaternions*. In eight dimensions, a nonassociative, noncommutative multiplication, with inverses, can be defined and the resulting structure is called the *octonions* or *Cayley numbers*. But no invertible multiplication can be constructed in the other dimensions.

13.5.2 Polynomial Equations

A field is the natural setting in which to study polynomial equations such as

$$a_n x^n + a_{n-1} x^{n-1} + \cdots + a_1 x + a_0 = 0 \quad (n \text{ a positive integer}). \tag{13.4}$$

Here we assume $\mathbb{F}$ is a field and that each of the coefficients a_j is an element of $\mathbb{F}$. The expression on the left-hand side of (13.4) is called a *polynomial*. The positive integer n is called the *degree* of the polynomial and of the equation. We also assume that $a_0 \neq 0$, because if $a_0 = 0$, then we can divide the equation by x and reduce the degree by 1.

The degree of the polynomial equation is often decisive for finding the solutions. Given a degree 1 equation $a_1 x + a_0 = 0$ (called a *linear equation*), we immediately find that $x = -a_0/a_1$. Degree 2 equations are called *quadratic equations*. Quadratic equations were considered in antiquity, and every student in elementary algebra must commit to memory the quadratic formula giving the solutions of $ax^2 + bx + c = 0$ as $x = (-b \pm \sqrt{b^2 - 4ac})/(2a)$. One might reasonably expect that the quadratic formula can be traced back at least a millennium and might well have been inscribed on some ancient Babylonian clay tablet, but though the solution method had long been known, the familiar form of the solution is due to Henry Heaton of Atlantic, Iowa, and made its first appearance in print in the *American Mathematical Monthly* in 1896.

Equations of degree 3 and 4 can be solved by formulas found in the 1500s. Equations of degree 5 and higher cannot be solved by general formulas involving only root extraction and field operations. This last fact is credited to Niels Henrik Abel (1802–1829).

We will not need to know much about solving an equation such as (13.4). We only need the following facts:

(1) An element r of $\mathbb{F}$ for which the Eq. (13.4) is true when x is replaced by r is called a *solution* of the equation and a *root* of the polynomial

$$a_n x^n + a_{n-1} x^{n-1} + \cdots + a_1 x + a_0.$$

13.5. Fields

(2) If $r_1, r_2, \ldots, r_k \in \mathbb{F}$ are roots of the polynomial, then the polynomial

$$(x - r_1)(x - r_2) \cdots (x - r_k)$$

divides the polynomial $a_n x^n + a_{n-1} x^{n-1} + \cdots + a_1 x + a_0$, meaning that there is a third polynomial, called the *quotient*,

$$b_{n-k} x^{n-k} + b_{n-k-1} x^{n-k-1} + \cdots + b_1 x + b_0$$

with its coefficients b_j in $\mathbb{F}$ such that

$$(x - r_1) \cdots (x - r_k)(b_{n-k} x^{n-k} + b_{n-k-1} x^{n-k-1} + \cdots + b_1 x + b_0)$$
$$= a_n x^n + a_{n-1} x^{n-1} + \cdots + a_1 x + a_0.$$

The fact that the degree of the quotient is $n - k$ is very important. For instance, it tells us that there can be at most n roots of the original polynomial.

(3) If we have $n-1$ roots $r_1, r_2, \ldots, r_{n-1} \in \mathbb{F}$, then we can apply the previous fact to obtain

$$(x - r_1)(x - r_2) \cdots (x - r_{n-1})(b_1 x + b_0) = a_n x^n + a_{n-1} x^{n-1} + \cdots + a_1 x + a_0,$$

so we see that the nth and last root r_n equals $-b_0/b_1$ and thus is also in $\mathbb{F}$.

A polynomial equation of degree n should have n roots, but to make that happen we need to count the roots with multiplicity and we may need to enlarge the field to find the "missing" roots.

A simple example helps to nail down the point. The polynomial equation $x^3 - 3x + 2$ has the multiple root 1 (the root occurs twice) and the simple root 2. Put in other words,

$$x^3 - 3x + 2 = (x - 1)(x - 1)(x + 2).$$

13.5.3 Finite Fields

The rational, real, and complex numbers are not the only examples of fields. There are also fields with only finitely many elements. You are already familiar with clock arithmetic, which is the underlying idea for finite fields. If it is 9 o'clock, then you know that in 7 hours it will be 4 o'clock, not 16 o'clock. In clock arithmetic, every time you pass 12, you start over—effectively, 12 is acting like 0. Finite fields are formed by using clock arithmetic, but with 12 replaced by a prime number p. We call the result *arithmetic modulo p*.

To perform an addition or multiplication modulo p, first do the operation in ordinary arithmetic, then divide that intermediate result by p; the remainder after division by p is the final result of the operation modulo p. Table 13.1 shows the addition and multiplication tables for arithmetic modulo 5.

Addition table
Row m, column n contains $m + n$

+	0	1	2	3	4
0	0	1	2	3	4
1	1	2	3	4	0
2	2	3	4	0	1
3	3	4	0	1	2
4	4	0	1	2	3

Multiplication table
Row m, column n contains $m \times n$

×	0	1	2	3	4
0	0	0	0	0	0
1	0	1	2	3	4
2	0	2	4	1	3
3	0	3	1	4	2
4	0	4	3	2	1

Table 13.1 Arithmetic modulo 5.

Every row of the addition table for arithmetic modulo 5 contains a 0. The element labeling that column is the additive inverse of the element labeling that row and vice versa. So for instance $-3 = 2$ modulo 5 and $-2 = 3$ modulo 5. We can use this information to set up the subtraction table for arithmetic modulo 5 shown in Table 13.2. The 0-row in the subtraction table shows the additive inverses of the numbers 0 through 4 modulo 5. For example, the entry 4 in the 0-row and 1-column tells us that 4 is the additive inverse of 1, that is, $-1 = 4$ modulo 5.

Having noted that, in arithmetic modulo 5, -1 equals 4, we should also take note of the fact that, modulo 5, -1 has the square roots 2 and 3. No imaginary number is needed to solve $x^2 + 1 = 0$ when we use arithmetic modulo 5.

Except for the 0-row, every row of the multiplication table for arithmetic modulo 5 contains a 1. The element labeling that column is the multiplicative inverse of the element labeling that row and vice versa. So for instance $1/3 = 2$ modulo 5 and $1/2 = 3$ modulo 5. If we did not use a prime number, there would be rows other than the 0-row without 1s; that is why we always endeavor to use a prime number. We can use this information to set up the division table for arithmetic modulo 5 shown in Table 13.2. Of course,

13.5. Fields

there is no 0-column in the division table, because we cannot divide by 0. The 1-row in the division table shows the multiplicative inverses of the numbers 1 through 4 modulo 5.

Subtraction table
Row m, column n contains $m - n$

−	0	1	2	3	4
0	0	4	3	2	1
1	1	0	4	3	2
2	2	1	0	4	3
3	3	2	1	0	4
4	4	3	2	1	0

Division table
Row m, column n contains m/n

÷	1	2	3	4
1	1	3	2	4
2	2	1	4	3
3	3	4	1	2
4	4	2	3	1

Table 13.2 Subtraction and division modulo 5.

Because the arithmetic operations modulo p are based on addition and multiplication in the integers and because addition and multiplication in the integers are commutative, associative, and distributive, it is not hard to show that arithmetic modulo p also satisfies the commutative, associative, and distributive laws. As a result, we see that the set of numbers $0, 1, 2, \ldots, p-1$ together with arithmetic modulo p form a field; we will denote this field by $\mathbb{F}_p$.

13.5.4 Polynomials Over Finite Fields

If a polynomial has integer coefficients and if p is a prime number, then we can treat the integer coefficients of the polynomial as elements of $\mathbb{F}_p$ as follows: If the coefficients are integers between 0 and $p-1$, then they already are elements of $\mathbb{F}_p$; otherwise, divide each coefficient by p and keep the remainder after division. This process is called *reducing the polynomial modulo* p.

As an example, we will consider the cubic polynomial

$$x^3 + 2x^2 + 3x + 4. \tag{13.5}$$

If we look for roots of (13.5) in the real numbers, we find that it has one simple real root, because the derivative of the polynomial, i.e., $3x^2 + 4x + 3$, is positive for real x. Over the complex numbers, (13.5) has three distinct roots: the real root already noted and two complex roots that are complex conjugates of each other, thus unequal. On the other hand, when we reduce the polynomial (13.5) modulo p, we will look for solutions of

$$x^3 + 2x^2 + 3x + 4 = 0 \tag{13.6}$$

in the field $\mathbb{F}_p$.

General theory tells us that provided we count solutions according to their multiplicities, (13.6) will have either no solution in $\mathbb{F}_p$, 1 solution in $\mathbb{F}_p$, or 3 solutions in $\mathbb{F}_p$.[6] With the aid of a computer it is not hard to generate Table 13.3 showing the solutions of (13.6) in $\mathbb{F}_p$ for the primes from 2 up to 53. Notice that for each field $\mathbb{F}_p$ listed in Table 13.3, there either is 1 solution, or 3 solutions, or no solution at all, just as theory tells us there should be.

Field	Solutions		
$\mathbb{F}_2$	0	1	1
$\mathbb{F}_3$	None		
$\mathbb{F}_5$	1	1	1
$\mathbb{F}_7$	4		
$\mathbb{F}_{11}$	None		
$\mathbb{F}_{13}$	2		
$\mathbb{F}_{17}$	None		
$\mathbb{F}_{19}$	None		

Field	Solutions		
$\mathbb{F}_{23}$	11		
$\mathbb{F}_{29}$	3		
$\mathbb{F}_{31}$	6		
$\mathbb{F}_{37}$	14		
$\mathbb{F}_{41}$	None		
$\mathbb{F}_{43}$	14	32	38
$\mathbb{F}_{47}$	17		
$\mathbb{F}_{53}$	24		

Table 13.3 Solutions of $x^3 + 2x^2 + 3x + 4 = 0$ in $\mathbb{F}_p$.

Table 13.3 shows us that reducing the equation $x^3 + 2x^2 + 3x + 4 = 0$ modulo p generates an enormous amount of information. Such a wealth of information can be organized and studied by using the data to define the coefficients of a function. In this case, we might consider the function

$$3z^2 + 3z^4 + z^7 + z^{13} + z^{23} + z^{29} + z^{31} + z^{37} + 3z^{43} + z^{47} + z^{53} + \cdots,$$

for which the coefficient of z^p is the number of solutions in $\mathbb{F}_p$, i.e., 0, 1, or 3. This sort of idea will play a significant role later.

---

[6] The reason there cannot be exactly two solutions is the following: If there are two solutions r_1 and r_2 to (13.6) in $\mathbb{F}_p$, then the polynomial on the left-hand side of (13.6) must be divisible by $(x - r_1)(x - r_2)$. The quotient is of degree one, i.e., of the form $x - r_3$, where $r_3 \in \mathbb{F}_p$. So we see that r_3 is the third solution of (13.6) in $\mathbb{F}_p$.

13.6 Algebraic Curves

A *Fermat triple* (a, b, c) is a set of three nonzero integers a, b c, without common factor, satisfying $a^n + b^n = c^n$, $n \geq 3$. The existence of a Fermat triple is equivalent to the existence of a pair of nonzero rational numbers x and y satisfying the equation

$$x^n + y^n = 1. \tag{13.7}$$

To see that this is the case, observe that given a Fermat triple (a, b, c), the rational numbers $x = a/c$ and $y = b/c$ are nonzero and satisfy (13.7), and conversely, given nonzero rational numbers x and y satisfying (13.7), we may let c be the least common denominator of x and y and then, setting $a = cx$, $b = cy$, we obtain the Fermat triple (a, b, c).

The Eq. (13.7) involves the two variables x and y. An equation in two variables involving only sums and products of constants and the two variables defines an *algebraic curve*. If the constants and variables are real, then the set of points satisfying the equation can be plotted in the Cartesian plane. If the constants are rational numbers, then we have particular interest in *rational points* on the curve, i.e., pairs of rational numbers x and y that satisfy the equation. The degree of the equation is the largest number of variables that are multiplied together. For example, if the term $x^3 y^5$ appears somewhere in an equation, then since that term has degree $3 + 5 = 8$, the degree of the equation will be 8 or larger.

As with equations in one variable, the degree is decisive. For the typical algebraic curve[7] with rational coefficients the following hold:

(1) A degree 1 rational algebraic curve contains infinitely many rational points.

(2) A degree 2 rational algebraic curve contains either no rational points or infinitely many rational points. This is a theorem of Legendre.

(3) A degree 3 rational algebraic curve contains either no rational points, finitely many rational points, or infinitely many rational points that can be generated from finitely many by the process of composition (which will be defined below). This is a version of Mordell's theorem of 1922.[8]

---

[7] Here "typical algebraic curve" technically means "smooth algebraic curve."
[8] Louis Joel Mordell (1888–1972).

(4) A rational algebraic curve of degree 4 or higher contains only finitely many rational points. This is Faltings' theorem of 1983.[9]

Faltings' theorem tells us that even if Fermat's last theorem were false, it would have only finitely many counterexamples in each degree. Perhaps more important than that finiteness result for Fermat's last theorem was the fact that Faltings used the methods of modern algebraic geometry to obtain results that impacted peoples' understanding of Fermat's last theorem.

13.7 Elliptic Curves

An *elliptic curve* is a curve defined by a cubic equation of the form

$$y^2 = x^3 + Ax^2 + Bx + C, \tag{13.8}$$

so an elliptic curve is a particular type of algebraic curve. We will always assume that A, B, and C are rational numbers, often integers. The name "elliptic" is traditional, but it is no help to one's intuition—don't look for an easy connection between an elliptic curve and one of the oval shaped figures of elementary geometry.

It is easiest to visualize the curve (13.8) when x and y are real variables, and the curve lies in the Cartesian (x, y)-plane. The appearance of the curve defined by (13.8) depends on the roots of the polynomial

$$x^3 + Ax^2 + Bx + C \tag{13.9}$$

on the right-hand side of (13.8). If the polynomial has three distinct roots, then either (1) the three roots are all real or (2) there is one real root and a pair of complex roots $a \pm bi$, where $b \neq 0$. Such curves are illustrated in Fig. 13.9. The curve defined by (13.8) is said to be *smooth* when the roots of the polynomial (13.9) are distinct, and in this case, the curve in $\mathbb{R}^2$ is in fact a smooth curve (as Fig. 13.9 leads us to believe). Even when the variables x and y are in a field other than the reals, the elliptic curve is still said to be smooth if the roots of (13.9) are distinct.

If the roots of (13.9) are not distinct, then there may be one triple root or there may be one double root and a third distinct simple root. These situations are illustrated in Fig. 13.10. The elliptic curve is said to be *singular* when the roots of (13.9) are not distinct and Fig. 13.10 shows that terminology is appropriate.

---

[9] Gerd Faltings (b. 1954).

13.7. Elliptic Curves

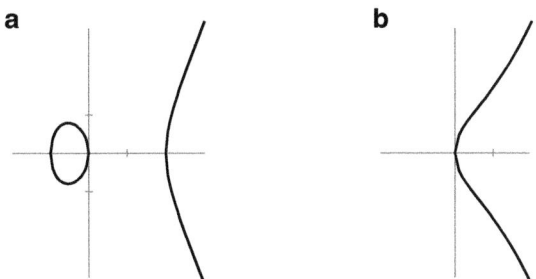

Figure 13.9 Smooth elliptic curves. (a) Three real roots. (b) One real root.

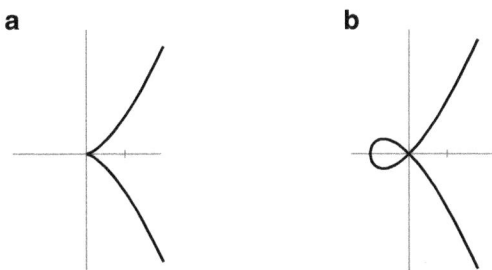

Figure 13.10 Singular elliptic curves. (a) Triple root. (b) Double root at the origin.

13.7.1 Composition

One part of the theory of elliptic curves is based on the idea of intersecting a line with the curve. Since an elliptic curve is given by an equation of degree 3, a line should intersect an elliptic curve at three points.[10] If two of the intersection points are known, then the third can be found by dividing one polynomial by another. The crucial feature of division of polynomials is that the coefficients of the quotient polynomial (and of the remainder polynomial, if there is one) are in the same field as the coefficients of the dividend and divisor. Thus, if the coordinates of the first two points are rational, then the coordinates of the third point will also be rational. This process of starting with two points on the elliptic curve and producing the third point is called *composition*. Composition is illustrated in Fig. 13.11.

---

[10]Here we are applying a theorem of Étienne Bézout (1730–1783).

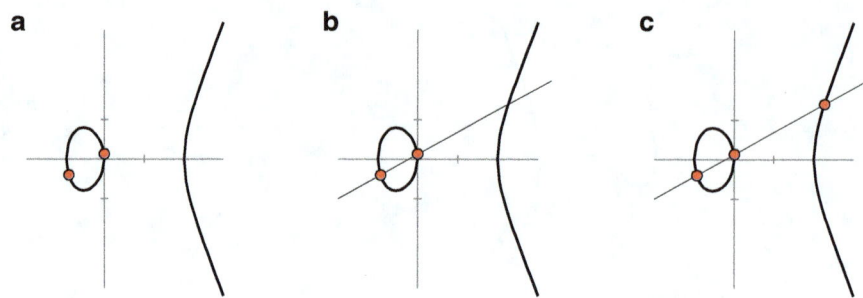

Figure 13.11 Composition of points on an elliptic curves. (a) Two points on the curve. (b) Line through two points. (c) Third intersection point.

13.7.2 Reduction

The existence of the operation of composition on elliptic curves is surprising, but composition is only the first hint of the remarkable structure that rational curves have. Mordell's theorem tells us that for a smooth rational elliptic curve, finitely many rational points on the curve can generate all the rational points via composition. Mordell's theorem is a second, major, indication of deep structure to elliptic curves.

To go further, we consider an elliptic curve with integer coefficients.

$$y^2 = x^3 + Ax^2 + Bx + C, \quad A,\, B,\, C \text{ integers.} \qquad (13.10)$$

Since (13.10) has integer coefficients, it will make sense to consider x and y to be elements of the finite field $\mathbb{F}_p$, where p is a prime. There are p^2 points of the form (x, y) with $x \in \mathbb{F}_p$ and $y \in \mathbb{F}_p$. Some of these points may satisfy the Eq. (13.10), but for $p \geq 5$ most will not.

For example, the curve

$$y^2 = x\,(x-2)\,(x+1),$$

contains only the three points $(0,0)$, $(2,0)$, and $(6,0)$ in $\mathbb{F}_7 \times \mathbb{F}_7$. If we count the number of points on the curve for other primes, we obtain the following data:

13.7. Elliptic Curves

Field	Points
$\mathbb{F}_2$	2
$\mathbb{F}_3$	4
$\mathbb{F}_5$	3
$\mathbb{F}_7$	3
$\mathbb{F}_{11}$	15
$\mathbb{F}_{13}$	15

Field	Points
$\mathbb{F}_{17}$	23
$\mathbb{F}_{19}$	15
$\mathbb{F}_{23}$	23
$\mathbb{F}_{29}$	27
$\mathbb{F}_{31}$	35
$\mathbb{F}_{37}$	39

Field	Points
$\mathbb{F}_{41}$	39
$\mathbb{F}_{43}$	47
$\mathbb{F}_{47}$	55
$\mathbb{F}_{53}$	43
$\mathbb{F}_{59}$	55
$\mathbb{F}_{61}$	55

Table 13.4 Points on the curve $y^2 = x(x-2)(x+1)$ in $\mathbb{F}_p \times \mathbb{F}_p$.

As suggested earlier, the thing to do with such a huge amount of data is to use it to define a function. In this case the function is complex-valued and is defined on the upper half-plane by the following series:[11]

$$\sum_{n=1}^{\infty} a_n \left[\cos(2n\pi(u+iv)) + i \sin(2n\pi(u+iv)) \right], \qquad (13.11)$$

where, when n is a prime and the reduced curve in $\mathbb{F}_n \times \mathbb{F}_n$ is smooth, the coefficient a_n is $n+1$ minus the number of points on the curve $\mathbb{F}_n \times \mathbb{F}_n$. When n is a composite number and when n is prime, but the reduced curve is not smooth, alternative definitions of a_n are needed, but we will not discuss those cases in any detail.

The behavior of the function in (13.11) is intimately tied to the elliptic curve itself. One might ask whether, as the curve is varied over the family of all elliptic curves, there might be some commonality of behavior of such functions. Based on examples and intuition, a conjecture concerning such a commonality did emerge. The conjecture, which originated with Yutaka Taniyama (1927–1958) in the mid 1950s, was that such functions would be modular, a notion we describe below. Taniyama inexplicably committed suicide, leaving the conjecture still vague. But Taniyama's close friend Goro Shimura (b. 1930) kept the idea going and made the conjecture more precise. Work of André Weil (1906–1998) was relevant to the conjecture—though Weil may have disbelieved the conjecture itself—so the conjecture came to be known as the Taniyama–Shimura–Weil Conjecture.

---

[11] In (13.11), u and v are real numbers and the series defines a function of the complex variable $u+iv$.

13.8 The Modular Group

The modular group is based on three geometric operations: Inversion in a circle, reflection across a line, and translation.

13.8.1 Inversion

Inversion in the unit circle maps a point at distance $r > 0$ from the center to the point on the same ray but whose distance from the center is $1/r$. Points on the unit circle stay fixed, points inside the unit circle (except for the center) are mapped to the outside of the unit circle, and points outside of the unit circle are mapped to the inside of the unit circle. Figure 13.12 shows the unit circle in red and a polygon and circle in black that will be inverted in the red circle. In Fig. 13.13, the inversion has been performed.

Inversion reverses orientations, but we would like to keep the original orientation. To accomplish that goal, inversion is followed by reflection across the vertical line through the center. The final result is shown in Fig. 13.14.

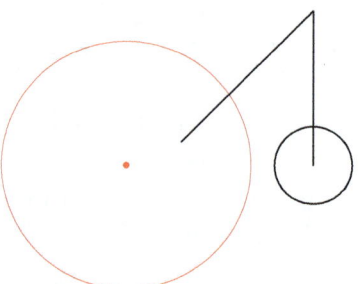

Figure 13.12 A figure to invert.

In Fig. 13.14 you will notice that the line segments have been mapped to arcs and the circle has been mapped to a circle. This is an instance of a general phenomenon: Under inversion in a circle, each line is mapped to a line or a circle and likewise each circle is mapped to a line or a circle.

We will only need inversion in one circle, namely, the unit circle centered at 0 in the complex plane, and we will restrict our attention to only the points in the upper half-plane. Figure 13.15 shows the inversion in the circle and reflection across the line restricted to the upper half-plane.

13.8. The Modular Group

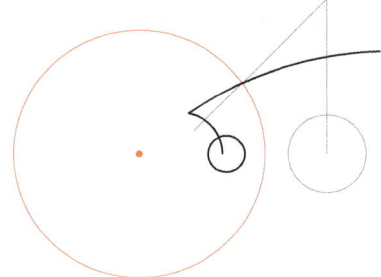

Figure 13.13 The figure inverted in the circle.

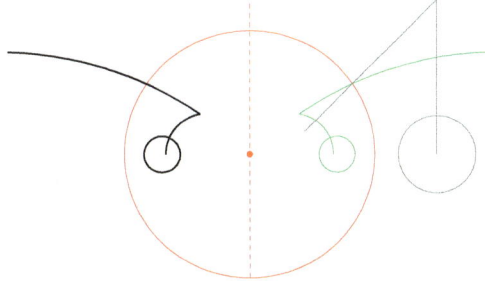

Figure 13.14 The figure inverted and reflected.

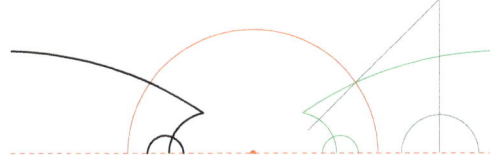

Figure 13.15 Inversion restricted to the upper half-plane.

13.8.2 Translation

A translation of the upper half-plane to the left or right maps the half-plane onto itself while preserving orientation. Using the Argand diagram, we can identify the upper half-plane with the set of complex numbers $u+iv$ for which v is positive. We will be interested in the translations that shift the upper half-plane left or right by an integer amount, that is, horizontal translations of the form

$$u + iv \longmapsto (u+m) + iv,$$

where m is an integer.

13.8.3 Modular Transformations

Any transformation of the upper half-plane that can be generated by repeated application of

- inversion in the unit circle followed by reflection across the imaginary axis
- horizontal translation by integers

is a *modular transformation*. The collection of all the modular transformations forms the *full modular group*.

13.8.4 The Congruence Subgroup

Fix a positive integer N. In the modular group there are many transformations for which the image of the upper half of the imaginary axis is a circular arc ending at a point of the form j/N, with j an integer. The collection of transformations generated by repeated application of

- transformations for which the image of the upper half of the imaginary axis is a circular arc ending at a point of the form j/N, with j an integer
- horizontal translation by integers

form the *congruence subgroup*, denoted $\Gamma_0(N)$.

13.9 The Taniyama–Shimura–Weil Conjecture

The Taniyama–Shimura–Weil Conjecture was that every rational elliptic curve is modular, where the meaning of "modular" can be made precise in various ways. For us "modular" means that when the function (13.11) is constructed from an elliptic curve, then there exists a positive integer N such that the function (13.11) behaves in a particularly nice way when the argument $z = u + iv$ is transformed by an element of the congruence subgroup $\Gamma_0(N)$ of the modular group.

13.9.1 Frey Curves

Given three integers A, B, C that satisfy

$$A + B + C = 0,$$

the elliptic curve

$$y^2 = x(x - A)(x + B) \tag{13.12}$$

is called a *Frey curve*.[12] The polynomial $x(x - A)(x + B)$ on the right-hand side of (13.12) has the roots 0, A, and $-B$. These three roots will all be different from each other if and only if none of A, B, and C equals 0 (because $A = -B$ is equivalent to $A + B = 0$ and we have assumed $A + B + C = 0$). Thus a Frey curve is a particular type of smooth elliptic curve. As an example, we take $A = 2$, $B = 1$, and $C = -3$. Figure 13.16 illustrates this Frey curve,

$$y^2 = x(x - 2)(x + 1),$$

in the usual Cartesian (x, y)-plane.

Of course, the really important Frey curve, the one that we would be most interested in seeing, is that curve associated with a Fermat triple (a, b, c):

$$\left.\begin{array}{c} y^2 = x(x - a^n)(x + b^n), \qquad a^n + b^n = c^n, \qquad n \geq 5 \text{ prime}, \\ a, \ b, \ c \text{ nonzero integers without common factor.} \end{array}\right\} \tag{13.13}$$

We now know that there are no Fermat triples, so no curve such as (13.13) can exist, but how was that proved?

Frey conjectured that the curve (13.13) could be understood well enough to show that it was not modular, and if Frey's conjecture were proved, then Fermat's Last Theorem would be a consequence of the Taniyama–Shimura–Weil Conjecture were the Taniyama–Shimura–Weil Conjecture to be proved. Jean-Pierre Serre (b. 1926) refined Frey's conjecture and Ken Ribet (b. 1947) proved Serre's refinement (called the *epsilon-conjecture*) in 1986.

---

[12] Though called Frey curves, these objects were introduced by Yves Hellegouarch about 15 years before Gerhard Frey (b. 1944) stimulated new interest in them.

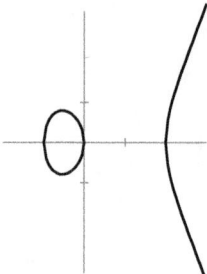

Figure 13.16 The curve $y^2 = x(x-2)(x+1)$.

13.9.2 Wiles's Proof of Fermat's Last Theorem

Andrew Wiles first learned of Fermat's last theorem when he was 10 years old, and even at that young age he decided that he would be the one to prove it. Of course, hundreds, maybe thousands, of other children must have had the same thought at some time, but unlike them Wiles was eventually successful.

The crucial motivating event for Wiles was Ken Ribet proving Serre's form of Frey's conjecture. When Wiles learned of Ribet's work, Wiles realized that if he proved the Taniyama–Shimura–Weil Conjecture, then he would also obtain Fermat's last theorem as a corollary. In fact, he would not need to prove the full Taniyama–Shimura–Weil Conjecture, but only enough of that conjecture to be applicable to the Frey curves.

At that time (1986), most mathematicians did not have the requisite background to make a realistic attempt at proving the Taniyama–Shimura–Weil Conjecture, but Wiles was well prepared for the task. When he learned of Ribet's work, Wiles dropped all else and concentrated on the Taniyama–Shimura–Weil Conjecture. He worked on it in secret for seven years, and only shared his work with one trusted colleague[13] during the last of those seven years.

When Wiles was finally convinced that he had a proof Fermat's last theorem in hand, he arranged to give a series of three talks during June 1993 at a number theory conference at the Newton Institute at Cambridge University. Cambridge was where Wiles earned his doctorate and the conference organizer was his former thesis advisor. The scene was perfectly set. The climactic words that ended Wiles series of lectures were, "And this proves Fermat's last theorem; I think I'll stop here."

---

[13] Nicholas Katz (b. 1943).

Wiles's triumphant presentation in Cambridge was followed by the anti-climatic refereeing process. Perhaps it is not obvious to the nonmathematician, but every mathematical paper of even modest length is likely to contain a few errors—such is the nature of our human endeavor. One hopes that the errors are inconsequential: typographical or at worst easily fixed. On the other hand, sometimes an error is found to be important indeed and a proof is thereby rendered invalid.

Wiles paper was 200 pages long—long enough to make at least minor errors inevitable. In fact, many errors were found and corrected. Unfortunately one error was found that stubbornly resisted correction. So stubbornly did it resist correction that in early December 1993, Wiles publicly announced the existence of a gap in his proof.

Every effort was made to fill the gap in Wiles's proof of Fermat's last theorem. In January 1994, as part of that effort, Wiles former doctoral student Richard Taylor came to Princeton to collaborate on the work. The pressure must have been intense. The International Congress of Mathematicians was to be held in Zurich early in August 1994, and Wiles was scheduled to be a keynote speaker.

It would have made a better ending to the story had Wiles been able to announce his successful completion of the proof of Fermat's last theorem during his keynote address at the 1994 International Congress, but that is not what happened. Instead he expressed his optimism that the gap would soon be filled. And so it was, using an important idea of Richard Taylor. The final good news did not happen until September 1994, a month after the International Congress. Today Wiles and Taylor are universally credited with having proved Fermat's last theorem.

A Look Back

As noted, Fermat's last problem was first formulated in 1637. Many mathematicians contributed to our understanding of the question. But one of the most remarkable ones was a woman working on her own—certainly one of the greatest and most remarkable woman mathematicians of all time.

Marie-Sophie Germain (1776–1831) was born in Paris the year of the American revolution. She was a young teenager during the time of the French revolution.

For her safety, Sophie's parents kept her at home, and away from school, during the most violent times of the French revolution. She diverted herself, and fought the tedium of being home alone by reading the books in her father's library. She was particularly fascinated by the story of Archimedes.

According to legend, when invading Roman troops marauded Syracuse, where Archimedes was living, he contented himself with his mathematics. Marcellus (268 BCE–208 BCE), the general who commanded the conquering troops, ordered that the great scientist Archimedes should be protected. Archimedes's first intimation that the city had been sacked was the shadow of a Roman soldier falling across his diagram in the soil. One version of the story is that the heathen stepped on Archimedes's diagram, causing the mild-mannered scholar to become angry and exclaim, "Don't disturb my circles!" Enraged, the soldier drew his sword and slew Archimedes.

Sophie Germain was fascinated with the notion that a person could be so absorbed in anything that he would ignore a soldier and then die as a result. She concluded that mathematics must be quite a worthwhile subject, and she was determined to study it.

With the largesse of her parents, Sophie Germain spent the time of the Reign of Terror studying differential calculus—and all without the aid of a tutor! Her father supported her financially throughout her life.

Because she was a woman, Sophie was not allowed to enroll at the prestigious École Polytechnique. But she managed to obtain copies of the notes from the courses. She studied them assiduously, and she was particularly enthralled with the teachings of Joseph-Louis Lagrange (1736–1813). She learned the name of a former student of Lagrange (Antoine-August LeBlanc), and at the end of the term submitted a paper on analysis to Lagrange using LeBlanc's name. Professor Lagrange was so impressed by the work that he demanded to meet the student who had written it. You can imagine his surprise to learn that said student was a young woman! Lagrange agreed to become Sophie's mentor.

After reading Adrien-Marie Legendre's *Essai sur la Théorie des Nombres*, she initiated a correspondence with him about some of the problems posed therein. The subsequent exchange of ideas can be considered as no less than a collaboration. In fact Legendre included some of her results in a supplement to the second edition of his *Essai*.

After Carl Friedrich Gauss published his book *Disquisitiones Arithmeticae* in 1804, Sophie Germain became fascinated by the subject of number theory. At the age of 28, she began corresponding with the great man—using the pseudonym "Monsieur LeBlanc." Between 1804 and 1809, she wrote a dozen letters to Gauss. In 1807 Gauss learned, only because of the French occupation of his hometown of Braunschweig, that his talented correspondent was a woman. Recalling the nasty fate that befell Archimedes, and fearing for Gauss's life, Sophie Germain contacted a French commander who was a friend of her family and asked for protection for Professor Gauss. When Gauss learned of Sophie's intervention on his behalf, he was lavishly grateful.

Gauss's letter to Germain, upon learning that she was a woman, read in part as follows:

> But how can I describe my astonishment and admiration on seeing my esteemed correspondent Monsieur LeBlanc metamorphosed into this celebrated person, yielding a copy so brilliant it is hard to believe? ...But when a woman, because of her sex, our customs and prejudices, encounters infinitely more obstacles than men, in familiarizing herself with their knotty problems, yet overcomes these fetters and penetrates that which is most hidden, she doubtless has the most noble courage, extraordinary talent, and superior genius.

The assertion that if p and $2p + 1$ are both prime, then any solution of $x^p + y^p = z^p$ must satisfy the conclusion that p divides at least one of x, y, z, is known as Sophie Germain's theorem.[14] Sophie's result was one of the first truly general results about Fermat's problem. Kummer later built on her work and generalized it. But she deserves credit for this fundamental insight.

REFERENCES AND FURTHER READING

[**Acz 96**] Aczek, A.D.: Fermat's Last Theorem: Unlocking the Secret of An Ancient Mathematical Problem. Four Walls Eight Windows, New York (1996)

---

[14]It must be noted that a prime number p such that $2p + 1$ is also prime is known to this day as a "Sophie Germain prime." This is an important idea in modern number theory.

[CSS 97] Cornell, G., Silverman, J.H., Stevens, G. (eds.): Modular Forms and Fermat's Last Theorem. Springer, New York (1997)

[DS 05] Diamond F., Shurman, J.: A First Course in Modular Forms. Springer, New York (2005)

[Dic 52] Dickson, L.E.: History of the Theory of Numbers. In: Diophantine Analysis, vol. II. Chelsea, New York (1952)

[Hel 02] Hellegouarch, Y.: Invitation to the Mathematics of Fermat–Wiles. Academic, San Diego (2002)

[Hus 87] Husemöller, D.: Elliptic Curves. Springer, New York (1987)

[Kle 00] Kleiner, I.: From fermat to wiles: Fermat's last theorem becomes a theorem. Elemente der Mathematik **55**, 19–37 (2000)

[Kob 93] Koblitz, N.I.: Introduction to Elliptic Curves and Modular Forms. 2nd edn. Springer, New York (1993)

[Miy 89] Miyake, T.: Modular Forms. Springer, New York (1989)

[Moz 00] Mozzochi, C.J.: The Fermat Diary. American Mathematical Society, Providence (2000)

[ST 92] Silverman, J.H., Tate, Jr. J.T.: Rational Points on Elliptic Curves. Springer, New York (1992)

[Sin 97] Singh, S: Fermat's Enigma. Walker and Company, New York (1997)

[Was 03] Washington, L.C.: Elliptic Curves: Number Theory and Cryptography. Chapman & Hall/CRC, Boca Raton (2003)

Chapter 14
Ricci Flow and the Poincaré Conjecture

14.1 Introduction

Jules Henri Poincaré (1854–1912) was one of the great geniuses of nineteenth and twentieth-century mathematics. Born into a distinguished family of academicians and public servants, he showed early talent in mathematics and science. Indeed, the entire country of France watched in awe as Poincaré developed from a child prodigy to a major leader in mathematics and physics.

Poincaré was born into a prominent, prosperous, bourgeois family. He had an idyllic youth and a fine education, though an early bout with diphtheria left him physically challenged during most of his young life. He was ambidextrous but nearsighted and uncoordinated. It was natural for him to avoid physical activity and devote himself to his studies. At that he excelled.

Poincaré graduated from the famous École Polytechnique in 1873 and then entered the equally famous École des Mines. He graduated with an Engineering degree. Although Poincaré's examiners were somewhat skeptical of his sloppily written thesis, he graduated with a first-class degree and assumed a position as Assistant Professor at the University of Caen. Poincaré rapidly worked his way up through the ranks and quickly arrived at the most important university in Paris, where he spent the rest of his career.

Poincaré spent most of his career as a professor at the Sorbonne and the École Polytechnique, and he taught a different graduate course every year. He made major contributions to algebraic topology, dynamical systems, differential equations, and differential geometry. But he was also an important theoretical physicist. At the 1904 World's Fair in St. Louis, Poincaré delivered an address in which he described *his* version of the theory of special relativity—this a full year before Einstein published his paper on the subject![1] Poincaré was subsequently nominated and promoted for the Nobel Prize in Physics, but prejudice against theoretical physics blocked his candidacy.

Poincaré is remembered today for many contributions to theoretical mathematics. But certainly the Poincaré conjecture has occupied central importance for over 100 years. The question posed by the Poincaré conjecture is whether a 3-dimensional surface that has the geometric characteristics of a sphere actually must *be* a sphere. The geometric characteristic that is of particular interest here is *homotopy*, which concerns the structural complexity of the surface. Specifically, can any rubber band configured on the surface be continuously deformed to a point? If you look at the surface in Fig. 14.1a (which is usually called a *torus*—it is the surface of a donut), you see a rubber band that *cannot* be continuously deformed to a point. By contrast, the surface in Fig. 14.1b (which is a 2-dimensional *sphere*) has the property that any rubber band on it *can* be continuously deformed to a point. The Poincaré conjecture posits that any 3-dimensional surface with the property that every curve (think rubber band) can be continuously deformed to a point actually is equivalent to the sphere, meaning that the 3-dimensional surface itself can be deformed continuously to a sphere. The conjecture was formulated by Poincaré in 1904, and while he had some ideas about proving it, they turned out to be inadequate to the task.

It is important to understand the subtlety in the Poincaré conjecture. Think of the usual 2-dimensional unit sphere in space (for instance, consider a basketball). We can think of creating it by taking two unit discs in the plane (Fig. 14.2a) and pasting them together along their boundaries (Fig. 14.2b). Similarly, one way to think about the 3-dimensional unit sphere (which would have to be embedded in 4-space, hence is impossible to envision in the usual way) as obtained by gluing two 3-dimensional balls together along their boundaries (Fig. 14.3). Now part of the fascination of the Poincaré conjecture is that our universe is 3-dimensional. That means that if we stand at any point in our universe, and look around us, it appears like a 3-dimensional space. But what is the global shape of our universe?

---

[1] See also our discussion of these matters in Chap. 7.

14.1. Introduction

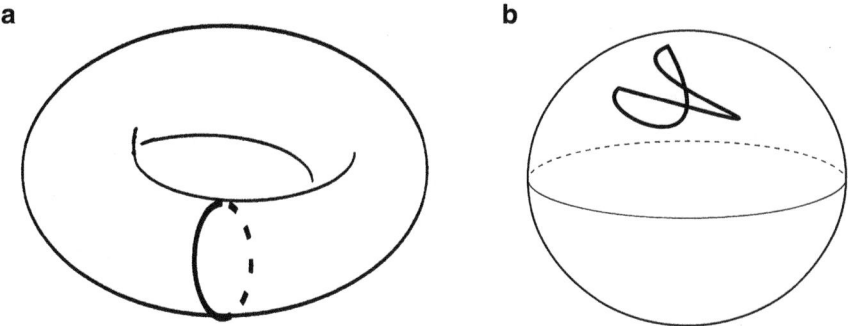

Figure 14.1 A rubber band on the torus versus a rubber band on the sphere.

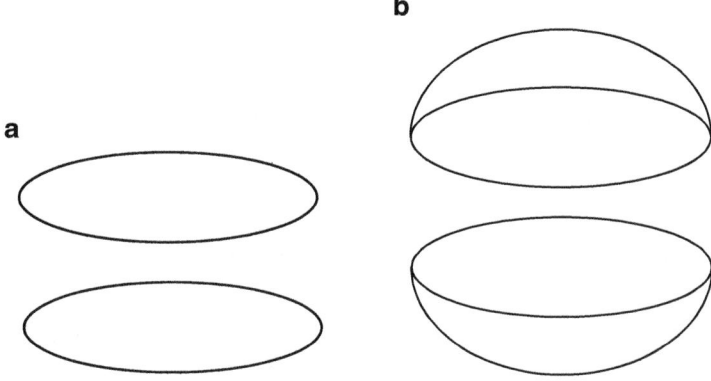

Figure 14.2 Gluing two 2-dimensional discs.

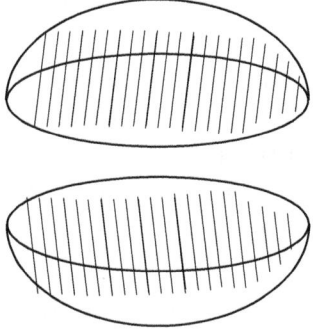

Figure 14.3 Gluing two 3-dimensional discs.

Most of us probably think of the universe as a big blob, but this is likely too naive. The universe could be in the shape of a torus (i.e., a donut), or a tube, or a box, or a sphere. Physical properties of our universe lend evidence more to some of these shapes rather than to others. The Poincaré conjecture would shed light on this question.

In general, we want to understand the structure of surfaces (or *manifolds*),[2] because almost everything that arises naturally in our world can be realized as a surface. The collection of all possible shapes for a ship's hull can be thought of as a 3-dimensional surface. The collection of all possible paths for a photon (necessarily traveling at the speed of light) can be thought of as a surface. Surfaces are essential to our analytical thinking.

Many twentieth-century mathematicians devoted their lives to proving the Poincaré conjecture. And many proofs have been announced. Until recently, they all turned out to be flawed. To be fair, these proofs contributed a number of important ideas and techniques to the flow of mathematical thought. But they didn't get the job done: they did *not* prove the Poincaré conjecture. One of the people who really opened up the modern approach to the problem is William P. Thurston(1946–2012). We shall discuss him now.

14.2 Thurston's Geometrization Program

Thurston was a graduate student at the University of California at Berkeley in the early 1970s. A student of Morris Hirsch, he rapidly distinguished himself as someone with new and original ideas about foliation theory. Here a foliation is a means of filling a surface or other geometric object with "leaves" as in Fig. 14.4. Here, by a "leaf," we mean a curve or surface.

Photo courtesy of George M. Bergman & M.F.Oberwolfach

Bill Thurston

In fact Thurston's thesis was so stunning, and so important, that he was appointed to a tenured full Professorship at Princeton University immediately on his completion of graduate school. He subsequently won the Fields Medal (the highest honor in the mathematics profession), the Waterman Prize, the Veblen Prize, and many other significant awards in mathematics. Thurston had many strong Ph.D. students, and disseminated his powerful ideas widely and effectively.

As noted, William P. Thurston was awarded the Fields Medal (in 1982 in Warsaw, Poland). He had done brilliant work on foliation theory (a branch

---

[2]Mathematicians often use the term "manifold" in place of "surface," especially when considering surfaces that are not 2-dimensional.

14.2. Thurston's Geometrization Program

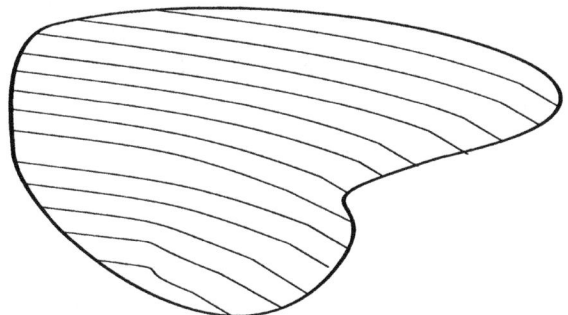

Figure 14.4 A foliation.

of topology, which is part of the modern theory of geometry)—work that was done *in his Ph.D. thesis!* These results completely revolutionized the field—solving many outstanding problems and opening up new doors. So the recognition was richly deserved. Thurston went on to do groundbreaking work in all aspects of low-dimensional topology. He became a cultural icon for mathematicians young and old.

In 1977, Thurston made a dramatic discovery (the first formal announcement was in 1982 in [Thu 82]). He had found a way to classify all 3-dimensional manifolds. Thurston's set of ideas was called the *geometrization program*. For a mathematician, a manifold is a surface with a specified dimension. This surface may or may not live in space; it could instead be an abstract construct. It was a classical fact from nineteenth-century mathematics that all 1-manifolds and 2-manifolds were completely understood and classified. The 1-manifolds are the circle and the line—see Fig. 14.5. Any 1-dimensional "surface" is, after some stretching and bending, equivalent to a circle or a line.

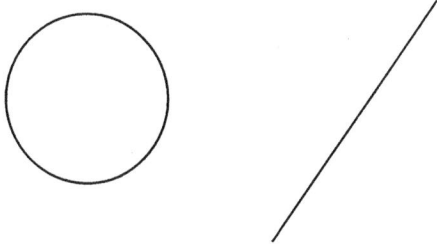

Figure 14.5 Classification of 1-manifolds.

Two-dimensional surfaces in 3-dimensional space are (thanks to nineteenth-century work of Camille Jordan (1838–1922) and August Möbius (1790–1868)) classified by the number of holes through the surface—that number is called the *genus* of the surface. A sphere has no hole through it, so it is of genus 0 (see Fig. 14.6a). The torus has 1 hole through it, so it is of genus 1 (see Fig. 14.6b). A surface of genus 2, that is, with 2 holes through it is often called the *2-torus* (see Fig. 14.6c). A surface of arbitrarily large genus is shown in Fig. 14.6d. Any 2-dimensional surface of genus $g > 0$ is, after some stretching and bending, equivalent to a sphere with g handles attached. Figure 14.7 shows how a sphere with 2 handles attached can be deformed into the genus 2 surface from Fig. 14.6c.

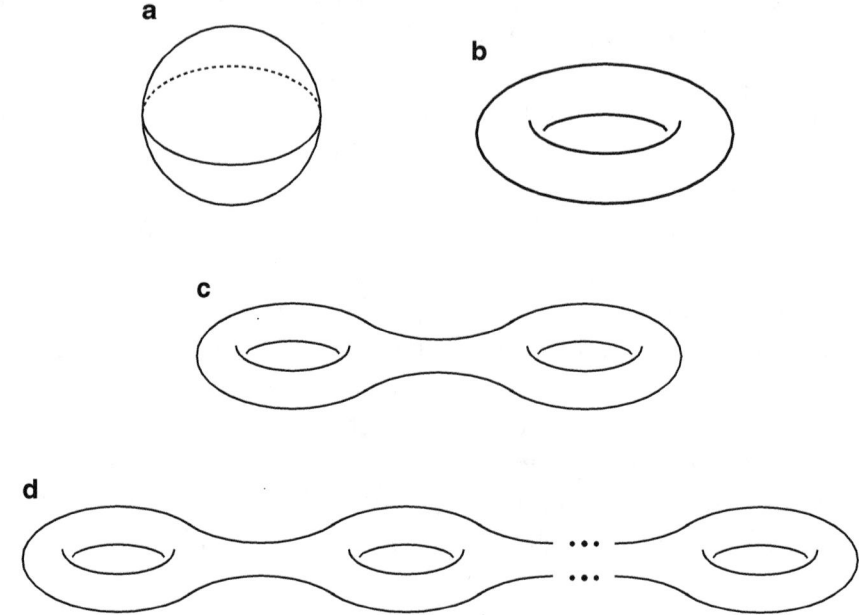

Figure 14.6 Classification of 2-manifolds.

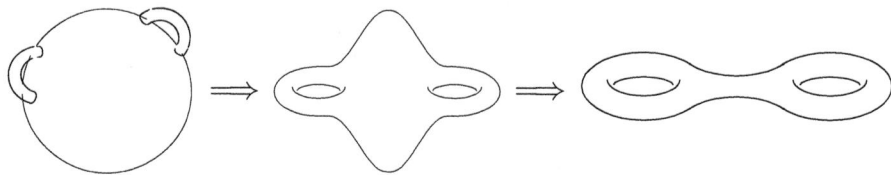

Figure 14.7 Deformation of a sphere with two handles.

14.2. Thurston's Geometrization Program

Thurston's daring idea was to break up any 3-dimensional manifold into pieces, each of which supports one of eight special classical geometries.[3] Thurston worked out in considerable detail what these eight fundamental geometries must be. His theorem would give a structure theory for 3-dimensional manifolds.

Of course 3-dimensional manifolds are much more difficult to envision, much less to classify, than are 1- and 2-dimensional manifolds. Prior to Thurston, almost nothing was known about this problem. Three-dimensional manifolds are of interest from the point of view of cosmology and general relativity—just because we live in a 3-dimensional space. For pure mathematicians, the interest of the question, and perhaps the driving force behind the question, was the celebrated Poincaré conjecture.

It seemed that every few years there would be an announcement—sometimes one that even made it into the popular press—of a new proof of the Poincaré conjecture. In 1986, Colin Rourke (b. 1943) of Warwick, a highly regarded mathematician, announced a proof; his proof survived until it was dissected in a seminar in Berkeley. In 2002, Martin J. Dunwoody (b. 1938) of the University of Southampton announced a proof. He even wrote a 5-page preprint. The effort quickly died.

Many mathematicians have tried and failed to prove the Poincaré conjecture. But, if Thurston's geometrization program were correct, then the Poincaré conjecture would follow as an easy corollary. Suffice it to say that there was considerable excitement in the air pursuant to Thurston's announcement. He had already enjoyed considerable success with his earlier work—he was arguably the greatest modern geometer—and he was rarely wrong. People were confident that a new chapter of mathematics had been opened for all of us.

But the problem was with the proof. The geometrization program is not something that one proves in a page or two. It is an enormous enterprise that reinvents an entire subject. This is what historian of science Thomas Kuhn [Kuh 70] would have called a "paradigm shift." Although Thurston was absolutely convinced that he could prove his new way of looking at low-dimensional geometry and topology (at least in certain key cases), he was having trouble communicating his proof to anyone else. There were so many new ideas, so many new constructs, so many unfamiliar artifacts, that

---

[3] In fact, mathematician Luigi Bianchi (1856–1928) first identified these eight fundamental geometries. But Thurston saw further than Bianchi, or anyone else, insofar as to how they could be used.

it was nearly impossible to write down the argument. After a period of time, Thurston produced a set of "notes" [Thu 80] (totaling 502 pages!) explaining the geometrization program.

It is important to understand what the word "notes" means to a mathematician. As this book explains and attests, mathematics has a long legacy—going back to Euclid and even earlier—of deriving ideas rigorously, using rigid rules of logic, and recording them according to a strict axiomatic method. Correctly recorded mathematics is crisp, precise, clear, and written according to a standard and time-honored model. As a counterpoint to this Platonic role model, modern mathematics is a fast-moving subject, with many new ideas surfacing every week. There are exciting new concepts and techniques springing forth at an alarming rate. Frequently a mathematician finds that he or she simply cannot take the time to write out his or her ideas in a linear, rigorous fashion. If the idea is a big and important one, it could take a couple or several years to get the recorded version just right. Frequently one just doesn't have time for that. So a commonly used alternative is "notes." What the mathematician does is give a set of lectures, or perhaps a course (at the advanced graduate level) and get some of the students to take careful notes. The professor then edits the notes (rather quickly) and then disseminates them. For example, the Princeton mathematics library has an elaborate and comprehensive set of mathematical notes—extending back 100 years or more—that is a valuable part of our literature. Many a mathematician has cut his or her teeth, and laid the basis for a strong mathematics education by studying those notes.

And this is where William Thurston found himself around 1980.[4] He had one of the most profound and exciting new ideas to come along in decades. It would take him a very long time to whip all these new ideas into shape and shoehorn them into the usual mathematical formalism. So he gave some

---

[4]The Poincaré conjecture is one of those problems that frequently finds its way into the popular press. It is one of the really big problems in mathematics, and when someone claims to have solved it that is news. In the mid-1990s, Valentin Poénaru (b. 1932) professed that he had a proof of the Poincaré conjecture. Poénaru is a professor at the University of Paris, and a man of considerable reputation. He produced a 1,200-page manuscript containing his thoughts. Unfortunately, none of the experts were able to battle their way through this weighty tract, and no definitive decision was ever reached on Poénaru's work. In 1999 he published an expository tract with a summation of his efforts. Poénaru has contributed a number of important ideas to the subject of low-dimensional topology. But the jury is still out on whether he proved the Poincaré conjecture.

14.2. Thurston's Geometrization Program

lectures and produced a set of notes. The Princeton University mathematics department, ever-supportive of its faculty, reproduced these notes and sold them to anyone—worldwide—who would send in a modest fee.

It is safe to say that these notes were a blockbuster. Many many copies were sold and distributed all over the world. There were so many beautiful new ideas here, and many a mathematician's research program was permanently changed because of the new directions that these notes charted. But the rub was that nobody believed that these notes constituted a proof of the geometrization program. Thurston found this very frustrating. He continued to travel all over the world and to give lectures on his ideas, and to produce Ph.D. students who would carry forth the torch into the mathematical firmament. But he felt that this was all he could do—given the time constraints and limitations of traditional mathematical language—to get his ideas recorded and disseminated. The catch was that the mathematical community—which in the end is *always* the arbiter of what is correct and accepted—was not ready to validate this work.

In fact, Thurston's pique with this matter was *not* transitory. In 1994, he published the paper *On proof and progress in mathematics* [Thu 94], which is a remarkable polemic about the nature of mathematical proof. It also, *sotto voce*, castigates the mathematical community for being a bit slow on the uptake in embracing his ideas. This article was met with a broad spectrum of emotions ranging from astonishment to anger to frustration.

Many years later Thurston published a more formal book [Thu 97]—in the prestigious Princeton University Press Mathematics Series—in which he began to systematically lay out the details of his geometrization program. In this tract he began at square one, and left no details to the imagination. He in effect invented a new way to look at geometry. His former Ph.D. student Silvio Levy (b. 1959) played a decisive role in developing that book. And it is a remarkable and seminal contribution to the mathematical literature. In fact, the book recently won the important AMS Book Prize. But it should be stressed that this book is the first step in a long journey. If the full saga of Thurston's proof of the geometrization program is to be revealed in this form, then a great many more volumes must be produced.[5]

---

[5]Unfortunately, William Thurston passed away on August 21, 2012, ending the hope for further volumes.

14.3 Ricci Flow

In 1982, mathematician Richard Hamilton (b. 1943) introduced a new technique into the subject of geometric analysis. Hamilton had established a strong reputation for himself as a geometer. His work on the Nash implicit function theorem (John Nash won the Nobel Prize in Economics a few years ago) was powerful and important. Hamilton's plan was to use differential equations in the study of geometry.

Before we go on, we should say a few words about what differential equations are. A *differential equation* is an equation that is to be solved for an unknown function. The term "differential" tells us that the equation under consideration involves not only the unknown function itself, but also its rates of change. (In calculus, rates of change are called *derivatives*.) Additionally, the equation may involve higher order rates of change of the unknown function (e.g., the rate of change of the rate of change of the function). Of course, the unknown function is a function and not a fixed numerical value, so it must be a function of something, that is, of one or more variables—those variables are called the *independent variables*. Certainly the independent variables themselves may appear explicitly in the equation being considered, as may other given functions of those independent variables.

If the unknown function in a differential equation depends on only one independent variable, then the equation is called an *ordinary differential equation*. If instead the unknown function depends on two or more independent variables, the equation is called a *partial differential equation*. The word "partial" is used not because the differential equation is in some way incomplete, but rather because, for a function of several variables, the rate of change of the function with respect to one variable is called a *partial derivative*. A partial derivative of a function provides only part of the information about how the function is changing.

Differential equations have practical importance because they can be used to model many physical phenomena: They describe the motions of the planets, the decay of a radioactive substance, the trajectory of a missile, the flow of a fluid, and countless other phenomena. The connection to physical problems works in both directions: The solution of a differential equation may inform us about the behavior of a physical system, or the behavior of a physical system may indicate the behavior to be expected from solutions of the corresponding differential equation.

14.3. Ricci Flow

There are certain differential equations for which the solution can be calculated and written down using a formula involving elementary functions. But more often a differential equation will be such that there is no formula for a solution. When there is no formula for a solution, one wants to prove that a solution exists—in some sense. It is at this point that the dichotomy between ordinary and partial differential equations becomes particularly relevant. While there are broadly applicable existence theorems available for ordinary differential equations, for partial differential equations the situation is much more complicated. Theorems concerning partial differential equations often require that the equation satisfy some very technical conditions.

Hamilton's new idea was called the method of *Ricci flow*; this is a means of mimicking the uniformizing behavior of heat flow in the context of geometry. The heat equation is a partial differential equation that models the physical fact that heat flows in response to a temperature difference, so temperature differences decrease as time passes. Asymptotically any temperature differences disappear (see Fig. 14.8). What Hamilton did was to write down a partial differential equation that mandates how the measure of distance between points on a manifold, what mathematicians call a *metric*, responds to the curvature of the manifold.

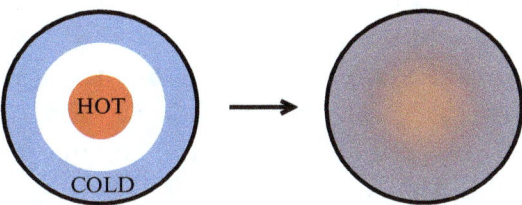

Figure 14.8 Heat flow.

The idea of changing the metric in response to the curvature is related to, and perhaps inspired by, what happens to a closed one-dimensional curve in the plane when it flows with each point moving perpendicularly to the curve in proportion to its curvature. As it evolves the curve shrinks, but it becomes more and more circular. See Fig. 14.9.

This is what is supposed to happen to a higher-dimensional manifold when it is subjected to the Ricci flow. The trouble is that things are much more complicated in higher dimensions. To start with, the curve evolution involves an ordinary differential equation, but the higher-dimensional cases involve a partial differential equation. It is difficult to prove existence of a solution to the partial differential equation. Also some nasty singularities

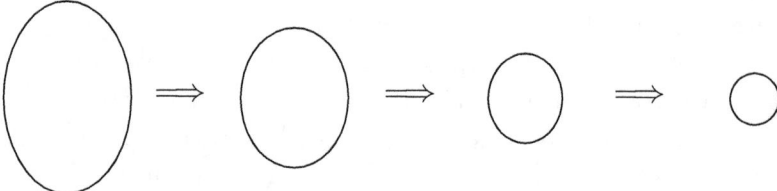

Figure 14.9 A curve evolving to a circle.

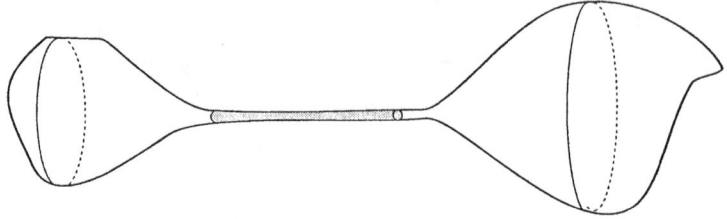

Figure 14.10 A "cigar" in the Ricci flow.

can arise during the deformation process. Particularly tricky are the so-called "cigars" that are like long thin tubes—see Fig. 14.10.

There is a sophisticated technique, developed at Princeton University by William Browder, John Milnor, Sergei Novikov, C.T.C. Wall, and others called *surgery theory* that allows one to cut out these nasty singularities—as with a knife—and then plug up the holes. The trouble was that, in the situation for the Poincaré conjecture, the singularities could run out of control. It was possible that one singularity could be removed and several others could spring up to take its place. (Grigori Perelman, about whom we will say more below, was able to show that the singularities evolved in finite time. This gave the means for controlling them and ultimately eliminating them.)

Hamilton could see that this was potentially a method for solidifying the method initiated by Thurston in his geometrization program. For one could start Ricci flows at different points of a surface and thereby generate the *geometric pieces* that Thurston predicted—refer to Fig. 14.11. The idea of each of these special pieces is that it contains an *ideal geometry*—one in which a minute creature, equipped with a tape measure for the special metric on that piece, could not tell one point from another.

Hamilton was able to harness the analytic techniques to completely carry out this idea only for surfaces of dimension 2. He made significant inroads in

14.4. Perelman's Three Papers

the case of dimension 3—enough to convince people that this was a potential program for proving the Poincaré conjecture—but some of the difficult estimates eluded him. And, as we have noted, 2-dimensional surfaces had already been classified in the mid-nineteenth century. Nothing new resulted from Hamilton's application of the Ricci flow to the study of 2-dimensional surfaces. He was able to obtain some partial results about the existence of solutions for Ricci flows in dimension 3, but only over very small time intervals. He could make some interesting assertions about Ricci curvature, but these were insufficient to resolve the Poincaré conjecture in 3 dimensions. And this is where Ricci flows had stood for more than twenty years.[6]

Figure 14.11 Ricci flow generates the geometrization decomposition.

14.4 Perelman's Three Papers

Enter Grigori (Grisha) Perelman. Born in 1966, Perelman exhibited his mathematical genius early on. But he never had any particular designs to be a mathematician. His father, an engineer, gave him problems and things to read, but Perelman entered the profession crablike.

He won the International Mathematics Olympiad with a perfect score in 1982. Soon after completing his Ph.D. at St. Petersburg State University, he landed a position at the Steklov Institute in St. Petersburg (the Steklov

---

[6]Hamilton has emerged as a hero in this story. He had an early idea that broke the Poincaré conjecture wide open. Although he could not himself bring the program to fruition, he was greatly admired both by Perelman and by Fields Medalist Shing-Tung Yau. He gave the keynote address at the International Congress of Mathematicians in August, 2006 about the status of the Poincaré conjecture.

Institutes are the most prestigious mathematics institutes in all of Russia). He accepted some fellowships at New York University and the State University of New York (SUNY) at Stony Brook in 1992, and made such an impression that he garnered several significant job offers. He turned them all down. He impressed people early on that he was an unusual person. He kept to himself, let his fingernails grow to 6 inches in length ("If they grow, why wouldn't I let them grow?" said Perelman), kept to a spartan diet of bread and cheese and milk, and maintained an eccentric profile.

In 1993 Perelman began a two-year fellowship at the University of California at Berkeley. At this time he was invited to be a speaker at the 1994 International Congress in Zurich. He was also invited to apply for jobs at Stanford, Princeton, the Institute for Advanced Study, and the University of Tel Aviv. He would have none of it. When asked to submit his curriculum vitae for the job application, Perelman said, "If they know my work, they don't need to see my CV. If they need my CV, then they don't know my work."

Grigori Perelman stopped publishing mathematics papers in 1994. After that he essentially dropped out of sight. Nobody was quite sure what he was up to.

In 1996 Perelman declined an important prize for young mathematicians from the European Mathematical Society. It is said that Perelman claimed that the judging committee was not qualified to appreciate his work. But at this point everyone knew that Grisha Perelman was a man who was going places.

After his time in the United States, Perelman returned to his position at the Steklov Institute in St. Petersburg. He was part of the Mathematical Physics group at that institution. The leading light of that group in those days was the brilliant and beautiful Olga Ladyzhenskaya (1922–2004). And it was a wellspring for studies of parabolic partial differential equations—the model for the Ricci flow that Perelman would later master and harness for the study of Poincaré's problem.

In the year 2002, on November 11 to be precise, Perelman wrote the groundbreaking paper *The entropy formula for the Ricci flow and its geometric applications* [Per(a)]. The fourth page of the introduction to this important paper contains the statement that in Sect. 13, he will provide a brief sketch of the proof of the Thurston geometrization conjecture. The "proof" was pretty sketchy indeed; and we are still unsure just what the paper [Per(a)] proves and does not prove. But the paper set the mathematical world aflame.

14.4. Perelman's Three Papers

On November 19, 2002, geometer Vitali Kapovitch of the University of California at Santa Barbara sent Perelman an e-mail that read

> Hi Grisha, Sorry to bother you but a lot of people are asking me about your preprint "The entropy formula for the Ricci..." Do I understand it correctly that while you cannot yet do all the steps in the Hamilton program you can do enough so that using some collapsing results you can prove geometrization? Vitali.

Perelman's reply the next day was, "That's correct. Grisha." Coming from a mathematician of the established ability and credentials of Perelman, this was a bombshell.

On March 10, 2003, Perelman released a second paper entitled *Ricci flow with surgery on three-manifolds* [Per(b)]. Among other things, this new paper filled in many of the details for the sketch provided in the first paper.

In April, 2003, Perelman gave a series of lectures at several high-level universities in the United States, including MIT (home of some of the most distinguished experts), SUNY at Stony Brook, New York University, and Columbia. These talks made a sufficiently strong impression that people began to take Perelman's program seriously. In July of that same year, Perelman released a third paper entitled *Finite extinction time for the solutions to the Ricci flow on certain three-manifolds* [Per(c)]. This paper provides a simplified version of a proof of a part of the geometrization program; that result is sufficient to imply the Poincaré conjecture.

One might cautiously compare Perelman's nine-month period—from November, 2002 to July, 2003—with Albert Einstein's "miracle year" (1905), in which he published four papers that completely changed the face of modern physics. In one of these papers Einstein introduced special relativity almost as an afterthought[7]—and it took several years for the idea really to catch on. But these were the four papers of Einstein's that really started everything. So too for Perelman. His three papers profess to be able to harness the Ricci flow in dimension three. He claims to be able to prove the assertions of Thurston's geometrization program, and therefore also to be able to prove the Poincaré conjecture.

It must be stressed that Perelman's three papers are full of original and exciting ideas. But they are written in a rather informal style. And Perelman has no plans to publish them. They are posted on the Internet on the preprint

---

[7] In the sense that relativity isn't even mentioned in the title!

server `arXiv`, and that is where they will remain. They will never be vetted or refereed, at least not in any formal sense. So it is up to the mathematical community to assess what these papers have to offer. A delightful discussion, on an informal level, of the Poincaré conjecture and Perelman's contributions, appears in [Strz 06].

14.5 Reaction to Perelman's Work

One rarely sees in mathematics the level of excitement and intense activity that these three papers by Perelman have generated. Conferences have been organized all over the world. Such an important institution as the Clay Mathematics Institute in Cambridge, Massachusetts funded Richard Hamilton to further his studies of the Ricci flow. It funded two mathematicians (Bruce Kleiner and John Lott) at the University of Michigan to develop Perelman's ideas and produce a detailed and verifiable proof of the geometrization program. Their resulting manuscript has proved to be a bellwether in the subject. Two other mathematicians (John Morgan of Columbia and Gang Tian of Princeton, also with support from the Clay Institute) have given time and effort to writing up all the details of Perelman's proof of the Poincaré conjecture; in fact their book [MT 07] has now appeared. Finally, Huai-Dong Cao and Xi-Ping Zhu published a 334-page paper in the *Asian Journal of Mathematics* in 2006 that purports to fill in all the details of Perelman's arguments and provide proofs both of the Poincaré conjecture and the geometrization conjecture. Cao and Zhu's paper has become somewhat controversial (see below), but the Kleiner–Lott work and the Morgan–Tian work are widely respected and generally believed.

Thus, even though Perelman himself has not played the usual academic game, even though he has not had his work formally refereed and published, others have done the job for him. We can honestly say that Perelman's work has been vetted and verified. And Perelman has been awarded both the Fields Medal and the Clay Prize (both of which he has declined).

There are a number of unusual threads to this story—some of which we introduced in the last paragraph—that require explication. The Clay Mathematics Institute has offered a $1 million dollar prize for the solution of each of seven seminal problems in modern mathematics. The Poincaré conjecture is one of them, and Perelman has in some sense proved it. He is the only living person who has a claim on one of the Clay Prizes. So that is very exciting.

14.5. Reaction to Perelman's Work

But there is the rub. For Perelman has not played by any of the rules of the Clay Math Prizes. First, he has not published any of his three papers (although others have published the details, and that may be sufficient). Perelman put his seminal work on the Poincaré conjecture and the Thurston geometrization program on `arXiv` and nowhere else. He has no intention of publishing his papers in the traditional fashion. Thus there has been no refereeing or reviewing. And the papers, sad to say, are not written in the usual tight, rigorous, take-no-prisoners fashion that is the custom in mathematics. They are rather loose and informal, with occasional great leaps of faith.

Perelman himself has—in spite of receiving a number of attractive offers from major university mathematics departments in the United States—returned to St. Petersburg so that he can care for his mother. He is more or less incommunicado, not answering most letters or e-mails. His view seems to be that he has made his contribution, he has displayed and disseminated his work, and that is all that he has to say. Perelman's position at Steklov pays less than $100 per month. But he lives an ascetic life. The jobs in the west that he declined pay quite prestigious (six-figure) salaries. Perelman claimed that he made enough money at his brief jobs in the west to support himself for the rest of his life.

But Perelman has left the rest of us holding the bag. The most recent information is that he has resigned his position at the Steklov Institute in St. Petersburg so that he can enjoy his solitude. He spends his time listening to opera and taking long walks. He no longer participates in mathematical life.

In 2008, Perelman was awarded the Fields Medal, but he declined the honor. In March, 2010 he was finally awarded the first Clay Millennium Prize (it was decided that others had published proofs of his work, and they had been vetted, and that was sufficient). He also declined that honor. Perelman will not speak to the press, and lives in seclusion.

Perelman now says that he is no longer a part of the mathematics profession. Because of a "competing" paper by Cao and Zhu, and because of vigorous campaigning by various highly-placed mathematicians, Perelman has concluded that the mathematics profession is deficient in its ethical code. He says that he has now quit mathematics to avoid making a fuss:

> As long as I was not conspicuous, I had a choice. Either to make some ugly thing or, if I didn't do this kind of thing, to be treated as a pet. Now, when I become a very conspicuous person, I cannot stay a pet and say nothing. That is why I had to quit.

All rather sad, and reminiscent of Fields Medalist Alexander Grothendieck (b. 1928). Grothendieck won the Fields Medal early on for his work (in functional analysis) on nuclear spaces. He later shifted interests, and worked intensely with his colleague and mentor Jean Dieudonné to develop the foundations of algebraic geometry. There is hardly any twentieth century mathematician who has received more honors, or more attention, than Grothendieck. He occupied a chair at the renowned Institute des Hautes Études Scientifiques for many years—indeed, he was a founding member (along with Dieudonné). But, at the age of forty, Grothendieck decided to quit mathematics. Part of his concern was over government funding, part of his concern was environmental issues, but an equally large concern was lack of ethics in the profession. Even in 1988 Grothendieck turned down the prestigious Crafoord Prize; his remarks at that time indicated that he was still disgusted with the lack of ethical standards among mathematicians. Today Grothendieck lives in the Pyrenees and is intensely introspective, to say the least. He believes that many humans are possessed by the devil.

In fact Grothendieck has written a remarkable tract called *Récoltes et Semailles*. This 2,000-page work is a personal reflection on Grothendieck's life and his mathematics and his role in the development of modern mathematics. Originally written in French, several people have dedicated themselves to translating the treatise into Japanese, English, and other languages. See

`http://www.fermentmagazine.org/rands/recoltes1.html`

for an account of the efforts of Roy Lisker in this regard. Although *Récoltes et Semailles* is a very personal document, it is also a remarkable record of the musings of a great mind on the meaning of his life.

14.6 Final Remarks on the Proof

Perelman was awarded the Fields Medal at the 2006 International Congress of Mathematicians in Madrid (the other recipients were Andrei Okounkov, Terence Tao, and Wendelin Werner). This is without question the highest honor in the profession. Perelman did not show up to accept his honor; he formally (in advance) declined the award. In fact, the Fields Committee determined at the end of May, 2006 to award Perelman (and three other mathematicians) the Fields Medal. President of the International Mathematical Union, Sir John M. Ball, traveled to St. Petersburg to endeavor to

14.6. Final Remarks on the Proof

convince Perelman to accept the medal. Perelman was quite gracious, and spent a lot of time with Ball, but was adamant that he would not accept the award. He made it plain that what was important was solving the problem, not winning the prize.

Before Perelman was awarded the Clay Prize, it was rumored that he did not want to publish his proof because then he would be a legitimate candidate for one of the seven Clay Prizes. And then, once he has $1 millon, Perelman feared that some Russian gangster may murder him. Since Perelman has declined the Clay Prize, this point may be moot.

At the International Congress of Mathematicians in 2006, held in Madrid, Richard Hamilton gave the keynote plenary lecture. His purpose was to announce, and to reassure the mathematical community, that the Poincaré conjecture had indeed been proved and all was well. As part of his talk, Hamilton also announced that he himself had an "alternative" proof of the Poincaré conjecture. That was in August of 2006. Late the following year, at another conference in China, Hamilton announced that he had written down about half of his proof, but he had run into problems. Nothing further has been heard in this matter.

A footnote to the fascinating Perelman story is that, in the thick of all the excitement over Perelman's work, Swiss mathematician Peter Mani-Levitska announced his own proof of the Poincaré conjecture. He wrote a self-contained, 20-page paper. And then retired from his academic position. His proof uses combinatorial techniques that harken back to some of the earliest ideas (due to Poincaré himself) about the Poincaré conjecture. And Mani-Levitska, although not a topologist by trade, is one of the ranking experts in these combinatorial techniques. The aforementioned 20-page paper does not seem to have appeared in print, but Mani-Levitska's proof is now in a 13-page paper that was posted on arXiv in January 2010.

Another development is that Huai-Dong Cao and Xi-Ping Zhu have a 334-page paper that has appeared in the *Asian Journal of Mathematics*—see [CZ 06]—which purports to prove *both* the geometrization program *and* the Poincaré conjecture. That is Fields Medalist S. T. Yau's journal, so this event carries some weight. It may be noted that the Cao–Zhu paper was published without the benefit of any refereeing or review. Yau obtained the approval of his board of editors, but *without* showing them the paper. On the other hand, Cao was a student of Yau. Yau is perhaps the premier expert in the application of methods of partial differential equations in geometry. He presumably read the paper carefully, and that counts for a lot.

Yau has aggressively promoted the Cao–Zhu work. Perelman himself has expressed some skepticism over what contribution this paper actually makes to the problem. He uses the paper as a touchstone to cast aspersions on the general ethical tenor in the mathematical profession.

Because of the very carefully written 473-page book of Morgan–Tian, it can now be said that Perelman's work has been carefully refereed and vetted. Two world-renowned experts have pronounced it (after some considerable ministrations of their own) to be correct. The entire Morgan–Tian book is available for purchase, and is also posted on the Web; it is therefore available for checking to the entire world. So it seems likely that the saga of the Poincaré conjecture has been brought to closure. Many experts say that we need to wait a while so that all the delicate, multidisciplinary aspects of the proof have time to gel and be fully understood. One wonders what Henri Poincaré himself might have thought of these developments that his problem inspired, or what he would say about the ultimate solution.

The American Mathematical Society (AMS) holds a large annual meeting, jointly with the Mathematical Association of America, each January. In 2007 that meeting was held in New Orleans, Louisiana. James Arthur, President of the AMS, had planned to have a celebration of the Poincaré conjecture at the gathering. A whole day of talks and discussions was planned. Fields Medalists John Milnor and William Thurston were to give background in the morning on the Poincaré conjecture and the geometrization problem respectively. In the afternoon Richard Hamilton and John Morgan and John Lott were to speak about their contributions to the program. Unfortunately, Hamilton backed out; he cited other commitments and general fatigue. After considerable negotiation and cogitation, it was decided to invite Zhu to replace him. At that point Lott said he would not share the stage with Zhu. Efforts were made to rescue the situation, but to no avail. The event was ultimately canceled. There are hopes that a new celebration may be organized in the future.

An interesting feature of the saga of the solution of the Poincaré conjecture is the role of teamwork in creating the solution of the problem. Up until World War II mathematicians tended to be single-combat warriors. The vast majority of mathematics papers were written by lone mathematicians working in isolation. This was partly a function of the fact that mathematicians did not have the resources or the means to travel (and hence to engender collaborations) and partly a matter of culture. There was a long legacy of mathematicians, and scientists in general, being rather secretive about

14.6. Final Remarks on the Proof

their work. University positions were few, and many mathematicians (such as Pierre de Fermat) had to pursue mathematics as a private avocation—really as a hobby. Other lucky ones, such as Leonhard Euler, found wealthy patrons. But it is easy to see that professional jealousies would often develop and be difficult to ignore. For many centuries there were no scholarly journals, and mathematicians published their work privately (at their own expense) if at all. In fact it is worth saying a few words about the genesis of scholarly journals.

In the mid-seventeenth century, Henry Oldenburg was active in the scientific societies. Because of his personality, and his connections, Oldenburg became something of a go-between among scientists of the day. If he knew that scientist A needed some ideas of scientist B, then he would arrange to approach B to ask him to share his ideas. Usually Oldenburg could arrange to offer a quid pro quo for this largesse. In those days books were rare and expensive, and Oldenburg could sometimes arrange to offer a scientific book in exchange for some ideas.

After some years of these activities, Oldenburg and his politicking became something of an institution. This led Oldenburg to create the first peer-reviewed scientific journal in 1665. He was the founding editor of *The Philosophical Transactions of the Royal Society of London*. At the time it was a daring but much-needed invention that supplanted a semi-secret informal method of scientific communication that was both counterproductive and unreliable. Today journals are part of the fabric of our professional life. Most scientific research is published in journals of some sort.

In modern times journals are the means of our professional survival. Scientists who want to establish their reputation must publish their ideas in scientific journals. If an assistant professor wants to get tenure, then it must be established that he or she has a substantial scholarly *gestalt*. This means that the individual will have created some substantial new ideas, and will have lectured about them and published some write-up of the development of this thought. This entire circle of considerations has led to the popular admonition of "publish or perish." The notion is perhaps worthy of some discussion.

For the past many years, and especially since the advent of National Science Foundation (NSF) grants, we have been living under the specter of "publish or perish." The meaning of this aphorism is that if you are an academic, and if you want to get tenured or promoted, or you want to get a grant, or you want an invitation to a conference, or you want a raise, or you want the respect and admiration of your colleagues, then you better publish original work in recognized, refereed journals or books. We might well wonder, "Who coined the phrase 'publish or perish'?"

One might think that it was a President of Harvard. Or perhaps a high-ranking officer at the NSF. Or some Dean at Caltech. One self-proclaimed expert on quotations suggested that it was Benjamin Franklin! But no, it was sociologist Logan Wilson (1907–1990) in his 1942 book *The Academic Man, A Study in the Sociology of a Profession* [Wil 42]. He wrote, "The prevailing pragmatism forced upon the academic group is that one must write something and get it into print. Situational imperatives dictate a 'publish or perish' credo within the ranks."

Wilson was President of the University of Texas and (earlier) a student at Harvard of the distinguished sociologist Robert K. Merton. So he no doubt knew whereof he spoke.

Marshall McLuhan has sometimes been credited with the phrase "publish or perish," and it is arguable that it was he who popularized it. In a June 22, 1951 letter to Ezra Pound he wrote (using Pound's favorite moniker "beaneries" to refer to the universities)

> The beaneries are on their knees to these gents [foundation administrators]. They regard them as Santa Claus. They will do "research on anything" that Santa Claus approves. They will think his thoughts as long as he will pay the bill for getting them before the public signed by the profesorry-rat. "Publish or perish" is the beanery motto.

It was not until the eighteenth and nineteenth centuries that mathematicians really began to disseminate their work. And it was not really until the twentieth century that mathematicians began to travel extensively and to openly share their ideas. Today there are many government agencies, both in the United States and internationally, that sponsor scientific research. These agencies, in particular, subvent travel and provide funding for conferences. It was considered quite astonishing when, in the first half of the twentieth century, British mathematicians John E. Littlewood and Godfrey H. Hardy wrote over 100 scholarly papers together. Still, in the days of Littlewood and Hardy, the vast majority of mathematics papers were written by single authors.

But today this has changed. Now the vast majority of papers are written collaboratively. Many papers have two authors, but many others will have three or more contributors. With the use of the Internet, there are now collaborative mathematical projects that involve upwards of 100 mathematicians. This has led to some remarkable developments in the way that mathematics is conceived and generated.

The solution of the Poincaré conjecture and the geometrization program follows in this new tradition. As we have noted, there have been many failed attempts to prove the Poincaré conjecture. Finally, inspired by the pioneering work of Richard Hamilton, Grigori Perelman wrote three rather unusual and sketchy papers—which he never published—providing a "proof" of the Poincaré conjecture and much more. Because of Perelman's track record doing original and pioneering work on other problems, the mathematical community took his efforts seriously. But nobody was sure that he really had a proof. It took a number of years, and the combined efforts of Kleiner, Lott, Morgan, Tian, Cao, Zhu, and many others to finally determine that the Perelman–Hamilton program really pays off and yields a solution of the Poincaré conjecture. As for the geometrization program, there are now four manuscripts purporting to prove that conjecture. These are authored by Cao/Zhu [CZ 06], Kleiner/Lott [KL 08], Morgan/Tian [MT 08], and of course Perelman. Taken as a group, these authors have considerable credibility, and it is a good bet that the geometrization program is valid. But it may be years before all the details are checked and the mathematical community is ready to embrace this theorem.

It is a tribute to the open nature of mathematical communication today, and to our group dedication to our subject, that everyone agrees that the Poincaré conjecture is Perelman's theorem. But there is no denying that the entire mathematical community, and in particular the talented geometers mentioned in the last paragraph, brought the proof to fruition, validated it, and confirmed it. No other science in today's world works in this fashion.

A Look Back

Certainly Henri Pioncaré was one of the geniuses of early twentieth century mathematics. His brilliance was recognized even when he was a small child.

Poincaré studied at the Lycée—today called the Lycée Henri Poincaré. He was the top student in every subject that he undertook. One of his instructors called him a "monster of mathematics." After the École Polytechnique, Poincaré spent some time as a mining engineer at the École des Mines. At the same time he studied mathematics under the direction of Charles Hermite. He earned his doctorate in 1879.

Poincaré was remarkable for his work habits. He engaged in mathematical research each day from 10:00 AM until noon and from 5:00 PM until 7:00 PM. He would read mathematical papers in the evening. Rather than build new ideas on earlier work, Poincaré preferred always to work from first principles. He operated in this fashion both in his lectures and in his writing.

Poincaré also believed that his best ideas would come when he stopped concentrating on a problem, when he was actually at rest:

> Poincaré proceeds by sudden blows, taking up and abandoning a subject. During intervals he assumes ... that his unconscious continues the work of reflection. Stopping the work is difficult if there is not a sufficiently strong distraction, especially when he judges that it is not complete ... For this reason Poincaré never does any important work in the evening in order not to trouble his sleep.[8]

In 1894 Poincaré published his important *Analysis Situs*. This seminal work laid the foundations for topology, especially algebraic topology. He defined the fundamental group—which is an important device for detecting holes of different dimensions in surfaces and other geometric objects. He proved the foundational result that a 2-dimensional surface having the same fundamental group as the sphere is in fact topologically equivalent to the sphere. He conjectured that a similar result is true in 3 dimensions, and ultimately in all dimensions. This is the Poincaré conjecture.

Poincaré is also remembered as the founder of the analytic theory of functions of several complex variables, and he did important work in number theory. Poincaré maintained his interest in physics, and made contributions to optics, electricity, telegraphy, capillarity, elasticity, thermodynamics, potential theory, quantum theory, relativity, and cosmology.

Of particular historical interest is Poincaré's participation in an 1887 competition that the King of Norway and Sweden initiated to celebrate the king's sixtieth birthday. Poincaré's paper on the 3-body problem[9] solved an important problem in celestial mechanics. Even though this paper was

---

[8]Translated from *Henri Poincaré*, Dr. Edouard Toulouse, Paris, 1910, page 146.

[9]The three-body problem asks about the long-term motion of objects (say the sun, the earth, and the earth's moon) acting on each other with the force of gravity. The French Academy first offered a prize question about the 3-body problem in the late eighteenth century, so the problem is well over 200 years old.

Today a great deal is known about what might happen to three planets in these circumstances, but the problem is far from being completely understood. The corresponding problem for n planets—called, appropriately enough, the "n-body problem"—is still open.

ultimately discovered to contain an error, it won the prize and has been very influential in twentieth century mathematics. In particular, the theory of dynamical systems (which ultimately led to fractal geometry and chaos) was founded in this paper. The paper was ultimately corrected and published in 1890.

A curious feature of Poincaré's career is that he never founded his own "school" since he never had any students. Poincaré's contemporaries used his results, but not his techniques. He certainly had considerable influence over the mathematics of his day (and on into the present day).

Poincaré is arguably the father of topology (popularly known as "rubber sheet geometry") and also of the currently very active area of dynamical systems. He made decisive contributions to differential equations, geometry, complex analysis, and many other central parts of mathematics.

References and Further Reading

[CZ 06] Cao, H.-D., Zhu, X.-P.: A complete proof of the Poincaré and geometrization conjectures—application of the Hamilton-Perelman theory of the Ricci flow. Asian Journal of Mathematics **10**, 165–498 (2006)

[KL 08] Kleiner, B., Lott, J.: Notes on Perelman's papers. Geometry & Topology **12**, 2587-2855 (2008)

[Kuh 70] Kuhn, T.S.: The Structure of Scientific Revolutions, 2nd edn. University of Chicago Press, Chicago (1970)

[MT 07] Morgan, J., Tian, G.: Ricci flow and the Poincaré conjecture. In: Clay Mathematics Monographs, American Mathematical Society, Providence (2007)

[MT 08] Morgan, J., Tian, G.: Completion of the proof of the geometrization conjecture. arXiv:0809.4040v1[math.DG]

[Per(a)] Perelman, G.: The entropy formula for the Ricci flow and its geometric applications (2008). arXiv:math.DG/0211159v1

[**Per(b)**] Perelman, G.: Ricci flow with surgery on three-manifolds (2008). arXiv:math.DG/0303109v1

[**Per(c)**] Perelman, G.: Finite extinction time for the solutions to the Ricci flow on certain three-manifolds (2008). arXiv:math.DG/0307245v1

[**Poi 04**] Poincaré, H.: Cinquième complément á l'analysus situs. In: *Œuvres de Poincaré*, pp. 435–498. Gauthier-Villars, Tome. VI, Paris (1953). Originally published in *Rendiconti de Circolo Matematica di Palermo*, **18** (1904), 45–110

[**Strz 06**] Strzelecki, P.: The Poincaré conjecture? American Mathematical Monthly **113**, 75–78 (2006)

[**Thu 80**] Thurston, W.P.: The Geometry and Topology of Three-Manifolds. Notes. Princeton University, Princeton, NJ (1980)

[**Thu 82**] Thurston, W.P.: 3-dimensional manifolds, Kleinian groups and hyperbolic geometry. Bulletin of the American Mathematical Society **6**, 357–381 (1982)

[**Thu 94**] Thurston, W.P.: On proof and progress in mathematics. Bulletin of the American Mathematical Society **30**, 161–177 (1994)

[**Thu 97**] Thurston, W.P.: 3-Dimensional Geometry and Topology, vol. 1. Princeton University Press, Princeton (1997)

[**Wil 42**] Wilson, L.: The Academic Man, A Study in the Sociology of a Profession. Oxford University Press, London (1942)

Epilogue

Many of us come away from school with the idea that mathematics is a fixed, crusty set of ideas cooked up three thousand years ago by a bunch of guys in togas. While we are indeed proud of the fact that mathematical ideas persist through the ages, and the mathematics of the ancient Greeks is as valid today as it was millenia ago, what we find exciting about math today is that there is always something new and different.

We have seen (Chap. 9 on cryptography) that mathematics plays a vital role in making and breaking secret codes—in ways that were undreamed of thirty years ago. We have seen (Chap. 2 on finance) that mathematics is not only good for counting your money, but for divising investment strategies and means of analyzing the market. We have seen (Chap. 14 on the Poincaré conjecture) that mathematics can be used to study the shape of the universe.

And on it goes. Mathematics is a powerful set of analytic tools and means of reasoning that enables us to understand and to analyze a wide variety of problems. What is important for you, the reader, to understand is that new mathematical ideas and techniques are being developed every day—by scholars at universities, by scientists at government institutions around the world, and by people at Google and Microsoft and other high tech companies.

The level of communication in mathematics today is very highly developed. With the Internet and email, and with the ease of world travel, mathematicians are collaborating like never before. Seventy-five years ago the vast majority of mathematical papers were written by lone individuals. Collaboration was rare. Today the significant majority of mathematical work is done by teams. Why is that so?

It is partly the ease of communication. But it is partly that mathematics has become much more sophisticated. Many of the problems that we think about today use ideas from many different parts of mathematics or science. It is difficult for just one person to be a master of all these different aspects. Thus it makes sense to have co-workers who bring in different kinds of expertise to attack the problem at hand.

It is also psychologically stimulating to have someone to bounce ideas off of, to have an interlocutor with whom to exchange new concepts, to have a reality check for whatever you are thinking about today. And it is also good to have someone to share the work.

Another relatively new development in mathematics is a resurgence of contact between mathematicians and researchers and practitioners in other disciplines. Fifty years ago or more, the typical mathematician sat alone in his or her office concentrating his or her attentions on some particular details of a very focused part of the discipline. Today mathematicians talk to everyone—physicists, engineers, computer scientists, and many others. And mathematicians collaborate with folks from all these applied disciplines. The result is a richer subject that is more fully grounded in the world that we live in. Certainly questions that arise from computer science have profoundly enriched many aspects of modern mathematics. Questions from engineering led to the theory of wavelets, which we treated in Chap. 8. Questions of finance led to the Black/Scholes option pricing scheme, which we studied in Chap. 2. Questions about the motions of the planets led to the theory of dynamical systems, which was treated in Chap. 4. Questions about the complexity of problems led to P/NP Conjecture which we learned about in Chap. 10.

We hope that this book has shown you the vast and varied panorama that is modern mathematics—where it comes from, how it has developed, and where it is going. We hope that you have a newfound appreciation for the variety and texture of the discipline, and for the many different approaches and techniques that are used. We hope that you have a new appreciation for the role that mathematical science plays in all of our lives—not just theoretically, but also in a variety of practical contexts. And we hope that you will continue to explore mathematics on your own, both as a source of pride and of pleasure.

Credits for Illustrations

Sec. (p.)	Title and credit/source for illustration.
1.2 (11)	*Percy Heawood.* Photograph courtesy of Robin J. Wilson.
2.1 (22)	*Clay accounting tokens. Susa, Uruk period (4000–3100 BCE).* ©Marie-Lan Nguyen / Wikimedia Commons.[2]
2.1 (23)	*Clay envelope and accounting tokens. Susa, Uruk period (4000–3100 BCE).* ©Marie-Lan Nguyen / Wikimedia Commons.[2]
2.1 (24)	*Clay accounting tablet. Susa, period III (3100–2850 BCE).* By Mbzt / Wikimedia Commons.[2]
2.3 (26)	*Fibonacci.* From *A Portfolio of Portraits of Eminent Mathematicians,* edited by David Eugene Smith, Open Court, Chicago, 1905.
2.5 (32)	*John Law.* From *Histoire de France,* edited by Ernest Lavisse, Librairie Hachette et Compagnie, Paris, 1909.
2.15 (54)	*Fischer S. Black.* Photograph courtesy of Black's daughter Alethea Black, author of *I Knew You'd Be Lovely.*
2.15 (54)	*Myron Scholes.* Nobellaureatesphotographer / Wikimedia Commons.[1]
2.15 (54)	*Robert C. Merton.* Digarnick / Wikimedia Commons.[4]
3.2 (64)	*Frank Ramsey.* Reproduced by kind permission of the Provost and Scholars of King's College, Cambridge.

3.3 (68) *Paul Erdős.*
By Kilian M. Heckrodt / Wikimedia Commons.[2]

4.2 (82) *The Mandelbrot set.*
By Wolfgang Beyer / Wikimedia Commons.[1]

4.2 (82) *Blown up portions of the Mandelbrot set.*
By Wolfgang Beyer / Wikimedia Commons.[1]

4.2 (85) *The Mandelbulb.*
By Ondřej Karlík / Wikimedia Commons.[1]

4.4 (97) *The Lorenz attractor.*
By Agarzago / Wikimedia Commons.[1]

5.2 (115) *The catenoid.*
By Krishnavedala / Wikimedia Commons.[1]

5.2 (115) *Leonhard Euler.* From *Leonhardi Euleri Opera Omnia,* series 1, volume 1, B. G. Teubner, Leipzig and Berlin, 1911.

5.2 (116) *Lagrange.* From *Œuvres de Lagrange,* volume 1, Gauthier-Villars, Paris, 1867.

5.5 (121) *A helicoid.*
By Krishnavedala / Wikimedia Commons.[1]

5.5 (122) *Meusnier's airship design (1784).* In *La Navigation Aérienne,* by Gaston Tissandier, Librairie Hachette et Compagnie, Paris, 1886.

5.6 (124) *Joseph Plateau.*
By Joseph Pelizzaro (1843) / Wikimedia Commons.[3]

5.8 (127) *The minimal surface of Costa, Hoffman, and Meeks.*
By Anders Sandberg / Wikimedia Commons.[1]

7.4 (177) *Henri Poincaré.* Photograph by Henri Manuel. Frontispiece of *Dernières Pensées,* Henri Poincaré, Flammarion, Paris, 1913.

9.1 (197) *Leonard Adleman.*
By Lenonard Adleman / Wikimedia Commons.[1]

10.2 (218) *A mechanical adding machine.*
By Roger McLassus / Wikimedia Commons.[1]

Credits for Illustrations

10.9 (250) *Stephen A. Cook.*
By Jiří Janíček / Wikimedia Commons.[1]

11.6 (262) *Fermat.* From *A Portfolio of Eminent Mathematicians,* edited by David Eugene Smith, Open Court, Chicago, 1905.

14.1 (344) *William Thurston.* Photograph courtesy of George M. Bergman and Mathematisches Forschungsinstitut Oberwolfach.

Wikimedia Licensing Information:

- **1:** Share, Remix, Attribute, Share Alike.
- **2:** Share, Remix, Attribute.
- **3:** Public domain: before 1922.
- **4:** Public domain: released by author.

Index

2-torus, 344
Guinness Book of World Records, 60, 75
Liber Abaci, 28
Principia Mathematica, 94
Tractatus Logico-Philosphicus, 64
"and", 296
"false", 288
"for all", 296–299
"if and only if", 295
"if–then", 291, 296
"iff", 296
"not", 291, 296
"or", 289, 296
"there exists", 296–299
"true", 288

a priori estimate, 132
A.W. Jones & Co., 44
Abel, Niels Henrik, 258, 320
acceptable in nondeterministic polynomial-time, 250
accepted, 225
acute angle, 153
addition modulo p, 322
adjacency matrix, 209
Adleman, Len, 197, 199
Agrawal, Manindra, 264, 271, 272
airship, 123
AKS primality test, 272
Al-Khujadi, Abu, 316

algebraic curve, 325
algebraic topology, 129
algorithm, 217
Almgren, Frederick, 134
American call option, 43
Apil-Sin, 24
Appel, Kenneth I., 12–14
Appel,Kenneth I., 15, 17, 18
Appell, Paul, 180
arbitrage, 42
arbitrage-free price, 42
Archbishop of Canterbury, 65
Archimedes, 336
Archimedes of Syracuse, 256
area-minimizing surface, 113
Argand diagram, 83, 319
Argand, Jean-Robert, 319
argument, 318
Aristotelian logic, 292
Aristotle, 25, 279, 304, 306
arithmetic modulo p, 322
Arthur, James, 358
associative law, 317
Atiyah–Singer Index Theorem, 13
atomic statement, 286
attractor, 97
automaton, 222
automaton, deterministic finite, 222
automaton, nondeterministic finite, 226

average return, 37
axiom, 279

Bézout, Étienne, 327
Babbage Difference Engine, 220
Babbage, Charles, 220
Bachelier, Louis, 37
Bachet, Claude, 265, 310
Ball, John, 356
base 3 notation, 102
basis vector, 172
Beaugrand, Jean, 265
Beethoven, Ludwig von, 285
Beltrami, Eugenio, 156
Bernays, Paul, 286
Bernoulli numbers, 316
Bernoulli, Daniel, 185
Bernoulli, Johann, 94
Besançon, 37
Bianchi, Luigi, 345
biconditional statement, 295
biconditional, material, 295
big "O", 238
binary digit, 233
Birkhoff, George D., 12
bit, 233
Black, Fischer, 54–56
Black–Scholes equation, 55
Black–Scholes option pricing, 366
Black/Scholes model, 57
Blu-ray, 195
Blu-ray disc, 186
Bolyai, Farkas, 154
Bolyai, János, 154–156, 160
bonds, 31
Boolos, George, 302
boundary of a current, 131
Bourbaki, Nicolas, 307

Brewster, David, 111
British East India Company, 33
Brooks, Robert, 85
Browder, William, 350
Bugia, 27
Butterfly Effect, 96

Caesar cipher, 197
Caesar, Julius, 197
Calabi conjecture, 161
Calderbank, Robert, 193
call option, 43
call option pricing, 45
Cantor set, 97, 98
Cantor ternary set, 100
Cao, Huai-Dong, 354, 355, 357, 358, 361
Carleson, Lennart, 82
Carmichael number, 268
Carolus guilder, 32
Cartan, Élie, 161
catenoid, 115
Catholic Church, 25
Cayley numbers, 193, 320
Cayley, Arthur, 2
CD, 186
celestial mechanics, 82, 117
central processing unit, 222
change of coordinates, 167
chaos, 82
Chern, Shing-Shen, 161
chromatic number, 8
Church–Turing thesis, 221
citizenship exam (Gödel), 284
Class **NP**, problems of, 218
Class **P**, problems of, 218
classification of surfaces, 344
clay envelope, 21

Clay Millennium Prize, 355
Clay Prizes, 354
clay tablet, 22
clay token, 21
clock arithmetic, 322
Coifman, Ronald R., 189, 194, 196
combinatorics, 76
commutative law, 317
compact disc, 186
Compagnie d'Occident, 33
Company of the West, 33
complete, 287
completeness of sentential logic, 301
completeness of the predicate calculus, 301
complex numbers, 318
complexity theory, 219
components of a vector, 172
composite number, 255
composition, 327
compound interest, 26
conditional statement, 292
conditional, material, 292
congruence of triangles, 141
congruence subgroup, 332
congruence, Angle-Side-Angle, 142
congruence, Side-Angle-Side, 142
congruence, Side-Side-Side, 141
conjunction, 288
connective, 290
consistency of the predicate calculus, 301
consistent, 287
consistent axiom system, 155
continuous compounding, 29
continuous dynamical system, 81
continuous map, 126
contrapositive, 296

convexity, 65
Cook, Stephen, 251
coordinate change, Lorentizian, 167
Costa, Celso José da, 128
Costa–Hoffman–Meeks surface, 128
Council of Aix, 25
Council of Lyons, 25
Council of Nicaea, 25
Council of Vienne, 25
counterpary risk, 39
CPU, 222
Craford Prize, 356
cryptography, 261
current, 131
curvature of a curve, 118
curve, algebraic, 325
curve, Frey, 333
curve, singular, 326
curve, smooth, 326

d'Espagnet, Étienne, 265
Dangerfield, Rodney, 60
Darboux, Gaston, 180
Daubechies, Ingrid, 192, 193
David, Guy, 194
De Giorgi, Ennio, 132, 133
de Morgan's laws, 292
De Morgan, Augustus, 2
decrementor, 239
definition, 279
definitions, 139
degree of a polynomial, 320
derivative, 38
design theory, 76
deterministic, 222
Dieudonné, Jean, 356
dike army, 32
Diophantus of Alexandria, 265, 310

Dirichlet, Johann Peter Gustav
        Lejeune, 59, 187, 258
dirigible, 123
disc-type surface, 126
discrete dynamical system, 81
disjunction, 289
distributive law, 317
division modulo p, 323
Doelalikar, Vinay, 253
Dolby noise reduction, 189
Douady, Adrien, 82
Douglas's solution of Plateau's
        problem, 129
Douglas, Jesse, 129, 133
dovetailing procedure, 249
Dunwoody, Martin, 345
Dutch East India Company, 31
DVD, 186, 195
dynamical system, continuous, 81
dynamical system, discrete, 81
Dyson, Freeman, 284

edge detection, 192
edge detection problem, 191
effective annual interest rate, 41
efficient market theory, 35
Einstein, Albert, 161, 163, 164, 167,
        171, 179, 283–285, 340, 353
elastic energy, minimizing, 132
elementary statement, 286
elliptic curve, 326
energy, 179
energy of a soap film, 114
Enneper, Alfred, 127
Enneper–Weierstrass formula, 127
epsilon-conjecture, 333
Eratosthenes of Cyrene, 256
Erdős , Paul, 63, 65, 68, 69

Erdős, Paul, 274
Euclid of Alexandria, 15, 16,
        137–140, 143, 152, 154, 256,
        257, 273, 279, 280, 311, 312,
        316
Euclid of Megara, 257
Euclid's common notions, 152
Euclid's postulates, 138
Euclid's Theorem, 255
Euclidean algorithm, 204, 262
Eudoxus of Cnidus, 256, 257, 278
Euler's phi function, 202
Euler's theorem, 202
Euler, Leonhard, 115, 116, 185, 202,
        316
euro, 32
European call option, 43
exclusive "or", 289
existential quantifier, 296, 297, 299
expiry date, 43

Faltings, Gerd, 326
Fama, Eugene, 35
feature detection, 192
Federer, Herbert, 132–134
Fermat test, 267
Fermat triple, 325
Fermat's Last Theorem, 265
Fermat's last theorem, 309, 310, 315,
        333–335
Fermat's little theorem, 266
Fermat, Pierre, 264, 265, 271, 309,
        310, 316, 359
Fermat, Samuel, 265
Fibonacci, 27, 28
field, 317
field, finite, 322
Fields Medal, 355

Index

final state, 225
finite field, 322
First Council of Carthage, 25
first-order logic, 300
Fleming, Wendell H., 132–134
Flemish pound, 32
Flexner, Abraham, 283
floor function, 262
Florence, 31
foliation theory, 342
formal language, 225
forward contract, 39
forward contract pricing, 40
Four-Color Problem, 1, 2, 13
Fourier analysis, 186
Fourier, Joseph, 184, 185, 187
fractal, 97
fractal dimension, 108
Franche-Comté, 37
Franklin, Benjamin, 360
Franklin, Philip, 12
Fredholm, Ivar, 180
Frege's theorem, 301
Frege, Gottlob, 286, 302, 303
French Revolution, 117
Frey curve, 333
Frey, Gerhard, 333
Frick, Henry Clay, 38
full modular group, 332
function, 184
Fundamental Theorem of Arithmetic, 255
Furtwängler, Phillip, 282

Gödel's incompleteness theorem, 304, 305
Gödel, Adele (née Porkert), 283
Gödel, Kurt, 281–287, 301, 303–305
Gödel, Marianne, 281
Gödel, Rudolf, 281
Gödel, Rudolf (brother of Kurt), 281
Gödel, Kurt, 283
Galois, Evariste, 258
Garey, Michael, 251
Garnier's solution of Plateau's problem, 127
Garnier, René, 127–129
Gauss, Carl Friedrich, 154, 155, 160, 258, 273
General Bank (France), 33
general relativity, 165
generalized surface, 131
Genoa, 31
genus, 8, 344
geometric Brownian motion, 37
geometric pieces, 350
geometric random walk, 35
geometrization program, 343
geometry, Riemannian, 160
Germain, Sophie, 335
Gingrich, Newt, 292
Glorious Revolution, 32
Goldstein, Rebecca, 284
Gonthier, Georges, 15
Graham number, 60, 75
Graham number, Knuth's notation for, 76
Graham, Ron, 60, 76
graph theory, 208
great circle, 149
greatest integer function, 9, 262
Grothendieck, Alexander, 356
guilder, 32
guru, 26
Guthrie, Francis W., 1
Guthrie, Frederick, 1, 2

Haar, Alfred, 194
Hadamard, Jacques, 262, 274
Haken, Wolfgang, 12–15, 17, 18
halted state, 231
Hamilton, Richard, 348, 350, 354, 357, 358, 361
Hamilton, Rowan, 2
hamiltonian circuit, 208
hang, 231
Hansen, Lars Peter, 35
happy end problem, 68
hardware, 219
Hardy, G. H., 360
Hawking, Stephen, 180
Heaton, Henry, 320
Heawood, Percy, 3, 6, 8–12
hedge fund, 44
Heesch, Heinrich, 12, 13
helicoid, 122
Hellegouarch, Yves, 333
Hilbert's program, 281
Hilbert, David, 138, 258, 281, 282, 305, 307
Hippocrates of Chios, 279
Hirsch, Morris, 342
Hitler, Adolf, 282
Hoffman, David Allen, 128
Hoffman, James T., 128
homotopy, 340
Hoogheemraadschap Lekdijk Bovendams, 32
Hooke's law, 91
Howarth, Jamie, 195
hyperbolic geometry, 155

ideal geometry, 350
Ilshu-bani, 24
image compression, 191
image denoising, 191
image enhancement, 192
image recognition, 192
imaginary axis, 319
imaginary part of a complex number, 83
Inanna of Elip, 24
inclusive "or", 289
incomplete, 287
incompleteness theorem, 304, 305
inconsistent, 287
incrementor, 233
independence of the parallel postulate, 149
industrial grade prime, 271
International Congress of Mathematicians, 335, 357
Internet routing, 76
inversion, 330

Jaffard, Stéphane, 194
James II (king), 32
Johnson, David, 251
joint-stock company, 31
Jones, Alfred Winslow, 44
Jordan, Camille, 7, 344
Jorge, Luquésio Petrola de Melo, 128
JPEG200, 195

Kahane, Jean-Pierre, 194
Kapovitch, Vitali, 353
Karlík, Ondřej, 85
Katz, Nicholas, 334
Kayal, Neeraj, 271, 272
Kempe chain, 4
Kempe, Alfred, 3, 4, 6
Kendall, Maurice, 34
Keynes, John Maynard, 65

Index

kinetic energy, 178
King's College, Cambridge, 64
Kleene, Stephen, 282
Klein, Esther, 65, 66, 68
Klein, Felix, 2
Kleiner, Bruce, 354, 361
Knuth, Donald, 76
Kuhn, Thomas, 345
Kummer's Criterion, 316
Kummer, Eduard, 316

Ladyzhenskaya, Olga, 352
Lagrange, Joseph-Louis, 116, 117, 133, 336
Lanford, Oscar, 13
Lanzone, Timothy, 253
LaVietes, Steve, 195
Law, John, 33, 34
LeBlanc, Antoine-August, 336, 337
Legendre, Adrien-Marie, 111, 325
length contraction, 171
length of a set, 98
Lenstra, Hendrk W., Jr., 272
Leonardo of Pisa, 27
Lesniewski, Stanislaw, 302
Levy, Silvio, 347
libration of the moon, 117
light cone, 166
light, speed of, 163, 164
lightlike vector, 174
Lisker, Roy, 356
Littlewood, John E., 194, 360
Lobachevsky, Nikoali Ivanovich, 154–156, 160
logical equivalence, 290, 291
logically equivalent, 295
logically equivalent statements, 293
long division, 204

Long-Term Capital Management, 44
Lorentz, Hendrik, 179
Lorentzian coordinate change, 167
Lorenz, Edward, 95, 96
Lott, John, 354, 358, 361
Louis XIV (king), 33
Louisiana, 33

Möbius, August, 7, 344
Mackay, Charles, 39
Mandelbrot set, 82, 108
Mandelbrot, Benoît, 82, 85, 108, 109
Mandelbulb, 85
Mani-Levitska, Peter, 357
manifold, 342
Marcellus, Marcus Claudius, 336
Marx, Karl, 69
Matelski, J. Peter, 85
material biconditional, 295
material conditional, 292
Maxwell's laws, 164
Maxwell, James Clerk, 164
Maytag-UK, 42
McLuhan, Marshall, 360
mean curvature, 120
mean curvature zero, 122
median, 144
Meeks, William Hamiton, III, 128
Merton, Robert C., 55
Merton, Robert K., 360
metric, 349
Meusnier, Jean-Baptiste, 122, 123
Meyer, Yves, 189, 194
Michelson–Morley experiment, 180
microchip structure, 76
Miller, Gary L., 268, 271
Milnor, John, 82, 350, 358
mina, 25

minimal surface, 114
minimal surface equation, 116
Mississippi Company, 34
Mittag-Leffler, Gösta, 94, 95, 180
model for a theory, 155
modular arithmetic, 200
modular group, 330
modular group, full, 332
modular transformation, 332
modulus, 318
modus ponendo ponens, 294
modus ponens, 294
modus tollendo tollens, 296
modus tollens, 296
molecular computation, 199
momentum, 171
Monge, Gaspard, 123
Mordell's theorem, 328
Mordell, Louis, 325
Morgan, Frank, 134
Morgan, John, 354, 358, 361
Morgenstern, Oskar, 283, 285
Mozart, Wolfgang, 285
multiplication modulo p, 322
multiplicity, 131

Nabi-ilishu, 24
Nash, John, 348
negation, 291
netting, 39
neutral geometry, 154
New York Central Railroad, 38
Newton's laws of motion, 171
Newton's second law, 92
Newton, Isaac, 94, 109, 161
Newtonian mechanics, 171
no-arbitrage price, 42
Nobel Prize, 55

nondeterministic acceptor, 244
nondeterministic finite automaton, 226
nondeterministic Turing machine, 243
nonlinear, 96
Novikov, Sergei, 350
NP-complete, 251
Nylander, Paul, 85

obtuse angle, 153
Occam's Razor, 140
Occam, William of, 140
octonions, 320
Okounkov, Andrei, 356
Oldenburg, Henry, 359
option, 39
orbit, 87
order of k, 202
orientation, 131
Oscar II (king), 94
osculating circle, 118

P/NP Conjecture, 250, 366
P/NP Problem, 219
Paley, Raymond, 194
parallel postulate, 139, 140
parallel postulate, Playfair's formulation of, 139
particle, 171
Penn Central, 38
Pennsylvania Railroad, 38
Perelman, Grigori, 252, 350–358, 361
perpetuity, 32
Petersen, Julius, 11
phi function, 202
Phragmén, Lars Edvard, 95
pigeonhole principle, 61

Pisa, 27
Plateau problem,
    higher-dimensional, 132
Plateau's problem, 125
Plateau's rules, 133
Plateau, Joseph, 113, 124, 125, 133
Plato, 256
Playfair, John, 138
Plimpton 322, 306
Poénaru, Valentin, 346
Poincaré conjecture, 252, 340, 345,
    353, 354, 357, 361, 365
Poincaré, Henri, 82, 94, 95, 339, 340,
    357, 358, 361–363
point mass, 171
polynomial, 320
polynomial equation, 320
polynomial time, 217
polynomial-time decidable, 249
Pomerance, Carl, 272
positive mass conjecture, 161
Pound, Ezra, 360
predicate calculus, 286
predicate calculus, completeness of,
    301
predicate calculus, consistency of,
    301
pricing, call option, 45
pricing, forward contract, 40
prime number, 255
Prime Number Theorem, 15, 262,
    273, 274
prime, industrial grade, 271
primitive Pythagorean triple, 313
principal curvatures, 120
principle of relativity, 164
Proclus Diadochus, 256
proofs, 139

propositional calculus, 285
provability, 301
prover, 209
pseudoprime, 267
pseudosphere, 156
Ptolemy, 256
publish or perish, 359
Puchberg am Schneeberg, Austria,
    64
pumping argument, 229
pumping theorem, 229
put option, 43
Pythagoras of Samos, 256
Pythagorean theorem, 311
Pythagorean triple, 311
Pythagorean triple, primitive, 313

quadratic equation, 320
quadratic formula, 320
quantifier, existential, 296, 297, 299
quantifier, universal, 296, 297, 299
quantifiers, 296
quaternions, 320
queueing theory, 76
quotient of polynomials, 321

Rabin, Michael O., 268, 271
Radó, Tibor, 129, 133
Rado's solution of Plateau's
    problem, 129
Ramsey number, 62, 73
Ramsey Theory, 65
Ramsey's Theorem, 62, 73
Ramsey, Arthur Michael, 65
Ramsey, Frank, 59, 64, 65
Ramsey, Lettice (née Baker), 64, 65
rational point, 314, 325
read/write head, 230

real numbers, 318
real part of a complex number, 83
Redford, Robert, 273
reduction modulo p, 323
reduction of an elliptic curve, 329
reflection, 330
regularity theory, 132
Reifenberg, Ernst, 132
relationship table, 69
relatively prime, 201
relatively prime integers, 203
relativistic momentum, 175, 179
relativity, general, 165
relativity, special, 164
replication method, 46
Ribet, Kenneth, 333, 334
Ricci flow, 349
Riemann, Bernhard, 160, 187, 273
Ringel, Gerhard, 12
risk-free interest rate, 41
Rivest, Ron, 197, 199
Robertson, G. Neil, 17
root of a polynomial, 320
Rorty, Richard, 284
Rosenthal, Steve, 195
Rourke, Colin, 345
Rouwenhorst, Geert K., 32
Royal Bank (France), 34
RSA encryption, 199, 204, 261
Russell's paradox, 303
Russell, Bertrand, 282, 286, 302, 305

Saccheri quadrilateral, 153
Saccheri quadrilateral, base of, 153
Saccheri quadrilateral, summit angle of, 153
Saccheri, Giovanni Girolamo, 153, 154

saddle-shaped surface, 120, 122
Sanders, Daniel P., 17
Saxena, Nitin, 271, 272
scalar component, 172
Scherk, Heinrich Ferdinand, 122
Schmidt, Ulrich, 14
Scholes, Myron, 54, 55
second-order logic, 300
section of a surface, 119
Selberg, Atle, 274
self-replicating set, 108
self-similarity, 105
selling short, 40
sentential logic, completeness of, 301
Serre, Jean-Pierre, 333
set, 302
set-theoretic difference, 98
Seymour, Paul, 15, 17
Shamash, 24
Shamir, Adi, 197, 199
shekel, 25
Shiller, Richard, 35
Shimura, Goro, 329
shortest path, 111
Sierpiński gasket, 105
Sierpiński triangle, 105
sieve of Eratosthenes, 258
signal compression, 192
signal design, 192
signal processing, 192
Simons, James H., 57
Sin-tajjar, 24
singular curve, 326
Smale, Stephen, 82, 252
smooth curve, 326
soap bubble, 124
soap film, 113, 122, 124
soap film, energy of, 114

Index 381

software, 219
Solovay, Robert M., 268
solution of an equation, 320
Sophie Germain prime, 337
Sophie Germain's theorem, 337
space-time diagram, 165
special relativity, 164
speed of light, 163, 164
sphere, 340
sphere with handles, 7
spherical geometry, 149
spring constant, 92
staircase representation, 87
Stalin, Joseph, 69
standard model (stock price), 35
state, 222
state, final, 225
state, halted, 231
statement, biconditional, 295
statement, conditional, 292
Steinman, Ralph, 55
Stichtse Rijnlanden, 32
stochastic partial differential equation, 55
straight line, 111
strange attractor, 97, 108
Strassen, Volker, 268
Strichartz, Robert, 16
strike price, 43
string theory, 161
Stromquist, Walter, 12
strong pseudoprimality test, 268
stuiver, 32
subtraction modulo p, 322
supplementary angles, 147
surface of least area, 113
surface spanning a curve, 125
surface, generalized, 131

surgery theory, 350
Szekeres, George, 65, 68

Tahales of Miletus, 137
Tait, Peter Guthrie, 11
tangent plane, 120
Taniyama, Yutaka, 329
Taniyama–Shimura–Weil Conjecture, 332, 334
Taniyama–Shimura–Weil conjecture, 329
Tao, Terence, 356
tautology, 300
Taylor, Jean E., 133, 134
Taylor, Richard, 310, 335
texture classification, 192
Thales of Miletus, 278, 306
Theaetetus of Athens, 256, 257, 278
theorem, 280
theorems, 139
Third Council of the Lateran, 25
third-order logic, 300
Thomas, Robin, 17
Thorp, Edward O., 57
three-body problem, 94
Thurston geometrization conjecture, 353
Thurston's geometrization conjecture, 354, 357, 361
Thurston, William, 342, 343, 345–347, 350, 358
Tian, Gang, 354, 358, 361
time dilation, 168
torus, 340
tractable, 218
transformation, modular, 332
transition, 222
transition rule, 222

translation, 331
transversal, 145
trial division, 264
triangle geometry, 140
truth table, 288, 291
tulipmania, 39
Turing acceptable, 232
Turing decidable, 232
Turing machine, 220
Turing machine, basic operations, 231
Turing machine, multitape, 243
Turing machine, nondeterministic, 243
Turing, Alan, 221, 230
twin paradox, 170
two-body problem, 94

underlying asset, 38
universal quantifier, 296, 297, 299
universal Turing machine, 221
universe, 298
upper half-plane model of hyperbolic geometry, 156

Vallée-Poussin, Charles de la, 262, 274
vector component, 172

Venice, 31
verifier, 209
volatility, 38
von Koch snowflake, 107
von Neumann, John, 283, 285, 304

Wall, C. T. C., 350
water board, Dutch, 32
wavelet theory, 189
Weierstrass, Karl, 127
Weil, André, 329
Weiss, Guido, 194
Werner, Wendelin, 356
White, Paul, 85
Whitehead, Alfred N., 282, 286, 305
Wickerhauser, Victor, 194, 195
Wiles, Andrew, 310, 334, 335
William of Orange, 32
Wilson, Logan, 360
Wittgenstein, Ludwig, 64
Woeginger, Gerhard J., 253
Wolfram, Stephen, 109
worldline, 165

Yau, Shing-Tung, 161, 351, 357, 358
Youngs, John W. T., 12

zero knowledge proof, 208
Zhu, Xi-Ping, 354, 355, 357, 358, 361